Lecture Notes in Mathematics

Edited by A. Dold and B. Eckr

1250

Stochastic Processes – Mathematics and Physics II

Proceedings of the 2nd BiBoS Symposium
held in Bielefeld, West Germany,
April 15–19, 1985

Edited by S. Albeverio, Ph. Blanchard and L. Streit

Springer-Verlag
Berlin Heidelberg New York London Paris Tokyo

Editors

Sergio Albeverio
Ruhr-Universität Bochum, Mathematisches Institut
Universitätsstr. 150, 4630 Bochum, Federal Republic of Germany

Philippe Blanchard
Ludwig Streit
Fakultät für Physik, Universität Bielefeld
Postfach 8640, 4800 Bielefeld, Federal Republic of Germany

Mathematics Subject Classification (1980): 22-XX, 28-XX, 31-XX, 34BXX, 35-XX, 35JXX, 46-XX, 58-XX, 60GXX, 60HXX, 60JXX, 73-XX, 76-XX, 81C20, 82-XX, 85-XX

ISBN 3-540-17797-3 Springer-Verlag Berlin Heidelberg New York
ISBN 0-387-17797-3 Springer-Verlag New York Berlin Heidelberg

Printed in Germany

Printing and binding: Druckhaus Beltz, Hemsbach/Bergstr.
2146/3140-543210

PREFACE

The Second Symposium on "Stochastic Processes: Mathematics and Physics" was held at the Center for Interdisciplinary Research, Bielefeld University, in April 1985. It was organized by the Bielefeld - Bochum Research Center Stochastics (BiBoS), sponsored by the Volkswagen Stiftung.

Our aim by choosing the topics of the conference was to present different aspects of stochastic methods and techniques concerning not only the mathematical development of the theory but also its applications to various problems in physics and other domains. The IInd BiBoS-Symposium was an attempt to provide an overview of these results, as well as of open problems. The success of the meeting was due first of all to the speakers: thanks to their efforts it was possible to take recent developments into account.

It is a pleasure to thank the staff of ZiF for their generous help in the organization of the Symposium, in particular Ms. M. Hoffmann. We are also grateful to Ms. B. Jahns, Ms. M.L. Jegerlehner and Dipl.-Phys. Tyll Krüger for preparing the manuscripts for publication.

S. Albeverio, Ph. Blanchard, L. Streit

Bielefeld and Bochum, February 1987

CONTENTS

JUMP PROCESSES RELATED TO THE TWO DIMENSIONAL DIRAC EQUATION

Ph. Blanchard — Theoretische Physik and BiBoS Universität Bielefeld

Ph. Combe — Université d'Aix Marseille II and CPT-CNRS, Marseille, and BiBoS

M. Sirugue — CPT-CNRS, Marseille

M. Sirugue-Collin — CPT-CNRS and Université de Provence, Marseille

I. Introduction

In four papers [1] written from 1963 to 1968, Symanzik indicated how to fit Feynman's formal approach [2] to Bose quantum field theory into a framework making possible mathematical control. The Euclidean strategy has become a central tool in the mathematical analysis of quantum field theory, which simplifies exact calculations and estimates of functional integrals (see e.g. [3] and the reference therein). The derivation of a path integral representation for the wave function of spin particles was solved by Feynman and Hibbs [4, ex. 2-6] and revisited many times [5], [6], [7], [8], [9], [10]. Recently, a probabilistic solution was derived for the Pauli equation [11], [12], [13] and an approach based on stochastic mechanics was formulated [14], [15].

There has been a revival of probabilistic representations of the solutions of the Dirac equation in the last few years with the work of Gaveau, Jacobson, Kac and Shulman [16] and the thesis of Jacobson [17] on the one hand, and a series of papers by Ichinose [18], [19], [20], [21] on the other hand. Ichinose proved the existence of a matrix valued countably additive path space measure on the Banach space of continuous paths for the Dirac equation in two dimensional space time. Gaveau et al. derived a probabilistic representation of the solution of the free Euclidean Dirac equation in two space time dimensions. However, they were not able to treat the Dirac equation in the presence of an external potential.

The purpose of this lecture is to present for real time a probabilistic representation for the solution of the two dimensional Dirac equation in terms of pure jump processes, even in the presence of an external electromagnetic field. We refer to [22] for a generalization for the 3- and 4-dimensional cases.

But briefly the deep origin for the derivation of such a probabilistic representation even for a real time is the existence of an underlying Poisson process, which allows to define Feynman's path integral as a bone fide integral. This is especially obvious in two space time dimensions. Indeed, using a time reversal the real time Dirac equation can be identified to the backward Kolmogorow equation of a jump process both in time and helicity.

The propagation of the nonrelativistic quantum mechanical Euclidean electron can be expressed in terms of the Wiener process or Brownian motion. In this lecture, we will show that other well-known stochastic processes, namely Markov jump processes, play the same role to describe the propagation of the relativistic electron even in the presence of external fields.

Moreover, taking advantage of this probabilistic representation the nonrelativistic limit $c \to \infty$ can be studied.

Finally, let us mention that lattice approximation in connection with path integral representation for the Dirac equation has been considered [17], [23].

II. Stochastic Models Related to Telegrapher's Equation and Euclidean Dirac Equation

In this section, we will first consider a strongly biased random walk, which leads not to a diffusion equation but to a hyperbolic one, the Telegrapher's equation. This model has been considered in [24]. We have a one dimensional lattice and a particle starting from the origin $x = 0$ which always move with speed c either in the positive direction or in the negative direction. Each step is of duration Δt and covers a distance ε (the lattice spacing). We have then $\varepsilon = c\Delta t$. At each lattice point we assume that $a\Delta t$ is the probability of reversal of direction, a being small. We introduce now the following dichotomic variable

$$\sigma = \begin{cases} +1 & \text{with probability } 1-a\Delta t \\ -1 & \text{with probability } a\Delta t \end{cases} \tag{2.1}$$

and consider a sequence $\sigma_1, \sigma_2, \ldots \sigma_{n-1}$ of such independent random variables. Let X_n be the displacement of the particle starting from the origin after n steps. If the particle starts in the positive direction then X_n will be

$$X_n = c\,\Delta t[1 + \sigma_1 + \sigma_1\sigma_2 + \ldots + \sigma_1\sigma_2 \ldots \sigma_{n-1}] \tag{2.2}$$

and if it starts in the negative direction this displacement will be $-X_n$. Let φ be a smooth function. We consider now the two following expectations

$$\varphi_n^{\pm}(X) = E[\varphi(X \pm X_n)] . \tag{2.3}$$

Writing

$$\varphi_n^+(x) = E\Big[\varphi[x + c\Delta t + c\Delta t\sigma_1(1 + \sigma_2 + \dots + \sigma_2\sigma_3\cdots\sigma_{n-1})]\Big]$$

and performing first the average over σ_1 we will obtain a recursion relation for the $\varphi_n^\pm$'s, namely

$$\varphi_n^+(x) = a\Delta t\ E\Big[\varphi[x + c\Delta t - c\Delta t(1 + \sigma_2 + \dots + \sigma_2\sigma_3\cdots\sigma_{n-1})]\Big]$$

$$+ (1 - a\Delta t)E\Big[\varphi[x + c\Delta t + c\Delta t(1 + \sigma_2 + \dots + \sigma_2\sigma_3\cdots\sigma_{n-1})]\Big]$$

$$= a\Delta t\ \varphi_{n-1}^-(x + c\Delta t) + (1 - a\Delta t)\varphi_{n-1}^+(x + c\Delta t). \tag{2.4}$$

In the same way we obtain for φ_n^- the following recursion relation

$$\varphi_n^-(x) = a\Delta t\ \varphi_{n-1}^+(x - c\Delta t) + (1 - a\Delta t)\varphi_{n-1}^-(x - c\Delta t). \tag{2.5}$$

The equation (2.4) can now be rewritten

$$\frac{\varphi_n^+(x) - \varphi_{n-1}^+(x)}{\Delta t} = \frac{\varphi_{n-1}^+(x + c\Delta t) - \varphi_{n-1}^+(x)}{\Delta t} - a[\varphi_{n-1}^+(x + c\Delta t) - \varphi_{n-1}^-(x + c\Delta t)].$$

Taking the limit $\Delta t \to 0$ in such a way that $n\Delta t = t$ we get

$$\frac{\partial\varphi^+}{\partial t} = c\,\frac{\partial\varphi^+}{\partial x} - a\,\varphi^+ + a\,\varphi^- \tag{2.6a}$$

$$\frac{\partial\varphi^-}{\partial t} = -c\,\frac{\partial\varphi^-}{\partial x} + a\varphi^+ - a\varphi^- \quad . \tag{2.6b}$$

Setting $f = \frac{1}{2}(\varphi^+ + \varphi^-)$ and $g = \frac{1}{2}(\varphi^+ - \varphi^-)$ we obtain

$$\frac{\partial f}{\partial t} = c\,\frac{\partial g}{\partial x} \tag{2.7}$$

$$\frac{\partial g}{\partial t} = c\,\frac{\partial f}{\partial x} - 2a\,g\,. \tag{2.8}$$

Differentiating (2.7) with respect to t and (2.8) with respect to x, we obtain for f the Telegrapher equation

$$\frac{1}{c}\,\frac{\partial^2 f}{\partial t^2} = c\,\frac{\partial^2 f}{\partial x^2} - \frac{2a}{c}\,\frac{\partial f}{\partial t}\,. \tag{2.9}$$

We will now discuss briefly limiting cases of this equation. If $a = 0$ then the probability of reversing direction is zero; it follows that

$$f(x,t) = \frac{1}{2}[\varphi(x+ct) + \varphi(x-ct)] \tag{2.10}$$

and we recognize the case of a vibrating string.

Letting now $a \to +\infty$ and $c \to +\infty$ in such a way that $\frac{2a}{c^2} = \frac{1}{D}$ stays constant, (2.9) becomes the diffusion equation

$$\frac{1}{D}\frac{\partial f}{\partial t} = \frac{\partial^2 f}{\partial x^2} . \tag{2.11}$$

In the model we consider the probabilities are either extremely small or extremely large. For the standard random walk model leading to Brownian motion the probability of a move to the right or to the left is $1/2$. Therefore, the only way to obtain $1/2$ is to let $a \to \infty$ as $\Delta t \to 0$. Moreover, in the limit of the random walk model the velocity of a Brownian particle is infinite: for this reason c must also go to infinity.

We will now briefly discuss the connection with Poisson process N_t and express the velocity $c(t)$ of the particle at time t . The number of reversals of direction up to time t is just N_t , so we have (assuming that the particle starts in the positive direction)

$$c(t) = c(-1)^{N_t}$$

and the position of the particle at time t , which is the continuous analogue of X_n , is given by

$$x(t) = c\int_0^t (-1)^{N_\tau}\, d\tau . \tag{2.12}$$

The solution of the Telegrapher's equation is simply in terms of Poisson process:

$$f(x,t) = \frac{1}{2} E\left[\varphi\left(x + c\int_0^t (-1)^{N_\tau}\, d\tau\right) + \varphi\left(x - c\int_0^t (-1)^{N_\tau} d\tau\right)\right] . \tag{2.13}$$

We notice a close analogy between (2.10) and (2.13). The time t in (2.10) is replaced by the random time $\int_0^t(-1)^{N_\tau}\, d\tau$ and the expectation with respect to the Poisson process is taken. This property is valid for all equations of this form in any numbers of dimensions. Consider for example the equation

$$\frac{1}{c^2}\frac{\partial^2 f}{\partial t^2}(t,x) + \frac{2a}{c^2}\frac{\partial f}{\partial t}(t,x) = \Delta f(t,x), \quad x \in \mathbb{R}^d . \tag{2.14}$$

To obtain a probabilistic representation of the solution of (2.14) it suffices to use a solution of the wave equation in d-dimensions, to replace the time t by the random time $\int_0^t (-1)^{N_\tau} d\tau$ and to take expectation with respect to the Poisson process.

Following [10] and [25], we will now describe the possibility to express the solution of the Euclidean Dirac equation using Poisson process. Setting

$$\phi = \binom{\varphi_+}{\varphi_-} \tag{2.15}$$

the Fokker-Planck equations (2.6a,b) can be rewritten

$$\frac{\partial\phi}{\partial t} = a(\sigma_x - \mathbb{1})\phi - c\,\sigma_z \frac{\partial\phi}{\partial x} \tag{2.16}$$

σ_x and σ_z being the Pauli matrices

$$\sigma_x = \begin{pmatrix} 0 & 1 \\ 1 & 0 \end{pmatrix} \qquad \sigma_z = \begin{pmatrix} 1 & 0 \\ 0 & -1 \end{pmatrix}. \tag{2.17}$$

Choosing $a = \frac{mc^2}{\hbar}$ and introducing a new spinor $\psi(t,x)$ by

$$\psi(t,x) = e^{+\frac{mc^2}{\hbar}t}\,\phi(x,t) \tag{2.18}$$

then ψ is a solution of the Euclidean Dirac equation in two space time dimensions, namely

$$\frac{\partial\psi}{\partial t} = -c\sigma_z \frac{\partial\psi}{\partial x}(x,t) + \frac{mc^2}{\hbar}\sigma_x\,\psi(t,x)\,. \tag{2.19}$$

This Euclidean equation is obtained from the relativistic Dirac equation by performing analytic continuations: $t \to it$ and $c \to -ic$.

This stochastic process describing the position at time t is given by the same expression as for Telegrapher's equation:

$$X(t) = x - \sigma c \int_0^t (-1)^{N_\tau}\, d\tau$$

where N_τ is the standard Poisson process with intensity $\frac{mc^2}{\hbar}$, $\hbar$ being the Planck constant divided by 2π

$$\mathrm{Prob}[N_\tau = k] = e^{-\frac{mc^2}{\hbar}t}\left(\frac{mc^2}{\hbar}\right)^k \frac{1}{k!}$$

and σ is the initial sign of the velocity.

III. The Relativistic Dirac Equation

It is well-known that the motion of the nonrelativistic quantum mechanical electron after analytical continuations to imaginary time can be described using the Wiener process or Brownian motion [3, 16]. The aim of this section is to show how Markov jump processes play an analogous role in the propagation of the relativistic electron, which is described by the Dirac equation.

The general strategy we use consists of choosing a representation of the relativistic Dirac equation such that we can identify this equation with a backward Kolmogorov equation for a Markov jump process. More precisely, the method consists of reducing the problem to an equation of the form

$$\frac{\partial f}{\partial t}(x,t) = (Af)(x,t) \tag{3.1}$$

where A is the generator of a pure jump process, namely of the form

$$(Af)(x,t) = a(x)\cdot\nabla f(x,t) + \int_{\mathbb{R}} d\mu(u)[f(x+c(x,u),t) - f(x,t)] \tag{3.2}$$

where a and c are smooth functions and μ is a (positive) bounded measure. It must be emphasized that the dimension of the space in which the infinitesimal generator A is defined can be bigger than the dimensional of the concrete problem we try to solve.

Classical results in probability theory ensure that the solution $f(x,t)$, $t \le T$, of such an integro-differential equation supplemented by the Cauchy (final) condition

$$f(x,T) = f_0(x) \tag{3.3}$$

is explicitly given by the following expectation

$$f(x,t) = E[f_0(X_t(\tau))] \tag{3.4}$$

where $X_t(s)$ for $t \le s \le T$ is the jump process solution of the stochastic differential equation

$$X_t(s) = x + \int_t^s a(X_t(\tau))d\tau + \int_t^s\int_{\mathbb{R}} c(X_t(\tau),u)\nu(d\tau,du). \tag{3.5}$$

In this last expression $\nu(d\tau,du)$ is a random Poisson measure such that

$$\nu([0,t),\ [a,b]) = N_t^{[a,b]} \tag{3.6}$$

$N_t^{[a,b]}$ being a Poisson process with intensity $\mu([a,b])$

$$E[N_t^{[a,b]}] = t\mu([a,b]) \quad . \tag{3.7}$$

We recall that if $[a,b] \cap [c,d] = \phi$ the corresponding processes N_t^I are independent. For more details we refer to [5].

Let us now consider the Dirac equation for a particle of mass m and charge q in the presence of a electromagnetic field in two space time dimensions.

Introducing the usual 2×2 Pauli matrices σ_x and σ_y

$$\sigma_x = \begin{pmatrix} 0 & 1 \\ 1 & 0 \end{pmatrix} \qquad \sigma_y = \begin{pmatrix} 0 & -i \\ i & 0 \end{pmatrix}$$

and denoting by c the velocity of the light and by $\hbar$ the Planck constant divided by 2π the Dirac equation can be written in the following form

$$\sigma_x(i\hbar\frac{\partial}{\partial t} - q\,A_o)\phi + i\sigma_y\, c(\frac{\hbar}{i} - q\,A_1)\sigma - mc^2\sigma = 0 \tag{3.8}$$

where $\phi(x,t)$ is a two-component spinor field and $A_o(x,t)$, $A_1(x,t)$ are real functions expressing the components of the external potential.

Associated with the two components of the spinor field ϕ let us introduce now a dichotomic variable $\sigma = 1$ which allows to rewrite (3.8) in the form

$$\begin{aligned} \partial_t\phi(x,\sigma,t) &= -\frac{imc^2}{\hbar}\phi(x,-\sigma,t) - c\,\sigma\frac{\partial}{\partial x}\phi(x,\sigma,t) \\ &\quad -i\frac{q}{\hbar}\Big[A_o(x,t) - \sigma c\,A_1(x,t)\Big]\phi(x,\sigma,t). \end{aligned} \tag{3.9}$$

Our goal is to find a probabilistic expression for the solution of (3.9) satisfying a given initial condition

$$\phi(x,\sigma,0) = \phi_o(x,\sigma) \ . \tag{3.10}$$

To make the connection between (3.9), (3.10) and the backward Kolmogorov equation (3.2), (3.3), we transform first, choosing a time $T > 0$, the initial condition (3.10) into a final condition and define for $t \le T$

$$\psi(x,\sigma,t) = \phi(x,\sigma,T-t) \tag{3.11}$$

$$\psi(x,\sigma,T) = \phi(x,\sigma,0) = \phi_o(x,\sigma) \ . \tag{3.12}$$

Finally let us introduce another supplementary real variable y by setting

$$\Phi(x,y,\ ,t) = e^{-\frac{mc^2}{\hbar}(T-t)}\, e^{-iy}\, \psi(x,\sigma,T-t)\ . \tag{3.13}$$

An easy computation shows that this function Φ satisfies now the following equation (with an obvious final condition)

$$\frac{\partial}{\partial t}\Phi(x,y,\sigma,t) + \frac{mc^2}{\hbar}\left[\phi(x,y+\frac{\pi}{2},-\sigma,t) - \phi(x,y,\sigma,t)\right]$$
$$- c\sigma\frac{\partial}{\partial x}\Phi(x,y,\sigma,t) + \frac{q}{\hbar}\left[A_o(x,T-t), -c\sigma A_1(x,T-t)\right]\frac{\partial}{\partial y}\phi(x,y,\sigma,t) = 0 \tag{3.14}$$

in which we recognize the canonical form of a backward Kolmogorov equation. We quote the following

Theorem: Let $X_t(s)$, $Y_t(s)$, $\Sigma_t(s)$ be the real valued stochastic processes defined by

$$X_t(s) = x - c\sigma\int_t^s (-1)^{N_\tau - N_t}\, d\tau$$

$$Y_t(s) = y + \frac{\pi}{2}(N_s - N_t)$$
$$+ \frac{q}{\hbar}\int_t^s \left[A_o(X_\tau(\tau),T-\tau) - c\tau(-1)^{N_\tau - N_t} A_1(X_t(\tau),T-\tau)\right] d\tau$$

$$\Sigma_t(s) = \sigma(-1)^{N_s - N_t} \qquad s \le t \le T$$

where N_t is the standard Poisson process with intensity $\frac{mc^2}{\hbar}$ and starting at 0 and A_o, A_1 are smooth real functions. Then the solution of (3.14) is given by

$$\Phi(x,y,\sigma,t) = E[\Phi(X_t(T),\, Y_t(T),\, \Sigma_t(T),T)]\ ,$$

E denoting the expectation with respect to the jump process (X,Y,Σ) with values in $\mathbb{R}\times\mathbb{R}\times\{-1, +1\}$.

□

Using (3.11) and (3.13) the above result implies the following corollary.

Corollary: The solution of the Dirac equation (3.7) admits the following probabilistic representation $(t=0)$

$$\phi(x,\sigma,T) = e^{\frac{mc^2}{\hbar}T}\, E\left[\phi_o\left(x - c\sigma\int_0^T (-1)^{N_\tau} d\tau,\ \sigma(-1)^{N_T}\right)\right.$$

$$\exp\left\{-i\frac{\pi}{2}N_T-\frac{iq}{\hbar}\int_0^T d\tau\,[A_o(X_o(\tau),T-\tau)-c\sigma(-1)^{N_\tau}A_1(X_o(\tau),T-\tau)]\right\}\Bigg]$$

The initial condition ϕ_o being a bounded continuous function with bounded first derivative. □

Remark 1: The above expression for ϕ is actually a generalized Feynman-Kac formula.

Remark 2: The typical paths of the process X_t are of the form

$$X_o(t)(\omega) = x - c\sigma(t-t_n) + c\sigma(t_n-t_{n-1}) + \dots$$

$$+ (-1)^n c\sigma(t_2-t_1) + (-1)^{n+1} c\sigma t_1 ,$$

the t_i being the times of jumps of ω up to time t. This agrees with the result obtained in [2] for the free equation.

Remark 3: For potentials A_o and A_1 which are Fourier transforms of bounded measures a similar result for the Dirac equation can be obtained in the momentum representation. For more details, see [22].

Remark 4: The method described in this section allows after obvious modifications to consider external potentials which are non-abelian.

Remark 5: The solution of the Euclidean Dirac equation (where $t \to it$) has no direct probabilistic representation. To obtain such a representation, the light velocity must also be analytically continued ($c \to ic$). See [10]. We will use this possibility in the next section.

IV. Nonrelativistic Limit of Euclidean Dirac Equation

In Section II, we have discussed a probabilistic representation for the solution of the free Euclidean two dimensional Dirac equation as well as the connection in a certain limit between the Telegrapher's equation (2.9) and the diffusion equation (2.11). It is a well-known fact that in the weakly nonrelativistic limit Dirac equation becomes in some sense the Pauli equation, for which a probabilistic interpretation for real time is not easy (see, however, [8]). For this reason we will consider in this section the Euclidean Dirac equation in which the light velocity is also taken imaginary. In this situation, the Dirac equation can be written after multiplyinn the spinor field by a factor $e^{-\frac{mc^2}{\hbar}t}$ as

$$\partial_t\psi(x,t,\sigma) = \frac{mc^2}{\hbar}[\psi(x,-\sigma)-\psi(x,\sigma)] - c\sigma\partial_x\psi(x,t,\sigma) + V(x,t,\sigma)\psi(x,t,\sigma) \qquad (4.1)$$

with

$$V(x,t,\sigma) = -\frac{q}{\hbar}[A_0(x,t) - ic\sigma A_1(x,t)] \quad .$$

If ψ_0 denotes the initial condition of (4.1) we obtain for the solution of (4.1) the following probabilistic representation

$$\psi_t(x,\sigma) \equiv \psi(x,t,\sigma) = E\left[\exp\left\{\int_0^t d\tau\, V(X_{x,\tau},\tau,(-1)^{N_\tau}\sigma)\right\}\psi_0(X_{x,t},(-1)^{N_t}\sigma)\right] \tag{4.2}$$

where N_t is the standard Poisson process with intensity $\frac{mc^2}{\hbar}$ and $X_{x,t}$ the process defined in terms of N_t by

$$X_{x,t} = x - c\sigma\int_0^t (-1)^{N_\tau}\, d\tau \quad . \tag{4.3}$$

We consider now the transition function $\Pi_t^c(x,x',\sigma,\sigma')$ associated with the process $(X_{x,t},(,1)^{N_t}\sigma)$ with values in $\mathbb{R}\times\{-1,+1\}$, which is solution of the following Chapman-Kolmogorov equation

$$\partial_t\,\Pi_t^c(x,x',\sigma,\sigma') = \frac{mc^2}{\hbar}\left[\Pi_t^c(x,x',-\sigma,\sigma') - \Pi_t^c(x,x',\sigma,\sigma')\right] - c\sigma\partial_x\,\Pi_t^c(x,x',\sigma,\sigma') \tag{4.4}$$

and satisfies the following initial condition:

$$\lim_{t\to 0}\Pi_t^c(x,x',\sigma,\sigma') = \delta(x-x')\delta_{\sigma\sigma'} \quad . \tag{4.5}$$

Denoting by E_p the quantity defined as

$$E_p = \frac{c}{\hbar}\,[m^2c^2 - p^2]^{1/2} \tag{4.6}$$

we obtain for Π_t^c the following explicit representation

$$\Pi_t^c(x,x',\sigma,\sigma') = \frac{1}{2\pi\hbar}\int_{\mathbb{R}} e^{i\frac{p}{\hbar}(x-x')}\, e^{-\frac{mc^2}{\hbar}t}\left[\left(\mathrm{ch}\,E_pt + \frac{icp\sigma}{\hbar E_p}\,\mathrm{sh}\,E_pt\right)\delta_{\sigma\sigma'} + \frac{mc}{\hbar E_p}\,\mathrm{sh}\,E_pt\,\delta_{\sigma,-\sigma'}\right]dp \quad . \tag{4.7}$$

If now $c\to+\infty$ the previous expression becomes

$$\lim_{c\to+\infty}\Pi_t^c(x,x',\sigma,\sigma') = \frac{1}{2}\,\frac{m}{\sqrt{2\pi\hbar t}}\,e^{-\frac{m}{2\hbar t}(x-x')^2} \tag{4.8}$$

where (4.8) involves the projection operator P

$$\frac{1}{2} = P_{\sigma,\sigma'} \qquad \forall\sigma,\sigma' = \pm 1 \quad . \tag{4.9}$$

Using (4.7) formula (4.2) can be written as

$$\psi_t(x,\sigma) = \sum_{n=0}^{+\infty} \int_0^t dt_{n-1} \cdots \int_0^{t_2} dt_1 \int_{\mathbb{R}} dx_n \cdots \int_{\mathbb{R}} dx_1 \sum_{\sigma_1,\ldots,\sigma_n}$$

$$\pi^c_{t-t_{n-1}}(x,x_n,\sigma,\sigma_n)\, V(x_n,t_{n-1},\sigma_n)\, \pi^c_{t_{n-1}-t_{n-2}}(x_n,x_{n-1},\sigma_n,\sigma_{n-1})$$

$$V(x_{n-1},t_{n-2},\sigma_{n-2}) \cdots\cdots \pi^c_{t_1}(x_2,x_1,\sigma_2,\sigma_1)\, \psi_0(x_1,\sigma_1) \quad . \tag{4.10}$$

Assuming that $V(x,t,\sigma)$ is integrable in x and t we obtain in the limit $c \to +\infty$

$$\lim_{c\to+\infty} \psi_t(x,\sigma) = \sum_{n=0}^{\infty} \left(-\frac{q}{\hbar}\right)^n \int_0^t dt_{n-1} \cdots \int_0^{t_2} dt_1 \int_{\mathbb{R}^n} dx_1 \ldots dx_n$$

$$P_{t-t_{n-1}}(x-x_n)\, A_0(x_n,t_{n-1})\, P_{t_{n-1}-t_{n-2}}(x_n-x_{n-1})\, A_0(x_{n-1},t_{n-2})$$

$$\cdots\cdot\; P_{t_1}(x_2-x_1)\, \frac{1}{2}\left[\psi_0(x_1, \;) + \psi_0(x_1,-\sigma)\right] \tag{4.11}$$

where $P_t(x) = \frac{m}{\sqrt{2\pi\hbar t}}\, e^{-\frac{m}{2\hbar t}x^2}$ is the transition probability of a Wiener process with intensity $\frac{\hbar}{m}$.

Formula (4.11) can be rewritten in a compact way using the projector P defined by $(P\psi)(x) = \frac{1}{2}\,[\psi(x,\sigma) + \psi(x,-\sigma)]$

$$\lim_{c\to+\infty} \psi_t(x,\sigma) = E_W\left[e^{-\frac{q}{\hbar}\int_0^t A_0(W_{x,\tau})d\tau}\, (P\psi)(W_{x,t})\right] \tag{4.12}$$

where the expectation in this Feynman-Kac formula is now taken with respect to the Wiener process starting at x and with intensity $\frac{\hbar}{m}$.

In the limit c going to infinity the mean number of jumps of the process N_τ in a given finite time interval increases to infinity, this increasing number of jumps acting as an average over σ. The result we obtain is a little bit more precise than the one obtained in [10] since it takes into account this averaging effect. However, it was already known on the basis of heuristic arguments.

References

[1] K. Symanzik, Euclidean quantum field theory in local quantum theory (Varenna 1968), R. Jost, ed., New York, Academic Press 1968

[2] R.P. Feyman, Space-time approach to nonrelativistic quantum mechanics, Rev. Mod. Phys. 20, 367-387 (1948)

[3] J. Glimm, A. Jaffe, Quantum Physics, New York, Springer Verlag (1981)

[4] R.P. Feyman, A.R. Hibbs, Quantum Mechanics and Path Integrals, MacGraw-Hill, New York (1965)

[5] G. Rosen, Formulation of classical and quantum dynamical theory, Academic Press, New York-London (1969)

[6] G.V. Riazanov, The Feyman Path Integral for the Dirac Equation, Soviet Phys. JETP 6 (33), 11o7-113 (1958)

[7] G.G. Papadopoulos, J.T. Devresse, Path Integral Solution of the Dirac Equation, Phys. Rev. D13. 2227-2234 (1975)

[8] A.O. Barut, N. Zanghi, Classical model of the Dirac electron, Phys. Rev. Lett., 52, 2oo9-2o12 (1984)

[9] M.A. Kayed, A. Inomata, Exact Path-Integral solution of the Dirac-Coulomb Problem, Phys. Rev. Lett. 53, 1o7-11o (1984)

[1o] T. Jacobson, L.S. Shulman, Quantum stochastics: the passage from a relativistic to a non-relativistic path integral, J. Phys. A17, 375-383 (1984)

[11] G.F. De Angelis, G. Jona-Lasinio, M. Sirugue, Probàbilistic solution of the Pauli type equation, J. Phys. A Math. Gen. 16, 2433-2444 (1983)

[12] G. Gaveau, J. Vauthier, Intégrales oscillantes stochastiques: l'équation de Pauli, J. Funct. Analysis 44, 388-4oo (1981)

[13] G.F. De Angelis, G. Jona-Lasinio, A stochastic description of spin 1/2 particles in a magnetic field, J. Phys. A15, 2o35-2o8o (1982)

[14] F. Guerra, R. Marra, Stochastic Mechanics of spin 1/2 Particles, Phys. Rev. D29, 1647 (1984)

[15] G.F. De Angelis, G. Jona-Lasinio, M. Serva, N. Zanghi, Stochastic description of the relativistic Dirac particle in two space time dimensions, to appear

[16] B. Gaveau, T. Jacobson, M. Kac, L.S. Shulman, Relativistic Extension of the Analogy between Quantum Mechanics and Brownian Motion, Phys. Rev. Lett. 53, 419-422 (1984)

[17] T. Jacobson, "Spinor chain path integral for the Dirac electron", Dissertation, University of Texas, Austin (1983)

[18] T. Ichinose, Path integral for the Dirac equation in two space time dimension, Proc. Jap. Acad. 58 A, 29o-293 (1982)

[19] T. Ichinose, Path integral formulation of the propagator for a two dimensional Dirac particle, Physica 124A, 419-426 (1984)

[2o] T. Ichinose, Path integral for Hyperbolic systems of the first order, Duke Math. Journal 51, 1-36 (1984)

[21] T. Ichinose, H. Tamara, Propagation of a Dirac particle. A path integral approach., Journal of Math. Phys. 25, 181o-1819 (1984)

[22] Ph. Blanchard, Ph. Combe, M. Sirugue, M, Sirugue-Collin, "Probabilistic expression of the solution of the Dirac equation in Fourier space", Proceedings of the Como Conference, Sept. 1985

[23] Ph. Blanchard, Ph. Combe, M. Sirugue, M. Sirugue-Collin, Stochastic Jump Processes associated with Dirac equation. Ascona Conference, 1985
Stochastic Processes in classical and quantum systems, Lectures Notes in Physics, 262, Springer 1986

[24] M. Kac, A stochastic model related to the telegrapher's equation, Rocky Mountain Journal of Math. 4 , 497-5o9 (1974)

A CONSTRUCTIVE CHARACTERIZATION OF RADON PROBABILITY MEASURES ON INFINITE DIMENSIONAL SPACES

E. BRÜNING

CENTRE DE PHYSIQUE THEORIQUE, CNRS, Luminy
Case 907, 13288 MARSEILLE, Cedex 09 and
UNIVERSITE DE PROVENCE, Dept. de Physique
Mathématique - Marseille - France

I. INTRODUCTION. NOTATIONS AND ASSUMPTIONS.

On one side a general characterization of Radon probability measures on various kinds of infinite dimensional vectorspaces is known since many years (see for instance [11] and the literature there). On the other side if more concrete properties are concerned our knowledge about such measures is still rather limited except in the case of Gaussian measures. But from various sides of applications (theory of generalized stochastic processes [7,8] and quantum field theory [9] one would like to know also more details about non-Gaussian measures on various infinite-dimensional spaces.

To be more concrete let us consider a real nuclear (LF)-space E (for example the Schwartz-space $\mathcal{D}(\mathbb{R}^n)$ and equip its topological dual E' with the weak * topology $\sigma=\sigma(E',E)$, denoted as E'_σ. The Fourier transform F maps each $\mu\in\mathcal{R}_1(E'_\sigma)$, the set of all Radon-probability measures on E'_σ, to a normalized positive definite function on E :

$$\hat{\mu}(x)=(F\mu)(x)=\int e^{i\langle\omega,x\rangle}\,d\mu(\omega) \qquad (1.1)$$

By well known inequalities for such functions [3]

$$|\hat{\mu}(x)-\hat{\mu}(y)|\leq C_0(1-\mathrm{Re}\,\hat{\mu}(x-y))^{1/2} \qquad (1.2)$$

one knows that $\hat{\mu}$ is continuous on E iff it is continuous at $x=0$.

If one knows that μ is "concentrated" on a certain class M of measurable subsets A of E'_σ, e.g. if for every $\delta>0$ there is $A\in M$ such that $\mu(A)\geq 1-\delta$, then one also knows some details about the continuity properties of $\hat{\mu}$: because then for all $x\in E$ (χ_A : characteristic function of A, $A^c=E'\setminus A$) we get

$$|\hat{\mu}(x) - \hat{\mu}(o)| \leq \int |e^{i<\omega,x>}-1|\chi_A(\omega)d\,\mu(\omega)+2\mu(A^c)$$

and thus for all $x\in \delta.A°$, $A°$ the absolute polar set of A in E,

$$|\hat{\mu}(x)-1| \leqslant \delta +2\mu(A^c) \leqslant 3\delta$$

as for all $\omega\in A$ and all $x\in \delta.A°$ $|e^{i<\omega,x>}-1|\leq|<\omega,x>|\leq\delta$.

Thus if $A°$ is a neighbourhood of zero in E then $\hat{\mu}: E\to\mathbb{C}$ is continuous.

If M is for instance a subclass of the class ε of all absolutely convex equicontinuous subsets of E' then $\hat{\mu}$ is continuous. A Radon measure on E'_σ is by definition concentrated on the (absolutely convex) compact subsets of E'_σ. Thus in the case of a (LF)-space the above argument shows $\hat{\mu}$ to be continuous (such a space is barreled and therefore its topology is identical to the Mackey-topology $\tau(E,E')$, the topology of uniform convergence on all absolutely convex compact subsets of E'_σ).
If we know that μ has at least a second moment T_2 such that

$$x \longmapsto p(x) = (T_2(x\otimes x))^{1/2} \tag{1.3}$$

is a continuous seminorm on E then we can estimate

$$|1-\hat{\mu}(x)| \leq p(x) \tag{1.4}$$

to get continuity of $\hat{\mu}$. Later in section 5 we will see explicitly on which subclass M of ε the Radon measure μ then is concentrated.

Conversely by a corollary of Minlos theorem, corollary 3 on p.237 of [11], we know for a nuclear space E that every normalized positive definite continuous function ϕ on E is the Fourier transform of a unique $\mu\in R_1(E'_\sigma)$ which is concentrated on some subclass M of ε. In the general case the relation between the concentration of $\mu\in R_1(E'_\sigma)$ and continuity of its Fourier transform $\hat{\mu}$ is established by the theorem of Sazonov-Badrikian-Schwartz (theorem 3 on p. 239 of [11]).

Thus by these general results the Fourier transform F is a bijection from $R_1(E'_\sigma)$ onto the set $C_{pd,1}(E)$ of continuous positive definite normalized functions on E, e.g. $\mu\in R_1(E'_\sigma)$ is characterized by $F\mu\in C_{pd,1}(E)$. But as the general form of $\phi\in C_{pd,1}(E)$ is not known (a recent source of information about positive definite functions is [3]) this characterization has proved till now to be very useful essentially only in the construction of Gaussian measures on E'_σ.

Furthermore, in most applications we know at best all moments or more realistically the first few moments but not the Fourier transform of the probability measure we are looking for. This leads to the problem to which extent a probability measure can be characterized in terms of its (first few) moments, e.g. to (truncated) moment problems over an infinite dimensional space.

To be more explicit note

$$\mathcal{R}_1(E'_\sigma) = \bigcup_{N=0}^{\infty} \mathcal{R}_1^N(E'_\sigma) \tag{1.5}$$

$\mathcal{R}_1^N(E'_\sigma)$ = set of all Radon-probability measures on E'_σ having (absolute) moments up to order $N\in\mathbb{N}$ or of every order for $N=+\infty$

and consider the <u>moment maps</u> $M_{(N)}$:

$$\mathcal{R}_1^N(E'_\sigma) \ni \mu \mapsto M^\mu_{(N)} = \{1, M_1^\mu \ldots, M_N^\mu\} \tag{1.6}$$

where M^μ_n is the linear map from the n-fold projective tensorproduct $E_n = \otimes^n_\pi E$ of E into $\mathbb{R}$ defined by linear extension of

$$M^\mu_n(x_1 \otimes \ldots \otimes x_n) = \int \langle\omega, x_1\rangle \ldots \langle\omega, x_n\rangle \, d\mu(\omega) \tag{1.7}$$

That is $M^\mu_{(N)}$ is a linear map

$$\mathcal{T}_N(E) = \bigoplus_{n=o}^{N} E_n \longrightarrow \mathbb{R} \qquad \text{defined by}$$

$$M^\mu_{(N)}(\underline{x}) = \int_{E'} \hat{\underline{x}}(\omega)\, d\mu(\omega) \;, \; \hat{\underline{x}}(\omega) = \sum_{n=o}^{N} \langle\omega^{\otimes n}, x_n\rangle \;, \; x_n \in E_n \tag{1.8}$$

If $N=+\infty$ we write $\mathcal{T}_\infty(E) = \mathcal{T}(E)$ and take the locally convex direct sum of the spaces E_n. $\mathcal{T}(E)$ is the tensor-algebra over E.

Therefore our problems mean that we are looking for a characterization of the range Ran $M_{(N)}$ of the moment map $M_{(N)}$ respectively of Ran $(M_{(N)} \restriction \mathcal{R}_1^L(E'_\sigma))$ for $N<L$.

A first important step is already known :

$$\text{Ran } M_{(N)} \subset \mathcal{T}_N(E)' = \text{the topological dual of } \mathcal{T}_N(E) \tag{1.9}$$

e.g. existence of the moments implies their continuity.
The proof is given by Dobrushin and Minlos [7,8].
Thus we consider the following problems : Given

$$T_{(N)} = \{1, T_1, \ldots, T_N\} \in \mathcal{T}_N(E)'$$

we ask for necessary and sufficient conditions on $T_{(N)}$ such that

a) $T_{(N)} \in$ Ran $M_{(N)}$, e.g. there is $\mu \in \mathcal{R}_1^N(E'_\sigma)$ such that $T_{(N)} = M^\mu_{(N)}$; (1.10)

and such measure is called a <u>representing measure</u> for $T_{(N)}$;

b) the representing measure is unique ;

c) $T_{(N)} \in \text{Ran}(M_{(N)} \restriction \mathcal{R}_1^L(E'_\sigma))$ for N L, e.g. there is a (unique) $\mu \in \mathcal{R}_1^L(E'_\sigma)$ such that (1.10) holds.

If we had "positive" answers to a)-c) we had a counter part to the Bochner-Minlos result in terms of moments. But in order to be useful in applications such a result should be supplemented by an answer to the question

d) what is the general form of functionals $T_{(N)}$ satisfying the necessary and sufficient conditions of a), b), respectively c) ?

Remarks :

ad a) For $N=+\infty$ an answer is known (Borchers and Yngvason, [4]).
Their proof does not work for $N \in \mathbb{N}$. Our method of proof treats the cases $N \in \mathbb{N}$ and $N=+\infty$ simultaneously and proceeds in its first part as Choquet [6] for the finite-dimensional moment problem for $N=+\infty$ and then uses instead of the Riesz' representation theorem the Bochner-Minlos theorem combined with some methods from the theory of ordered topological vector spaces.
Two kinds of problems arise :

i) As the underlying space is now not assumed to be locally compact various classical arguments needed in this context do not apply.

ii) In the case of truncated moment problems and for $N \in \mathbb{N}$ a difficulty arises which does not occur in the finite-dimensional truncated moment problems [1,3,12]. In the infinite dimensional case an additional continuity property of $T_{(N)}$ has to be required.

ad b) In 1972 Berezanskij [2] presented a set of sufficient conditions for the uniqueness of the measure for the case $N=+\infty$.
Whereas the classical approach to the finite dimensional moment problems usually proceeds via operator -and spectral theory to prove uniqueness by showing that the commuting operators involved are essentially self-adjoint on a suitable domain so that they have exactly one self-adjoint extension we treat this part quite elementary and directly by locating exactly in which part of the existence proof the possibility of non-uniqueness enters and control this step precisely.
The result is :
(i) necessary and sufficient conditions for uniqueness
(ii) convenient sufficient conditions for uniqueness.

The approach to uniqueness in the one dimensional moment problem for $N=+\infty$ via analytic functions [1,3,12] involves quantities which seem to have no meaning in the infinite-dimensional case.

ad c) For this very hard problem we only have some special results but which nevertheless are interesting as they lead to non-Gaussian measures.

ad d) To answer this question means to analyse the positivity condition of a) for continuous functionals on $\mathcal{T}_N(E)$. A detailed answer to this is not only important for the construction of measures in terms of their moments but also would allow to prove various " correlation-inequalities".

In order to answer question d), using the results of Minlos, Sazonov, and Schwartz mentioned above, we will show that and in which way the second moment (more precisely the covariance $T_2 - T_1 \otimes T_1$ if T_1 is the mean of μ) determines the concentration of μ. The result is a kind of parametrization of μ which shows this concentration quite explicitly. Furthermore a detailed relation between the higher moments and the "type" of the measure becomes apparent. Thus we arrive at a set of necessary (and sufficient) conditions on the form of $T_{(N)}$.

Canonically associated with such a measure $\mu \in \mathcal{R}_1^N(E'_\sigma)$ there is an equivalence class of generalized stochastic processes $\phi \in \mathcal{L}(E, \mathrm{Meas}(\Omega, m, \mathbb{R}))$ ((Ω, m): a probability space ; $\mathrm{Meas}(\Omega, m; \mathbb{R})$: the space of equivalence classes of real measurable functions on (Ω, m)) such that the moments of these processes equal the given moments :

$$\int \phi_{x_1} \cdots \phi_{x_n} \, dm = T_n(x_1 \otimes \ldots \otimes x_n) \quad ; \quad x_i \in E$$

In applications one is interested in such processes if they appear as solutions of a differential equation. This then amounts to the fact that the moments of higher order appear as functions of those of lower order. In section 6 we give some additional comments on this problem.
The details of the proofs of the statements of section 2-4 are given in part I of [5] For section 5 we refer to part II.

II. Positive linear extensions

For (1.10) to hold it is obviously necessary that

$$0 \leq T_{(N)}(\underline{x}) \quad \text{for all } \underline{x} \in \mathcal{P}_N \tag{2.1}$$

where $\mathcal{P}_N = \{ \underline{x} \in \mathcal{T}_N(E) \mid \hat{\underline{x}}(\omega) \geq 0 \text{ for all } \omega \in E' \}$ (2.2)

A functional satisfying (2.1) is called positive.

For a positive functional $T_{(N)}$ we have the following equation :

$$T_{(N)}(\underline{x}) = T_{(N)}(\underline{y}) \quad \text{whenever } \underline{x}, \underline{y} \in \mathcal{T}_N(E) \text{ satisfy } \hat{\underline{x}} = \hat{\underline{y}} . \tag{2.3}$$

Now $\hat{\underline{x}}$ is a polynomial function $E'_\sigma \longrightarrow \mathbb{R}$. Therefore by (2.3) we can define a linear functional $\hat{T}_{(N)}$ on the polynomial functions on E'_σ by

$$\hat{T}_{(N)}(\hat{\underline{x}}) = T_{(N)}(\underline{x}) \quad \text{where } \underline{x} \in \mathcal{T}_N(E) \text{ is any representative of } \hat{\underline{x}} . \tag{2.4}$$

By (2.1) this functional $\hat{T}_{(N)}$ is positive on positive polynomial functions. Thus we are in the frame of the ordered real vector space of continuous functions on E'_σ
In order to proceed we introduce for $N \geq 2$

$$
\begin{aligned}
\chi &= C(E'_\sigma, \mathbb{R}) , \chi_+ = \{ f \in \chi \mid f(\omega) \geq 0 \quad \text{for all } \omega \in E' \} \\
V_N &= V_N^+ - V_N^+ , \quad V_N^+ = V_N \cap \chi_+ = \{ \hat{\underline{x}} \mid \underline{x} \in P_N \}
\end{aligned} \tag{2.5}
$$

and then

$$
\begin{aligned}
\hat{V}_N &= [V_N]_{\chi_+} = (V_N + \chi_+) \cap (V_N - \chi_+) = \{ f \in \chi \mid \exists\, \underline{x} \in P_N : |f| \leq \hat{\underline{x}} \} \\
\hat{V}_N^+ &= \hat{V}_N \cap \chi_+ = \{ f \in \chi_+ \mid \exists\, \underline{x} \in P_N : 0 \leq f \leq \hat{\underline{x}} \} .
\end{aligned} \tag{2.6}
$$

Thus $\hat{V}_N$ is the vector space of real continuous functions on E'_σ which are bounded by some polynomial of degree $\leq N$. In particular $\hat{V}_N$ contains all bounded continuous real functions on E'_σ for any $N \geq 0$.

A slight extension of known results on positive linear extensions [6] yields

Theorem 2.1

Suppose $T_{(N)} = \{ 1, T_1 \ldots, T_N \} \in T_N(E)'$ is positive. Then

a) The set of positive linear extensions of $T_{(N)}$,

$$Ex_p(T_{(N)}) = \{ F \in L^+(\hat{V}_N, \mathbb{R}) \mid F(\hat{\underline{x}}) = T_{(N)}(\underline{x}) \quad \text{for all } \underline{x} \in T_N(E) \}$$

is a non-empty weak $*$-compact convex subset of

$$\hat{V}_N^* = L(\hat{V}_N, \mathbb{R}) = \{ F : \hat{V}_N \longrightarrow \mathbb{R} \mid F \ \text{linear} \} .$$

b) $Ex_p(T_{(N)})$ contains exactly one element $\Leftrightarrow$ $1^+ = 1^-$ (2.7)

where $1^\pm = 1^\pm(T_{(N)})$ are defined by

$$
\begin{aligned}
1^+(f) &= \inf \{ T_{(N)}(\underline{y}) \mid \underline{y} \in T_N(E) , \hat{\underline{y}} \geq f \} \\
1^-(f) &= \sup \{ T_{(N)}(\underline{x}) \mid \underline{x} \in T_N(E) , \hat{\underline{x}} \leq f \}
\end{aligned} \tag{2.8}
$$

c) If U is a neighbourhood of zero in $T_N(E)$ and if $T_{(N)}$ is bounded on U by one then every positive linear extension F of $T_{(N)}$ is also bounded by one on

$$
\begin{aligned}
[\tilde{U}] &= (\tilde{U} + \chi_+) \cap (\tilde{U} - \chi_+) \quad \text{where} \\
\tilde{U} &= \{ \hat{\underline{x}} - \hat{\underline{y}} \mid \underline{x} , \underline{y} \in U \cap P_N \}
\end{aligned} \tag{2.9}
$$

This theorem answers the main questions about positive linear extensions : Part a) assures existence, part b) controls uniqueness, and part c) determines continuity of positive linear extensions.

III. Solution of the (truncated) moment problems

If for $N \geq 2$ a functional $0 \leq T_{(N)} = \{1, T_1, \dots, T_N\} \in \mathcal{T}_N(E)'$ is given we know by theorem 2.1 that the set of positive linear extensions of $T_{(N)}$ is not empty. For $F \in \mathrm{Ex}_p(T_{(N)})$ define a function $\phi_F : E \longrightarrow \mathbb{C}$ by

$$\phi_F(x) = F(C_x) + i\, F(S_x) \qquad \text{for } x \in E \tag{3.1}$$

where $C_x, S_x \in \hat{V}_N$ are given by

$$C_x(\omega) = \cos \langle \omega, x \rangle \quad , \; S_x(\omega) = \sin \langle \omega, x \rangle \qquad \text{for all} \quad \omega \in E'. \tag{3.2}$$

If $V_N^c = \hat{V}_N + i\,\hat{V}_N$ denotes the complexification of $\hat{V}_N$ we can write $\phi_F(x) = F^c(e_x)$ with the $\mathbb{C}$-linear functional F^c associated with F and $e_x \in V_N^c$, $e_x(\omega) = e^{i\langle \omega, x \rangle}$.

Proposition 3.1

If $0 \leq T_{(N)} \in \mathcal{T}_N(E)'$, $N \geq 2$, is given define for $F \in \mathrm{Exp}(T_{(N)})$ ϕ_F by eq. (3.1). Then

a) $\phi_F \in C_{pd,1}(E)$ and for all $x, y \in E$

$$|\phi_F(x) - \phi_F(y)| \leq C_0 (T_2((x-y)^{\otimes 2}))^{1/2} = C_0 p(x-y) \tag{3.3}$$

b) There is exactly one $\mu_F \in \mathcal{R}_1(E'_\sigma)$ such that for all $x \in E$

$$F^c(e_x) = \int_{E'} e_x(\omega)\, d\mu_F(\omega) = (F\,\mu_F)(x) = \phi_F(x) \tag{3.4}$$

Proof : The main part of the argument has already been given in section 1 (corollary 3 of Minlos' theorem). Furthermore we use

$$1\text{-Re}\, \phi_F(x) = F(1 - C_x) \leq F(\tfrac{1}{2}\hat{x}^2) = \tfrac{1}{2} T_2(x \otimes x) \ .$$

In order to decide whether the measure $\mu_F \in \mathcal{R}_1(E'_\sigma)$ given by proposition 3.2 has moments up to order N and to answer the question "what are these moments ?" we use the following :

Lemma 3.2

If $G : V_N^c \rightarrow \mathbb{C}$ is a positive linear functional for $N = 2M$, $M \in \mathbb{N}$, and if μ is a Radon probability measure on E'_σ such that for all $x \in E$ $\hat{\mu}(x) = G(e_x)$ then

a) $\mu \in \mathcal{R}_1^N(E'_\sigma)$, e.g. μ has moments up to order N

b) $\int_{E'} \hat{x}^n(\omega)\, d\mu(\omega) = G(\hat{x}^n)$ for all $x \in E$ and $n \leq N-1$

c) $\int_{E'} \hat{x}^N(\omega)\, d\mu(\omega) \leq G(\hat{x}^N)$.

The proof is done by induction on N, using Taylors formula for C_x and S_x and many details of it concerning positivity together with Fatou's lemma and Lebesgue's dominated convergence theorem.

If lemma 3.2 is applied to an extension F of $T_{(N)}$ and the measure μ_F determined by proposition 3.1 we see that the highest moment of this measure is not, for $N \in \mathbb{N}$, automatically equal to T_N. In order to assure this equality an extra continuity condition for the extension is needed. By the proof of lemma 3.2 one realizes that indeed a rather weak continuity condition, "continuity along a special path in $\hat{V}_N$", suffices. For fixed $x \in E$ the path in question is defined by $t \longrightarrow r_m(t\hat{x})$ where for all $\omega \in E'_\sigma$

$$r_m(t\hat{x})(\omega) = (m!)^{-1} \int_0^1 (1-s)^m . s\ (s\langle\omega, tx\rangle)^{-1} \sin(s\langle\omega, tx\rangle)\, ds \tag{3.5}$$

Theorem 3.3

For $N = 2M, M \in \mathbb{N}$, suppose $T_{(N)} = \{1, T_1, \ldots, T_N\} \in \mathcal{T}_N(E)'$ to be positive.

a) For every $F \in Ex_p(T_{(N)})$ there is exactly one $\mu_F \in \mathcal{R}_1^N(E'_\sigma)$ such that

(i) $\int \underline{\hat{x}}(\omega)\, d\mu_F(\omega) = T_{(N)}(\underline{x})$ for all $\underline{x} \in \mathcal{T}_{N-1}(E)$

(ii) $\int \hat{x}^N(\omega)\, d\mu_F(\omega) \leq T_N(x^{\otimes N})$ for all $x \in E$.

b) There is $\mu \in \mathcal{R}_1^N(E'_\sigma)$ such that for all $\underline{x} \in \mathcal{T}_N(E)$

$$\int \underline{\hat{x}}(\omega)\, d\mu\ (\omega) = T_{(N)}(\underline{x}) \tag{3.6}$$

iff there is $F \in Ex_p(T_{(N)})$ such that

$$\lim_{t \to 0} F(\hat{x}^N r_{N-2}(t\hat{x})) = F(\hat{x}^N)\, r_{N-2}(0). \tag{3.7}$$

c) If $T = \{1, T_1, T_2, \ldots\} \in \mathcal{T}(E)'$ is positive then for all $F \in Ex_p(T_{(N)})$ there is exactly one $\mu_F \in \mathcal{R}_1^N(E'_\sigma)$ such that $T(\underline{x}) = \int \underline{\hat{x}}(\omega)\, d\mu_F(\omega)$.

Remark : There are at least two ways to realize the continuity condition (3.7) :

(i) $T_{(N)}$ has an extension $F \in L^+(\hat{V}_{N+2}, \mathbb{R})$.

(ii) To apply part c) of theorem 2.1.

We conclude this section with a comment on question c) in section 1. If $N < L$ then $\mathcal{R}_1^L(E'_\sigma) \subset \mathcal{R}_1^N(E'_\sigma)$ and thus for $T_{(N)} \in \mathcal{T}_N(E)'$ to be in the range of $M_{(N)} \upharpoonright \mathcal{R}_1^L(E'_\sigma)$ it is necessary an sufficient that

(i) $T_{(N)}$ has an extension $T_{(L)}$ to $\mathcal{T}_L(E)$ which is positive there ;

(ii) this extension $T_{(L)}$ satisfies the necessary and sufficient conditions of theorem 3.3 to belong to Ran $M_{(L)}$.

The main difficulty is (i). This problem of positive linear extension is much harder than that of section 2 and no general solution is known because the space to which we

want to extend is not the order convex hull of the space where the functional is given as it was the case in section 2. In part I of [5] an example of this extension is treated for the case $L = +\infty$. There this extension is achieved by reducing it to a "monotone" continuous linear extension and applying recent results on this kind of extensions.
(A functional $T_{(2M)} \in \mathcal{T}_{2M}(E)'$ is called "monotone" iff $T(\underline{x}^* \cdot \underline{x}) \geq 0$ for all $\underline{x} \in \mathcal{T}_M(E)$).

IV. Uniqueness

According to proposition 3.1 and theorem 3.3 $T_{(N)}$ has exactly one representing measure if and only if all the functions ϕ_F, $F \in Ex_p(T_{(N)})$, defined by eq. (3.1), are equal.

Another way of expressing this is that all positive linear extensions F of $T_{(N)}$ agree on the subspace

$$H = \mathrm{Lin}\{ C_x, S_x \mid x \in E \} \subset \hat{V}_N \tag{4.1}$$

By part b) of theorem 2.1 this is equivalent to

$$1^+(T_{(N)}) \restriction H = 1^-(T_{(N)}) \restriction H \tag{4.2}$$

A first reduction of this condition is given by the following

Lemma 4.1
Suppose $T_{(N)} \in \mathcal{T}_N(E)'$ is positive and $N = 2M$, $M \in \mathbb{N}$, or $N = +\infty$.
Then equation (4.2) holds iff

$$\bigwedge_{x \in E} \; \bigwedge_{\gamma \in \{0,-\frac{\pi}{2}\}} \; \bigwedge_{\varepsilon > 0} \; \bigvee_{\underline{x},\underline{y} \in \mathcal{T}_N(E)} \quad \begin{array}{l} \text{i) } \hat{\underline{y}} \leq C_{x+\gamma} \leq \hat{\underline{x}} \\ \text{ii) } T_{(N)}(\underline{x} - \underline{y}) \leq \varepsilon \; . \end{array} \tag{4.3}$$

The proof of this lemma uses some simple calculations in ordered vector spaces in order to realize (4.2) in terms of inequalities.

In order to get more convenient sufficient conditions for uniqueness we choose in (4.3) i) special elements $\underline{y},\underline{x}$ and realize condition ii) of (4.3) as a condition on $T_{(N)}$.

Given $x \in E$, $\gamma \in \{0, -\pi/2\}$ introduce

$$\begin{aligned} \underline{y} = \underline{y}(x,\gamma,m) &= \sum_{j=0}^{2m-1} (-1)^j \, [(2j)!]^{-1} \, (x + \gamma 1)^{\times 2j} \\ \underline{x} = \underline{x}(x,\gamma,m) &= \sum_{j=0}^{2m} (-1)^j \, [(2j)!]^{-1} \, (x + \gamma 1)^{\times 2j} \end{aligned} \tag{4.4}$$

then $\underline{x} - \underline{y} = [(4m)!]^{-1} \, (x + \gamma 1)^{\times 4m}$ and for all $\omega \in E'_\sigma$

$$\hat{y}(\omega) \leqslant \hat{y}(\omega) + (-1)^{2m} [\hat{x}(\omega)+\gamma]^{4m} r_{4m-2}(\hat{x}(\omega)+\gamma) = C_{x+\gamma}(\omega) =$$

$$= \hat{x}(\omega) - (-1)^{2m} [\hat{x}(\omega)+\gamma]^{4m+2} r_{4m}(\hat{x}(\omega)+\gamma) \leqslant \hat{x}(\omega).$$

Thus for this choice of $\underline{x}$, $\underline{y}$ we get :

$$T_{(N)}(\underline{x}-\underline{y}) = ((4m)!)^{-1} \sum_{k=0}^{4m} \binom{4m}{k} \gamma^{4m-k} T_k(x^{\otimes k})$$

and therefore

<u>Theorem 4.2</u>

Suppose that $T_{(N)} \in \mathcal{T}_N(E)'$ is positive, $N = 2M$, $M \in \mathbb{N}$, or $N = +\infty$

a) There is at most one representing measure for $T_{(N)}$ iff condition (4.3) holds.

b) If the following condition (4.5) holds

$$\bigwedge_{x\in E} \bigwedge_{\gamma\in\{0,-\pi/2\}} \bigwedge_{\varepsilon>0} \bigvee_{\substack{m\in\mathbb{N}\\ 4m\leqslant N}} 0 \leqslant ((4m)!)^{-1} \sum_{k=0}^{4m} \binom{4m}{k} \gamma^{4m-k} T_k(x^{\otimes k}) \leq \varepsilon \tag{4.5}$$

then there is at most one representing measure for $T_{(N)}$.

c) If in the case $N = +\infty$ condition (4.5) holds (without '$4m \leqslant N$') then $T = \{1, T_1, T_2, ..\}$ has exactly one representing measure $\mu \in \mathcal{R}_1^\infty(E'_\sigma)$

<u>Corollary 4.3</u>

If $T = \{1, T_1, T_2, ...\} \in \mathcal{T}(E)'$ is positive and if for all $x \in E$

$$\lim_{m\to\infty} ((4m)!)^{-1} T_{4m}(x^{\otimes 4m}) = 0 \tag{4.6}$$

then there is exactly one $\mu \in \mathcal{R}_1^\infty(E'_\sigma)$ such that $T = M^\mu$.

Proof : In order to show that (4.6) implies (4.5) use the estimate

$$|T_{jm+k}(x^{\otimes(jm+k)})| \leqslant T_{4m}(x^{\otimes 4m})^{j/4} \cdot T_{4k}(x^{\otimes 4k})^{1/4} \tag{4.7}$$

for $j = 0,1,2,3,$ $k = 1,...,m$ and $x \in E$ which follows from positivity.

<u>Remarks</u> :

a) For $N \in \mathbb{N}$ conditions (4.3) and (4.5) seem to have only a rate chance to be satisfied. This indicates that it is hard for the truncated moment problem to have a unique solution (and this is not surprising).

b) According to proposition 3.1 the functions $\phi_F, F \in Ex_p(T_{(N)})$, are continuous with respect to the seminorm p on E. Thus these functions are uniquely determined by their restrictions to a p-dense subspace $E_\circ$ of E. Therefore we might relax conditions (4.3) - (4.6) by replacing E by $E_\circ$.

Example :

If $T=\{1,T_1,T_2,\dots\} \in \mathcal{T}(E)'$ is positive and if there are constants α and C, $0\leq \alpha <1$, $0\leq C$, and a (continuous) seminorm q on E such that for all $x\in E$ and all $n\in \mathbb{N}$

$$|T_n(x^{\otimes n})| \leq C^n (n!)^\alpha \, q(x)^n \tag{4.8}$$

then there is exactly one $\mu\in \mathcal{R}_1^\infty(E'_\sigma)$ such that $T=M^\mu$. Clearly (4.8) implies (4.6). In the Euclidean approach to quantum field theory estimates of the type (4.8) are extensively used for the sequence of Schwinger functions [9]. Thus in those cases where this sequence is positive there is exactly one Radon probability measure on E'_σ ($E = \mathcal{D}(\mathbb{R}^d)$ for instance) such that the moments of this measure are the Schwinger functions.

If a functional $T_{(N)}$ satisfies the hypotheses of theorem 3.3 then the set $\mathcal{M}(T_{(N)})$ of representing measures for $T_{(N)}$ is not empty. The results of this section indicate that quite often, as expected, $\mathcal{M}(T_{(N)})$ will contain many elements. Thus further investigations are necessary which impose additional constraints in order to determine the measure uniquely by $T_{(N)}$ and these constraints.

V. Concentration of the measure and form of its moments

In this section we suppose that $T_{(N)}$ has at least one representing measure and then ask : what can be said about the concentration of any representing measure $\mu\in\mathcal{M}(T_{(N)})$ in terms of properties of $T_{(N)}$? This then will also answer question d).
We begin by analysing the positivity condition for a continuous linear functional $T_2\in E'_2$ when E is a real nuclear barreled space.
As a preparation we note a consequence of barreledness and nuclearity [14]:

Lemma 5.1
Let E be a real barreled space and $T_2 : E\otimes_\pi E \longrightarrow \mathbb{R}$ a continuous linear functional. Then if $T_2(x\otimes x) \geq 0$ for all $x\in E$:

a) $p(x):=(T_2(x\otimes x))^{1/2}$ defines a continuous Hilbertian seminorm on E.

b) there is a continuous Hilbertian seminorm $q \geq p$ on E such that the canonical map $I_{pq}:E_q \longrightarrow E_p$ between the associated Hilbert spaces is an injective Hilbert-Schmidt map.

Using the wellknown representation for Hilbert-Schmidt-maps and a slight generalization of the Gram-Schmidt orthogonalization procedure we get

Lemma 5.2
Under the assumptions of Lemma 5.1 there exist
(i) an equicontinuous sequence $\{t_j\}$ of linearly independent element $t_j\in E'$
(ii) a sequence $\{e_j\}$ of linearly independent elements $e_j\in E$
(iii) a sequence $\{\lambda_j\}$ of positive numbers $\lambda_j > 0$ such that for all $x,y \in E$

a) $T_2(x \otimes y) = \sum_{j=1}^{\infty} \lambda_j^2 t_j(x) t_j(y)$ (5.1)

b) $q(x)^2 = \sum_{j=1}^{\infty} t_j(x)^2$

c) $t_j(e_i) = \delta_{ij}$

d) $\sum_{j=1}^{\infty} \lambda_j^2 < \infty$

Associated with the seminorm p there is a neighbourhood U of zero in E, $U=\{x\in E \mid p(x)\leq 1\}$, and its absolute polar U° in E', $U^\circ = \{\omega\in E' \mid |<\omega, x>| \leq 1$ for all $x\in U\}$. The subspace of E' generated by U°, e.g. Lin U°, is a Hilbert space when equipped with the dual norm p',

$$p'(\omega) = \sup \{ |<\omega, x>| \mid p(x) \leq 1 \}$$

This subspace is denoted by $E'_{p'}$. Similarly we get a Hilbert space $E'_{q'} \subset E'$ (See for instance section 0.11 of [14]).

Lemma 5.3

Under the hypotheses and in the notation of lemma 5.1 and 5.2 we have

a) $E'_{q'} = \{\omega\in E' \mid \sum_{j=1}^{\infty} <\omega, e_j>^2 < \infty\}$, $q'(\omega) = (\sum_{j=1}^{\infty} \omega(e_j)^2)^{1/2}$ (5.2)

b) $E'_{p'}$ is a subspace of $E'_{q'}$ and the identical injection $I'_{q'p'}: E'_{p'} \longrightarrow E'_{q'}$ is a Hilbert-Schmidt map.

c) $f_{\underline{t}} : l^2(\mathbb{R}) \longrightarrow E'_{q'}$, defined by $f_{\underline{t}}(\underline{\sigma}) = \sum_{j=1}^{\infty} \sigma_j t_j$ (5.3)

is a unitary map from the Hilbert space of square summable sequences of real numbers onto $E'_{q'}$ with inverse

$$f_{\underline{t}}^{-1}(\omega) = (\omega(e_j)_{j\in\mathbb{N}}) \quad (5.4)$$

Now suppose that for $N \geq 2$ $0 \leq T_{(N)} \in \mathcal{T}_N(E)'$ satisfies the hypotheses of theorem 3.3 such that the set $\mathcal{M}(T_{(N)})$ of representing measures for $T_{(N)}$ is not empty. For convenience assume $T_1 = 0$ and then apply Lemma 5.1 - 5.3 to T_2. As indicated in the introduction the Fourier transform $\hat{\mu}$ of any $\mu\in\mathcal{M}(T_{(N)})$ can be estimated according to (1.4) or (3.3). Thus using first theorem 1, p.193 of [11] and then Sazonov's theorem (for instance theorem 2, p. 215 of [11]) in connection with Lemma 5.3 we get

Theorem 5.4

Suppose $0 \leq T_{(N)} \in \mathcal{T}_N(E)'$ satisfies the hypotheses of theorem 3.3. Assume furthermore $N \geq 2$ and $T_1 = 0$. Associate with T_2 the seminorms p and q according to lemma 5.1 and the Hilbert spaces $E'_{p'}$ and $E'_{q'}$ according to lemma 5.3.
Then every representing measure μ for $T_{(N)}$ is actually a Radon probability measure

on E'_q, having moments of order N :

$$M(T_{(N)}) \subseteq R_1^N (E'_\sigma) \tag{5.5}$$

Corollary 5.5

Under the hypotheses of theorem 5.4 every $\mu \in M(T_{(N)})$ is the image of a Radon probability measure ν on $l^2(\mathbb{R})$ under the map $f_{\underline{t}}$ of lemma 5.3

$$\mu = f_{\underline{t}}(\nu) \tag{5.6}$$

$\nu \in R_1^N(l^2(\mathbb{R}))$ has the following properties :

a) For $n \leq N$ and $j_1 \dots, j_n \in \mathbb{N}$

$$\int |\sigma_{j_1} \dots \sigma_{j_n}| \; d\nu(\underline{\sigma}) < \infty \;, \quad \int \| \underline{\sigma} \|^2 \, d\nu(\underline{\sigma}) = \sum_{j=1}^{\infty} \lambda_j^2 < \infty \tag{5.7}$$

b) $$\int \sigma_{j_1} \dots \sigma_{j_n} \; d\nu(\underline{\sigma}) = T_n(e_{j_1} \otimes \dots \otimes e_{j_n}) \tag{5.8}$$

While lemma 5.2 provides a complete analysis of the second moment theorem 5.4 and corollary 5.5 present the implications of this analysis for the underlying measure. Now we want to discuss what one can say about the general form of the higher moments. Thus we assume $N \geq 2$.

To this end let us denote the real numbers defined by equation (5.8) by

$$C_{j_1 \dots j_n} \;, \quad j_i \in N \;, \; n \leqslant N \tag{5.9}$$

The representation (5.8) of these numbers as moments of ν (or the positivity condition for $T_{(N)}$) immediately implies the following positivity condition :

Whenever $a^\circ, a^n_{j_1 \dots j_n}$ is a finite set of real numbers such that for all $\sigma_i \in \mathbb{R}$

$$a^\circ + \sum_n \sum_{j_1 \dots j_n} a^n_{j_1 \dots j_n} \cdot \sigma_{j_1} \dots \sigma_{j_n} \geqslant 0 \qquad \text{(a)}$$

then (5.10)

$$a^\circ + \sum_n \sum_{j_1 \dots j_n} a^n_{j_1 \dots j_n} \cdot C_{j_1 \dots j_n} \geqslant 0 \qquad \text{(b)}$$

This positivity condition just says that the sequence of coefficient-tensors

$$C^{(n)} = ((C_{j_1 \dots j_n}) \;, \quad j_i \in \mathbb{N})$$

has to belong to the bidual cone of the cone of sequences generated by $(1, \sigma^{\otimes n}, n \in \mathbb{N})$ where the pairing is given by (5.10).

The sequence $\{e_j\}$ of lemma 5.2 generates a subspace F of E. Then a simple calculation shows (assumptions of theorem 5.4)

Lemma 5.6

a) $T_n \restriction \otimes^n F = \sum_{j_1 \dots j_n} C_{j_1 \dots j_n} \; t_{j_1} \otimes \dots \otimes t_{j_n} \restriction \otimes^n F \;, \; n \leqslant N$ (5.11)

b) If $q(x_i)=0$ for one $i \leqslant n$, $n \leqslant N$, then for all $x_j \in E, j \neq i$,
$T_n(x_1 \otimes \dots \otimes x_i \otimes \dots \otimes x_n) = 0$

The second statement of this lemma admits a stronger version. By symmetry and positivity the following estimate holds for $2n \leqslant N+2$:

$$T_n(x_1 \otimes \dots \otimes x_i \otimes \dots \otimes x_n)^2 \leqslant T_2(x_i \otimes x_i)\; T_{2n-2}(y^* \otimes y) \qquad (5.12)$$

with $y = x_1 \otimes \dots \hat{x}_i \otimes \dots \otimes x_n$, x_i omitted. Therefore:

If $p(x_i) = 0$ for one $i \leqslant n$, $2n \leqslant N+2$, then for all $x_j \in E$, $j \neq i$
$T_n(x_1 \otimes \dots \otimes x_i \otimes \dots \otimes x_n) = 0$ (5.13)

Clearly one would like to extend equality (5.11) to all of E_n.
For the moments of order n, $2n \leqslant N+2$, inequality (5.12) provides "enough continuity" to do so. For the other moments the fact that μ has moments up to order N represents a continuity condition which allows to prove a somewhat weaker statement.
First we define a sequence of continuous projections $\pi_m : E \longrightarrow F$ by

$$\pi_m(x) = \sum_{j=1}^{m} t_j(x) e_j \quad \text{for all } x \in E, \; m \in \mathbb{N}. \qquad (5.14)$$

Lemma 5.2 implies $\lim_{m \to \infty} q(x - \pi_m(x)) = 0$ (5.15)

for all $x \in E$. For an absolutely convex compact subset k of $E'_{q'}$ denote $\mu_k = \chi_k \cdot \mu$ and then

$$T_n^k = M_n^{\mu_k} \;, \quad n \leq N \,. \qquad (5.16)$$

The following proposition shows now that and in which way the moments T_n are determined by their restriction to $\otimes^n F$.

Proposition 5.7

a) For $2n \leqslant N+2$ and all $x_i \in E$

$$T_n(x_1 \otimes \dots \otimes x_n) = \lim_{m_1 \to \infty} (\dots (\lim_{m_n \to \infty} T_n(\pi_{m_1}(x_1) \otimes \dots \otimes \pi_{m_n}(x_n))) \dots) \qquad (5.17)$$

b) For all $n \leqslant N$, all $x_n \in E_n$, and all compact subsets k of $E'_{q'}$

$$T_n^k(x_n) = \lim_{m \to \infty} T_n^k \circ \pi_m^{\otimes n}(x_n) \qquad (5.18)$$

and $T_n(x_n) = \lim_{k \nearrow E'_{q'}} T_n^k(x_n)$

c) If $N = +\infty$ then (5.17) holds for all $n \in \mathbb{N}$.

Remarks

a) Though corollary 5.5 only is a reformulation of theorem 5.4 we think it to be interesting for the following reasons :

i) It shows clearly that and in which way the concentration of the measure is determined by its second moment (more precisely by its covariance).

ii) It separates two aspects of the problem :

 α) the aspect of concentration, in eq. (5.6) expressed by the map $f_{\underline{t}}$ given by the second moment ;

 β) the "type" of the measure which, given T_2, is determined by the higher moments expressed in eq. (5.6) by ν .

iii) It provides a reduction of the problem of determining $\mu \in R_1^N(E'_\sigma)$ such that $T_{(N)} = M_{(N)}^{\mu}$ to that of determining $\nu \in R_1^N(l^2(\mathbb{R}))$ satisfying (5.7) and (5.8).

b) Corollary 5.5 can be used to construct many Radon probability measures $\mu \in R_1^N(E'_\sigma)$ with prescribed concentrations and at the same time with some knowledge about the form of its moments :

i) Choose an equicontinuous sequence $\{t_j\}$ of linearly independent elements $t_j \in E'$ and $\nu \in R_1^N(l^2(R))$ with vanishing mean such that for all $x \in E$ and $i,j \in N$:

$$\sum_{j=1}^{\infty} t_j(x)^2 < \infty, \qquad \int \|\sigma\|^2 d\nu(\underline{\sigma}) < \infty$$
$$\int \sigma_i \sigma_j d\nu(\underline{\sigma}) = 0 \ \text{ if } i \neq j \ \text{ and } \int \sigma_j^2 d\nu(\underline{\sigma}) > 0 \tag{5.19}$$

By equation (5.6) we get $\mu \in R_1^N(E'_\sigma)$ concentrated on the compact subsets of $E'_{q'} := f_{\underline{t}}(l^2(\mathbb{R}))$. The moments of this measure can be calculated according to equations (5.1), (5.8), (5.11) and (5.16) - (5.18).

ii) On a separable Hilbert space it is not too hard to construct cylindrical measures. Their images under Hilbert-Schmidt-maps then are Radon-probability measures. Thus an approach of how to construct $\nu \in R_1^N(l^2(\mathbb{R}))$ satisfying (5.14) (or (5.7) and (5.8)) is known.

6. Conclusion

We think that our results clearly show: it is not hopeless to construct Radon probability measures on E'_σ, E a real nuclear (LF)-space, in terms of their moments : These results provide a general basis.
But obviously they have to be supplemented by various details in order to be interesting for (more concrete) applications. This is under investigation. For instance as indicated in section 1 one would like to construct such a measure μ on E'_σ such that the (generalized) stochastic process canonically associated with μ satisfies an appropriate

differential equation. Then one would prefer to take $E = \mathcal{D}(\mathbb{R}^d)$, $d \in \mathbb{N}$, the Schwartz space of C^∞- functions with compact support. A first step in this program would be to express this fact in terms of the moments. The idea to do this is as follows :
First one determines the "positivity-irreducible part" T_n^i of each moment T_n :

$$T_n = F_n(T_1,\ldots,T_{n-1}) + T_n^i$$

(for instance $T_1^i = T_1$ and $T_2^i = T_2 - T_1 \otimes T_1$ = covariance).
Then depending on the case at hand one has to fix T_n^i for $n > n_0$ as a function of $T_1^i,\ldots,T_{n_0}^i$, expressing the "dynamics" eventually nonlinear.
In the well-known case of a Gauss-measure on E'_σ one has $n_0=2$ and then [7]

$$T_n = \sum_{k=0}^{m} T_1^{\otimes(n-2k)} \circledS T_2^{i \circledS k} \quad , \; m = \left[\frac{n}{2}\right] .$$

Another application which is under investigation is to the Euclidean approach to quantum field theory. This is basically a special case of the above case but in addition one has to incorporate various concrete properties for the moments and has some ideas about the form of the differential equation.

After having finished this article I learned by the author that related questions have been considered in

H.Zessin: a) Moments of states over nuclear LSF-spaces, in 'L.Arnold, P.Kotelenez (eds. eds.), Stochastic Space-Time Models and Limit Theorems, 249 - 261, D.Reidel Publishing Company, 1985

b) The Methods of Moments for Random Measures, Z.Wahrscheinlichkeitstheorie verw. Gebiete, 62, 395 - 409 (1983)

I thank Prof. Zessin for this information and some discussions about this subject.

REFERENCES

1 Akhieser, N.I. The Classical Moment Problem and some Related Questions in Analysis ; Translated by N. Kemmer, Olive and Bryol, Edinburg & London, 1965.

2 Berezanskiy, Ju.M. : Generalized Power Moment Problems, in : Colloquia Mathematica Societatis Janos Bolyai, 5. Hilbert Space Operators, Tihany (Hungary) 1970, Ed.B.SZ. Nagy, North-Holland, Amsterdam London 1972.

3 Berg, C.,Christensen, J.P.R., Ressel, P.: Harmonic Analysis on Semigroups : Theory of Positive Definite and Related Functions. Graduate Texts in Mathematics Springer-Verlag New York Berlin Heidelberg Tokyo 1984

4 Borchers, H.J. Yngvason, J. : Integral Representations for Schwinger Functionals and the Moment Problem over Nuclear Space, Commun. Math. Phys. 43,255-271 1975

5 Brüning, E. On the Construction of Radon-Probability Measures on Infinite Dimensional Spaces.
I. : Preprint N.110, Project N.2, mathematics+physics, Zentrum für interdisziplinere Forschung, Universität Bielefeld, 1984.
II.: in preparation.

6 Choquet, G. : Lectures on Analysis, Volume I and II, W.A. Benjamin, Inc. New York 1969.

7 Dobrushin, R.I., Minlos, R.A. : Polynomials in Linear Random Functions, Russian Math. Survey 32.2 (1981), 71-127 ; from : Uspekhi Mat. Nauk. 32.2 (1977), 67-122

8 Dobrushin, R.I. Minlos, R.A. : The Moments and Polynomials of a Generalized Random Field ; Theory of Probability and its Applications, Vol. XXIII, 1978, N.4, 686-699.

9 Glimm, J., Jaffe, A. : Quantum Physics. A Functional Integral Point of View. Springer-Verlag, New York Heidelberg Berlin 1981

10 Hegerfeldt, G.C. : Extremal Decomposition of Wightman Functions and States on Nuclear *-Algebra by Choquet Theory ; Commun. Math. Phys. 45,133-135 (1975)

11 Schwartz, L. : Radon Measures on Arbitrary Topological Spaces and Cylindrical Measures ; Oxford University Press, London W1 , 1973.

12 Skohal,J.A., Tamarkin, J.D. : The Problem of Moments, Math. Survey 1, Amer. Math. Soc. Providence, Rhode Islands 1943

13 Robertson, A.P., Robertson,W. : Topological Vector Spaces, Cambridge University Press 1973

14 Pietsch, A. : Nuclear Locally Convex Spaces, Springer-Verlag New York Heidelberg Berlin 1972

A "Brownian motion" with constant speed.

Ian M. Davies
Department of Mathematics and Computer Science,
University College of Swansea,
Singleton Park,
Swansea,
SA2 8PP,
United Kingdom.

Abstract

We consider the time integral of a Brownian motion on a sphere in $\mathbb{R}^3$ and show that in some cases it could be mistaken for a Brownian motion in $\mathbb{R}^3$.

Introduction

When one studies the motion of a particle on a manifold due to some acting potential one often changes from using the natural metric to using the action metric. The benefit of this is that the classical motions are now along the geodesics and are of constant speed. If the manifold were very rough then we might observe the motion of the particle to be very erratic although it actually had constant speed.

We are going to consider the very simple example of a particle moving in $\mathbb{R}^3$ whose velocity is generated by a Brownian motion on a sphere of constant radius. The position of the particle will be given by the time integral of the velocity process. Since one would determine the behaviour of the position process by measurement we will calculate some of the lower order moments of the position process. We will show that in a specific case the process, with finite speed, strongly resembles a Brownian motion.

The paper of M. Yor (in these proceedings) gives the law of position process in the case of infinite speed.

The position process

Let V_t be Brownian motion on a sphere of radius R in $\mathbb{R}^3$. Following the scheme of Lewis and van den Berg we have the following Itô stochastic differential equation for V_t

$$dV_t = n(V_t) \wedge dB_t + \tfrac{1}{2}[(n(V_t) \wedge \nabla) \wedge n(V_t)]\, dt$$

where $n(V_t)$ is the unit normal to the sphere at V_t. Since $n(V_t) = R^{-1}V_t$ this stochastic differential equation reduces to

$$dV_t = R^{-1}V_t \wedge dB_t - R^{-2}\, V_t\, dt.$$

One could also use Strook's form of the s.d.e. for a Brownian motion on a sphere and still obtain the same results as we will do. B_t is a standard Brownian motion in $\mathbb{R}^3$. To give us a little more scope we introduce the time changed motion on the sphere $v_t = V_{t/c}$ where c is a real positive constant. This new process is a solution of

$$dv_t = \alpha v_t \wedge dB_t - \alpha^2 v_t\, dt$$

where $\alpha^2 = R^2/c$ for convenience.

Let X_t denote the position of the particle at time t and without loss of generality set $X_o = 0$. Define X_t by

$$X_t = \int_0^t v_s\, ds.$$

If we denote expectations by $E(\cdot)$ and notice that $\exp\{\alpha^2 t\}v_t$ is a martingale we can quickly compute $E(X_t)$ and $E(X_t . X_t)$.

$$\begin{aligned} E(X_t) &= \int_0^t E(v_s)\, ds \\ &= \int_0^t e^{-\alpha^2 s} E(e^{\alpha^2 s}\, v_s)\, ds \\ &= \alpha^{-2}(1 - e^{-\alpha^2 t})\, v_o . \end{aligned} \qquad (1)$$

$$\begin{aligned} E(X_t . X_t) &= \int_0^t ds \int_0^t dr\, E(v_s . v_r) \\ &= 2\int_0^t ds \int_0^s dr\, e^{-\alpha^2(s+r)} E(e^{2\alpha^2 r}\, v_r . v_r) \\ &= 2R^2\alpha^{-2}(t + \alpha^{-2}(e^{-\alpha^2 t} - 1)) \end{aligned} \qquad (2)$$

since $E(v_r . v_r) = R^2$.

These expected values will enable us to demonstrate the drasticly differing behaviour of X_t for various values of c. We assume that R is large.

$c = 1$.

For this value the Brownian motion on the sphere is running at the standard rate and $\alpha^2 = R^{-2}$. We have that

$$E(X_t) \sim v_o t, \qquad E(X_t.X_t) \sim R^2 t^2.$$

This is the behaviour which one would intuitively expect.

$c = R^{-2}$.

The Brownian motion on the sphere is now running quite fast and correspondingly $\alpha^2 = 1$. We have that

$$E(X_t) \sim v_o t, \qquad E(X_t.X_t) \sim R^2 t^2 \qquad \text{for small } t,$$

$$E(X_t) \sim v_o, \qquad E(X_t.X_t) \sim R^2(t-1) \qquad \text{for large } t.$$

$c = R^{-4}$.

The velocity process is now running extremely fast and $\alpha^2 = R^2$. We have

$$|E(X_t)| < R^{-1}, \qquad \text{for all } t,$$

$$E(X_t.X_t) = 2t + O(R^{-2}), \qquad \text{for } t \geq O(R^{-1}).$$

We see that for this case the process X_t resembles a Brownian motion in $\mathbb{R}^3$. If we set $\alpha^2 = 2R^2/3$ then $E(X_t.X_t) \sim 3t = E(B_t.B_t)$ where B_t is the standard Brownian motion in $\mathbb{R}^3$.

Returning to the general case, one can show that

$$E(X_t.X_s) = E(X_s.X_s) + R^2\alpha^{-4}(1 - e^{-\alpha^2 s})(1 - e^{-\alpha^2(t-s)}), \quad t \geq s, \qquad (3)$$

and so when $\alpha^2 = O(R^2)$ we have

$$E(X_t.X_s) = E(X_s.X_s) + O(R^{-2}).$$

The process X_t appears to be a martingale.

The fourth moment of $|X_t|$ for $\alpha^2 = O(R^2)$

As there is the possibility of not being able to distinguish X_t from a Brownian motion in $\mathbb{R}^3$ when $\alpha^2 = O(R^2)$ we will calculate $E((X_t.X_t)^2)$. The calculation is straightforward but extremely tedious and so we only list the important details herein.

We express $(X_T.X_T)^2$ as a fourth order iterated integral and so we must compute the expected values of terms such as $v_1(t)v_1(s)v_1(r)v_1(q)$

and $v_1(t)v_1(s)v_2(r)v_2(q)$ where there is no given ordering of t,s,r and q. Once we impose an ordering t,s,r and q we are required to calculate the expectations of

$$v_1(t)v_1(s)v_1(r)v_1(q),$$
$$v_1(t)v_1(s)v_2(r)v_2(q),$$
$$v_1(t)v_2(s)v_1(r)v_2(q),$$
$$v_1(t)v_2(s)v_2(r)v_1(q),$$

where $t \geq s \geq r \geq q$, the other expectations being obtained by observation. To ease the calculations one requires to know the following: if $y_t = \exp\{\alpha^2 t\}v_t$ then

$$y_1, \quad e^{\alpha^2 t}y_1^2 - \tfrac{1}{3}R^2 e^{3\alpha^2 t}, \quad e^{3\alpha^2 t}y_1^3 - \tfrac{3}{5}R^2 e^{5\alpha^2 t}y_1,$$

$$e^{\alpha^2 t}y_1y_2, \quad e^{3\alpha^2 t}y_1^2y_2 - \tfrac{1}{5}R^2 e^{5\alpha^2 t}y_2$$

are all martingales, and

$$E(y_1^4(t)) = \tfrac{1}{5}R^4 e^{4\alpha^2 t} + \tfrac{6}{7}R^2(y_1^2(0) - \tfrac{1}{3}R^2)e^{\alpha^2 t} + a_o e^{-6\alpha^2 t},$$

$$E(y_1^2(t)y_2^2(t)) = \tfrac{1}{15}R^4 e^{4\alpha^2 t} - \tfrac{1}{7}R^2(y_3^2(0) - \tfrac{1}{3}R^2)e^{\alpha^2 t} + a_1 e^{-6\alpha^2 t}.$$

a_o and a_1 are constants which one need not evaluate. We are left with the following functions to integrate

$$e^{\alpha^2(6q - 3r - 2s - t)}, \quad e^{\alpha^2(-4q - 3r - 2s - t)},$$

$$e^{\alpha^2(q - r + s - t)}, \quad e^{\alpha^2(q + 2r - 2s - t)}.$$

Since $\alpha^2 = O(R^2)$ we need only concern ourselves with the $O(R^{-4})$ terms which come from the integrations, the only one being $T^2/2\alpha^4$. Now taking care of all the constants and multiplicities introduced by using an ordered iterated integral we get

$$E((X_T.X_T)^2) = \frac{20R^4T^2}{3\alpha^4} + O(R^{-2}). \qquad (4)$$

If $\alpha^2 = 2R^2/3$ we have $E((X_T.X_T)^2) = 15T^2 + O(R^{-2})$. If B_T is a standard Brownian motion in $\mathbb{R}^3$ then $E((B_T.B_T)^2) = 15T^2$. The value of $E((X_T.X_T)^2)$ reinforces our earlier statement that for $\alpha^2 = O(R^2)$ it is easy to confuse X_T with a Brownian motion.

Conclusions

We have shown that the process X_t is remarkably similar to a Brownian motion if the velocity process on the sphere of radius R is running very fast and if we are unable to make accurate measurements down to $O(R^{-2})$. While we have only considered moments of orders less than five the resemblance will persist in the higher moments (see the result on the law of X_t in the paper of M. Yor). The expression given in (3) can be used to show that

$$h^{-1}E((X_{t+h} - X_t).X_t - X_t.(X_t - X_{t-h})) \sim \text{constant for } h \geq O(R^{-1})$$

and so the process X_t appears to have non-differentiable continuous paths if $O(R^{-2})$ is at the limit of our ability to measure. We conclude by emphasising that for $\alpha^2 = O(R^2)$ the process X_t looks just like Brownian motion if one can only make precise observations down to a scale of $O(R^{-1})$.

Acknowledgements

The author would like to thank Chris Rogers for a short tutorial which convinced him that he had the ability to carry out the calculations contained herein. The author also wishes to thank Marc Yor for a most enlightening conversation. The author wishes to thank S. Albeverio, Ph. Blanchard and L. Streit for their invitation to Z.I.F. and for their hospitality and support.

Reference

'Brownian Motion on a Surface in $\mathbb{R}^3$', J. Lewis and M. van den Berg, BiBoS Preprint.

THE SEMI-MARTINGALE APPROACH TO THE OPTIMAL RESOURCE ALLOCATION IN THE CONTROLLED LABOUR-SURPLUS ECONOMY

M.H.A. DAVIS
Dept. of Electrical Eng.
Imperial College
London SW7, 2BT, UK

Guillermo L. GÓMEZ M.
Department of Mathematics
University of Erlangen-Nürnberg
D-8520 Erlangen, FRG

Abstract

This paper deals with the characterization of an economic policy that over the long-run leads to the elimination of labour-surplus, i.e. one which directs the labour force from less productive or idleness to more highly productive employment.

Crucial is here the assumption that the rational response of individuals to economic development policies is uncertain and only partially known. Therefore, the dynamics of the labour participation in the more productive activities is stochastic and described by a stochastic differential equation of the diffusion type. We solve the associated stochastic optimization problem using semi-martingale techniques which we introduce in some detail with the purpose of making this paper self-contained. Our presentation is, however, intuitive and heuristic, since our emphasis lies in applications rather than in theories.

Contents

1. Introduction and Background

Economic development is a very complex process which involves social, political, technological and cultural evolutionary components.

For the purpose of the present study we let the most essential social and cultural features of development be represented in a utility function by means of which the control board reveals the prevailing system of preferences and value judgements. Further, we consider the economic, technological and institutional components as the growth aspects of development. The interrelation and interdependence of these components make sure that economic growth and (utility) welfare decisions influence the paths of economic development effectively.

Therefore, our main concern is to introduce elements that counter the effects of factors inhibiting economic growth by increasing the degree of utilization and improving the productivity of the available human and physical resources. This in turn furthers performance motivation and social awareness, increases the level of individual expectations and social aspirations, etc., breaking vicious circles typically attached to dual systems in less developed countries.

Thus, our primary objective is to work out development paths along which surplus-labour can be absorbed into production. The facts that a capital cost is entailed in training labour to a level of skill adequate for participation in production and that it costs to settle people where they can enter production and to endow unemployed labour with working tools at the existing standards provide barriers to the elimination of unemployment.

At this point we are not interested in debating whether there is any reasonable argument why responsible governments tolerate the continuous existence of surplus-labour, whether there is any economic logical basis to argue that may be optimal to leave substantial amount of unemployment with income redistribution undertaken to compensate those unemployed, whether profitability criteria should bear higher priorities than social benefits in matters of economic policy making, and so on.

For the time being, we limit ourselves just to forward our view that democratic principles like individual freedom and equal rights, and concepts like national sovereignty and peaceful international coexistence acquire real meaning only by adhering to each other and by interacting to generate world wide social and economic development, cultural and technical progress from which no nation should be deprived for any ideological, political or strategic reason or whatsoever.

In a set of theoretical studies on dualistic economic systems we have designed a highly aggregate and simplified dynamic model of neoclassic-post Keynesian mixed type that captures, we believe, essential features of the development process of less developed countries. That model consists of three sectors obtained by the mode of production criterion and contains enough structure to allow exploring short-term and long-term implications of intertemporal decisions and development strategies.

The present paper deals with one of these intertemporal decisions: the choice of the optimal consumption-investment path or equivalently the choice of the optimal rates of labour participation and savings in the advanced sector. Let us briefly recall results obtained in Gômez (1981, 1983) upon which the present investigation relies. First, we have the following long-run system dynamics:

$$dx = a(t,\Gamma)\, x\, dt + \sigma(t,x)dB + \sum_{j=1}^{n} \delta(t-t_j)\theta_j \tag{1-a}$$

$$x(t_0^+) = x_0 \quad \text{with} \quad x,x_0 \in \mathbb{R}_+^d\, ;\; t,t_0 \in \mathbb{R}_+\, ;\; B \in \mathbb{R}^{d_1} \tag{1-b}$$

$$d,\, d_1 \in \mathbb{N}\, , \quad \theta_j \in \mathbb{R}_+^n \tag{1-c}$$

where $x(t)$ and $B(t)$ are d- and d_1-dimensional column random real-valued vectors defined for any $t \in [t_0,T]$, $T \in \overline{\mathbb{R}}_+$, d and d_1 are integers. We omit the time variable unless any ambiguity may arise. Moreover, $B = (B(t))_{t\geq 0}$ represents a $\mathbb{R}^d$-valued Brownian motion defined on an abstract probability space $(\Omega, \underline{\underline{F}}, \mathbb{P})$ with the usual properties and endowed with a filtration $\underline{F} = (\underline{\underline{F}}_t)_{t\geq t_1}$ of sub-σ field of $\underline{\underline{F}}$. The variable x represents the state of the economy and consists here of eight variables L_i, Y_i and q_i denoting respectively production, labour and real price of the i-th sector where i takes on the values 1,2 and 3. The product of the secondary sector has been taken as numéraire. Γ stands for the structure of the economy and is characterized by the following structural parameter: β_i elasticity of production with respect to labour, b_i rate of technical progress, α_i income and η_i price elasticity of the demand of the product of the i-th sector and δ_i the corresponding capital depreciation.

Finally, $a(t,\Gamma)$ and $\sigma(t,x(t))$ are $\mathbb{R}^d \times \mathbb{R}^d$-valued diagonal matrices which depend respectively on (t,Γ) and on $(t,x(t))$. The time horizon $\bar{T} := [t_0,T]$ consists of an exhaustive sequence of non-overlapping time intervals $\bar{T}_j = [t_{j-1},t_j)$ that we call the j-th time period, where the t_j are impulsion times with j taking on the values $1,2,3,\ldots,n$. $\delta(t-t_j)$ is a Dirac function which equals 1 when $t = t_j$ and equals 0 (zero) elsewhere. At the j-th impulsion time t_j the state variable is given by

$$x(t_j) = x(t_j-0) + \theta_{j+1}\,. \tag{1-d}$$

During the j-th time period $\bar{T}_j$ the structural parameters remain constant but change instantaneously at the impulsion time t_j as a consequence of the (capital) impulse θ_{j+1}. We refer the interested reader to Gômez (1984a) where we formulate the model and obtain the dynamics, and to Gômez (1984b) where we introduce impulsion (or stopping) times and deal with the short-time optimization problem based on the interpretation of the development process as a learning process. We like to point out that x is a short-hand for $x = (x(t), \underline{\underline{F}}_t^x)$, $t \in \bar{T}_j$, with $\underline{\underline{F}}_t^x = \sigma(x(s) : t_o \leq s \leq t)$ the σ-field generated by the time-path of x until t. The same holds for B.

In order to make this paper self-contained, we describe roughly a few features of our controlled labour-surplus economy.
We consider a backwarder sector and call it S_1, primary sector, characterized by a small role for wage-labour, low capital but higher labour intensities and a predominating survival motivation. Handicraft manufacture is allocated to the primary sector regardless of how closely its products may approximate the product of the capitalist industry in physical terms. We consider an advanced sector and call it S_2, secondary sector, characterized by a mode of production in which wage-labour predominates. It is relatively well-developed and operates with high capital intensities. The capitalist agriculture is allocated to the secondary sector by the mode of production criterion notwithstanding the homogeneity of the products of the capitalistic and traditional agriculture.

Finally, we consider a sector S_3, tertiary sector that includes services, market and official activities characterized by low capital intensities.

Since our characterization of the dual economy stresses the differences in the mode of production and in the form of economic organization rather than in the diversity of the products of the various sectors we like to highlight briefly the effects of some sectorial differentials.

Usually advanced sectors exhibit capital-labour ratios and as a consequence of better trained and skilled workers higher productivities and wage rates.
A rational response to positive wage differentials may be migration towards advanced sectors. Since typically dual economies do not enjoy adequate growth potentials, migration acts upon the system as a disequilibrium factor.

To help the system to overcome the resulting impasses we attempt to weaken those sectorial differentials and to enhance the system growth potential. We achieve that injecting certain amount of capitál (impulse). With the help of stopping and impulse control methods we determine the impulsion time t_j and the associated impulse θ_j.
These capital impulses enable the economy to make a more suitable choice of techniques thereby inducing instantaneously, for the sake of simplicity, structural changes through capital deepening and widening as well as through enhancing capital efficiency (both physical and human) and a better allocation of labour. As a consequence, these capital injections bring about an increase in the producitivity, in the rate of technical progress and in the capability of the economy to absorb the labour-surplus.

For purposes of economic policy and in order to give more clarity to the model conception, we find it convenient to split the matter at issue, i.e. the optimal development of the economy, into three highly interrelated optimization problems. Namely,

(P1a) Here, one determines the capital impulses θ_j, $j = 1,2,3,\ldots,$ taking into account technological and institutional constraints.

(P2) Given the impulses θ_j as well as two functionals representing an average subjective evaluation of benefits and social costs of the development path, one determines the time $t_j = t(\theta_j)$ at which the net benefits associated with the j-th time period are optimal.

(P3) Now, taking t_j as given, one determines the optimal paths of the rate of capital accumulation and of the level of employment according to a short-run macro-welfare function.

(P1b) Finally, one adjusts the impulses θ_j so that a given long range welfare function gets an optimum.

The present paper deals with a stochastic version of the optimization problem (P3).

2. Formulation of the Optimization Problem

First of all, we owe to point out that our optimization approach involves feedback and adjustment mechanisms that rely heavily on the interpretation of the process of economic development as a learning process, see Gômez (1984b), and on the attempt to control migration by means of changes of the minimum wage rate of the secondary sector.

Consequently, we do not mean that individuals are really engaged in a conscious optimization and planning process but instead that they learn from and adapt to their socio-economic environment. Thus, using as a guide their own as well as others success and failure experiences they develop behavioural patterns, following which they expect to get the highest possible pay-off.

Once more our modelling rationale directs attention to the idea of approximation which relies upon our belief that invidividuals learn and that under adaptive pressures they come to behave in a way that can be closely described as optimal. However, this hints to a further source of uncertainty which has to be taken into consideration.

As we have already mentioned, migration acts upon the system as a disequilibrium factor and may prevent it from reaching equilibrium. The interpretation of migration as a rational response of individuals to differentials or changes in their socio-economic and cultural environment leads us to rely on concepts, latent variables, which are determinants of human behaviour but that cannot be observed directly and in consequence cannot be measured properly, i.e. motivation, aspiration level, need for achievement, social awareness, etc., ... see Jöreskog and Wold (1982).

Even, if we assume that individuals respond to signals from their economic environment, a wage differential for instance, we may not know to what extent does it happen, neither do we know how they build their expectations nor how they formulate subjective probabilities of future events even making enough information available. These points suggest again uncertainty that we like to capture assuming that the growth of the labour force, in the secondary sector, satisfies the following diffusion equation

$$dL_2 = a^{(3)}(t)L_2dt + b^{(3)}(t)L_2dB^{(3)} \qquad (2)$$

where $a^{(3)}(t)$ and $b^{(3)}(t)$ are real-valued coefficients that depend on structural parameters and remain constant within the j-th time period as we have already pointed out. The upper index (3) points out that we are delaying with a coefficient or component related ot the third component of the state variable x, i.e. with L_2. Hence $B^{(3)}$ stands for the component of the Brownian motion corresponding to L_2. For the sake of simplicity we suppress the upper indices unless any ambiguity may arise. We assume that all our random functions are appropriately defined on the abstract probability space $(\Omega, \underline{\underline{F}}, P)$, introduced in Section 1. For details and a derivation of this equation we refer to Gômez (1985c). Since our analysis here refers only to the j-th time period, we set $a(t) := a$ and $b(t) := b$, hence eq. (2) becomes

$$dL_2 = aL_2dt + bL_2dB \,. \qquad (2')$$

The designation "diffusion" refers to the fact that under the conditions of uncertainty just mentioned a change ΔL_2 in the labour participation during the incremental time Δt can be described, given that at time t it assumes the value $L_2(t)$, as the sum of two changes $aL_2\Delta t + \delta L_2 + o(\Delta t)$, where δ is a constant and aL_2 is the macroscopic change (drift) due to the system dynamics and δL_2 is a random change caused by collisions and conflicting goals of interacting individuals given their limited information.

Furthermore, $\mathbb{M}(\delta L_2)^2 = b\Delta t + o(\Delta t)$ i.e. b is in a certain sense proportional to the mean uncertainty of the individuals reacting to the guess or belief that the labour supply at time t was $L_2(t)$. $\mathbb{M}$ stands for the expectation operator.

Let us now obtain the equation of per capita capital accumulation. Let K_2, I_2, Y_2, k_2 and y_2 denote respectively, capital stocks, gross investments, output, capital-labour and output-labour ratio in the secondary sector. Further, let us reacall that

$$Y_2 = F(t, K_2, L_2) \tag{3-a}$$

$$I_2 = s_{K_2}(Y_2 - W_2 L_2) + s_{L_2} W_2 L_2 \tag{3-b}$$

$$C_2 = (1 - s_{K_2})(Y_2 - W_2 L_2) + (1 - s_{L_2}) W_2 L_2 \tag{3-c}$$

$$0 < s_{K_2}, s_{L_2} < 1 ; \; s_{K_2}, s_{L_2} \in \mathbb{R} \tag{3-d}$$

$$\dot{K}_2 = I_2 - \delta_2 K_2 \tag{3-e}$$

$$\delta_2 > 0 , \; \delta_2 \in \mathbb{R}_+ \tag{3-f}$$

$$Y_2 > 0 , \; {}^0K_2 > 0 ; \; Y_2 , {}^0K_2 \in \mathbb{R}_+ \tag{3-g}$$

where s_{K_2} and s_{L_2} denote the fraction of capital and wage income saved (savings := investments) in the secondary sector, and δ_2 stands for the rate of capital depreciation. $F(t;K_2,L_2)$ is a neoclassic production function, i.e. is a smooth function of class $C^{1,2,3}(\mathbb{R}^3_+ \to \mathbb{R}_+)$ of flows of capital K_2 and labour services L_2, homogeneous of degree 1 with respect to K_2 and L_2, the marginal productivities of both factors are non-negative but diminishing with successive increments of inputs, etc. Before we go over to formulate the control problem, we like to introduce some additional notations.

Let us recall that the economy lives the j-th time period $\bar{T}_j$ which is the real interval $[t_{j-1}, t_j)$ where t_j is given by the function $\theta_j \to t(\theta_j)$ whose characterization is part of otimization problem (P2). Thus, for the time being, we consider t_j as given and set $t^* = t_j$. We denote the initial capital endowment $K_2(t_{j-1})$ by ${}^0K_2$ and assume that it is fully utilized. Let C_2 denote aggregate consumption, ${}^0L_2$ the total labour force available to the secondary sector during the j-th time period, $\bar{L}_2$ the employment level at which the wage bill exhausts output and ${}^0\bar{L}_2 = \min\{{}^0L_2, \bar{L}_2\}$. We assume that once full employment is reached the control board is able to require investors to save any fraction s_{k_2} of capital income as long as it remains below the upper bound $\bar{s}_{k_2}$. Further, we assume that the capitalists' control of the disposition of their income is limited to the ability to impose or, to put it less drastic, to negotiate with the control board the upper bound $\bar{s}_{K_2}$ on the saving rate. Moreover, we assume a minimum wage rate $\bar{W}_2$, exogeneously fixed, which remains constant over the time period. Finally, we introduce two strict concave functionals $U(\cdot)$ and $g(\cdot)$ of class $C^1_b(\mathbb{R}_+ \to \mathbb{R}_-)$. U measures aggregate consumption utility by means of which the control board expresses its preferences and judgements with respect to social welfare. The functional g is a terminal pay-off and measures the utility value in terms of consumption of the capital bequeathed to the (j+1)-th time period. We shall understand U and g as a set of socio-economic alternatives and social utility values resulting from the government program by means of which the ruling party should have come to power and to its fulfillment it should have committed itself.

Now we have all the economic elements we need in order to obtain an equation for the per capita capital accumulation. Let us recall a simple version of the Ito Rule. Let τ be a Markov time, let $\bar{\sigma}(s)$ be a process with values in a set of matrices of dimension dxd_1, and let $\bar{a}(s)$ be a d-dimensional process. We assume that $\chi_{s\leq\tau}\,\bar{\sigma}(s)$, $\chi_{s\leq\tau}\,a(s)$ are progressively measurable. $\chi_{s\leq\tau}$ is a characteristic function, i.e. $\chi_{s\leq\tau} = 1$ if $s \leq \tau$, elsewhere it is 0. If the process $\xi(t)$ satisfies the relation

$$\mathbb{P}\left\{ \sup_{t\leq\tau}\left|\xi(t) - \int_0^t \chi_{s\leq\tau}\,\bar{\sigma}(s)dB(s) + \int_0^t \chi_{s\leq\tau}\,\bar{a}(s)ds\right| = 0\right\} = 1 \qquad (4)$$

it is convenient to write

$$d\xi(t) = \bar{\sigma}(t)dB(t) + \bar{a}(t)dt \qquad (5)$$

$$t \leq \tau \qquad \xi(t_0) = x\ .$$

Eq. (5) is only a short-hand representation of eq. (4).

<u>Itô Lemma</u>: Assume the process $\xi(t)$ satisfies the stochastic differential equation (SDE), eq. (5), and that the functional $h(t,\xi(t))$ is of class $C^{1,2}(\mathbb{R}^{d+1} \to \mathbb{R}^d)$. Then the process $h(t,\xi(t))$ satisfies the following SDE

$$dh(t,\xi(t)) = \frac{\partial h(t,\xi(t))}{\partial t}\,dt + \nabla_x h(t,\xi(t))d\xi(t)$$

$$+ \frac{1}{2}\sum_{i,j=1}^{d} h_{x^{(i)}x^{(j)}}(t,\xi(t))d\xi^{(i)}(t)d\xi^{(j)}(t) \qquad (6)$$

$$t \leq \tau \qquad h(t_0,\xi(t_0)) = h(t_0,x)$$

where $h_{x^{(i)}} = \frac{\partial h}{\partial x^{(i)}}$, $h_{x^{(i)}x^{(j)}} = \frac{\partial^2 h}{\partial x^{(i)}x^{(j)}}$ and $\nabla_x = (h_{x^{(1)}}, h_{x^{(2)}},\ldots,h_{x^{(d)}})$ the gradient which is to be regarded as a row vector. Eq. (6) is an analog of Taylor's formula with two terms and is a <u>chain rule for stochastic differentials</u>. Applying the usual rules for removing parentheses and for the product of stochastic differentials, i.e. $(dB^{(i)}(t))^2 = dt$, $dB^{(i)}(t)dB^{(j)}(t) = 0$ for $i \neq j$, $dB^{(i)}(t)dt = 0$ and $(dt)^2 = 0$, one obtains for the third term in eq. (6)

$$d\xi^{(i)}(t)\, d\xi^{(j)}(t) = \frac{1}{2} \sum_{k=1}^{d_1} \bar{\sigma}^{(ik)}(t)\, \bar{\sigma}^{(jk)}(t)dt = S^{(ij)}(t)dt$$

with $$S^{(ij)}(t) = \frac{1}{2} \sum_{k=1}^{d_1} \bar{\sigma}^{(ik)}(t)\, \bar{\sigma}^{(jk)}(t)\,.$$

Introducing the operator $\mathbb{L}^{\bar{\sigma},\bar{a}}$

$$\mathbb{L}^{\bar{\sigma},\bar{a}} = \sum_{i,j=1}^{d} S^{(ij)}(t) \frac{\partial^2}{\partial x^{(i)} \partial x^{(j)}} + \sum_{i=1}^{n} \bar{a}^{(i)}(t) \frac{\partial}{\partial x^{(i)}}\,. \tag{7}$$

Eq. (6) becomes in its integral version

$$h(t,\xi(t)) = h(t_0,x) + \int_0^t \chi_{x\leq\tau}\ \nabla_x h(s,\xi(s))\bar{\sigma}(s)dB(s)$$
$$+ \int_0^t \chi_{x\leq\tau}\ (\frac{\partial}{\partial t} + \mathbb{L}^{\bar{\sigma},\bar{a}})h(s,\xi(s))ds,\ t \leq \tau \tag{8}$$

and is known as the <u>change of variable formula</u>.

We refer the interested reader to Gihman and Skorohod (1972, 1979), Elliott (1982) and Ikeda and Watanabe (1981) for a full and modern treatment of SDE's. Schuss (1980) and Arnold (1974) are excellent works covering various and interesting application fields.

After this short incursion on the Itô calculus we return to our economic problem.
The capital-labour ratio $k_2 = \frac{K_2}{L_2}$ is a function of labour L_2 and time t, a fact that we stress setting $k_2 := h(t,L_2)$. On the other hand, L_2 satisfies the SDE (2) and also eq. (5) with $\bar{a}(t) := aL_2(t)$ and $\bar{\sigma}(t) := bL_2(t)$. Therefore, simple application of Itô's Lemma gives

$$dk_2 = \left[\frac{\partial}{\partial t}(\frac{K_2}{L_2}) + aL_2 \frac{\partial}{\partial L_2}(\frac{K_2}{L_2}) + \frac{1}{2}(bL_2)^2 \frac{\partial^2}{\partial L_2^2}(\frac{K_2}{L_2})\right] dt + bL_2 \frac{\partial}{\partial L_2}(\frac{K_2}{L_2})dB(t)$$

and further

$$dk_2 = \left[\frac{I_2}{L_2} - (a + \delta_2 - b^2)k_2\right]dt - bk_2 dB(t) \tag{9}$$

which we rewrite in the following compact form to which we shall refer as the <u>stochastic equation of the accumulation of capital</u>

$$dk_2 = f(t,k_2,\alpha(t))dt + \sigma(t,k_2)dB(t) \tag{9'}$$

where $f(t,k_2,\alpha(t)) := s(t)y_2 - (a + \delta_2 - b^2)k_2$, $y_2 = \frac{Y_2}{L_2}$,
$s(t) = s_{K_2}(t)\pi_K + s_{L_2}(t)\pi_L$ with π_K and π_L representing the relative income share of capital and labour respectively, $\sigma(t,k_2) := -bk_2$ and $\alpha(t) := \alpha(s_{k_2}(t), L_2(t))$ is a function of class $C_b^{1,1}(\mathbb{R}_+ \to \mathbb{R}_+)$ which stresses the dependence of the first term in the right-hand side of eq. (9) on the time paths of s_{k_2} and L_2. In our framework $s_{k_2}(t)$ and $L_2(t)$ are choice variables, therefore $\alpha(t)$ is also one and we use it here as a control variable. The function f is obviously of class $C^{1,2,1}(\mathbb{R}_+ \times \mathbb{R}_+ \times \mathbb{R}_+ \to \mathbb{R}_+)$.

Bourguignon (1974), Merton (1975) study eq. (9) and properties of its solution. Malliaris (1981) gives a full account of applications of Itô calculus to economic problems.

Now, we are in a position to formulate our optimization problem as the following completely observable stochastic control problem:

Find admissible time path $\{\alpha(t)\}$ or, equivalently, find admissible time paths $\{s_{k_2}(t)\}$, $\{L_2(t)\}$ such that the mathematical expectation of the functional

$$\int_0^{t^*} U(k_2(t), \alpha(t))dt + g(k_2(t^*)) \tag{10}$$

reaches its supremum subject to eq. (9) and the constraints related to eq. (3). For notational simplicity we set $t_{j-1} = 0$. Shortly, consider the completely observable stochastic control problem

$$\sup_\alpha \mathbb{M} \left\{\int_0^{t^*} U(k_2,\alpha)dt + g(k_2(t^*))\right\} \tag{10'}$$

subject to eqs. (9) and (3).

As a matter of fact, the integrand in eq. (10) should read $U(t,C_2(t))$. However, we do not consider the time explicitly for the sake of simplicity. Furthermore, the optimization problem is more convenient to deal with in terms of the pair $(s_{K_2}(t),L_2(t))$ than in terms of the investment-consumption mix $(I_2(t),C_2(t))$. On the other hand, substituting Y_2 given by (3-a) into C_2 one obtains the control $\alpha(t) = \alpha(s_{K_2}(t),L_2(t))$ and the objective functional, see eq. (10), turns out in terms of the fraction of profit income invested s_{K_2} and the labour-participation L_2. Moreover, the constraints reduce to the inequalities

$$\begin{aligned} 0 < L_2(t) &\le {}^0\bar{L}_2 \\ 0 < s_{K_2}(t) &\le \bar{s}_{K_2} . \end{aligned} \tag{11}$$

As a matter of fact, the choice of a couple $(s_{K_2}(t),L_2(t))$ amounts to the choice of a couple $(I_2(t),C_2(t))$ as one easily recognizes looking at eqs. (3).

Those readers willing only to see the solution of the stochastic optimization problem just stated, may go directly to paragraph 5 where we construct the optimal saving-employment development strategy according to the tools introduced and the results obtained in paragraphs 3 and 4.

In the two following paragraphs we present some basic concepts of stochastic control theory and survey the most essential steps of the semi-martingale formulation of the problem. Since we are mainly interested in the application of control techniques that help us constructing a development strategy aimed at eliminating unemployment, we proceed in an intuitive and heuristic way and refer to the received literature those readers interested in a rigorous formulation.

3. Control of Diffusion Processes

3.1. Heuristic Formulation

Let us consider in $\mathbb{R}^d$ a random (state) process $x(t)$ governed by the (Itô) SDE

$$x(t) = \xi_o + \int_0^t f(s,x(s),\alpha(s))ds + \int_0^t \sigma(s,x(s))dB(s) \tag{12}$$

where $\sigma(t,y)$, $f(t,y,\alpha)$ are given functions of $y \in \mathbb{R}^d$, time t and of a control parameter α, ξ_o is the initial value of the process $x(t)$, $B(t)$ is a d_1-dimensional real-valued Brownian motion and d, d_1 are integers.
Further $f(t,y,\alpha)$ is a d-dimensional vector-valued function, $\sigma(t,y)$ is a dxd_1-dimensional matrix-valued function whose components are C_b^1 functions of y and α, and of y respectively. The control $\alpha(t)$ is a feedback of the current state, i.e. $\alpha(t) = \alpha(t,x(t))$ for some given functions $\alpha(t,y)$ taking values in the control set A.
The object of control theory is to select the control $\alpha(t)$ so that the corresponding process $x(t)$ possesses some desired properties. We steer thereby the process $x(t)$ considered. This gives rise to the question as to whether there exists a solution of eq. (12) for the random process $\{\alpha(t)\}$ chosen and if this is the case whether it is unique.
An extensive treatment of these and related problems can be found in Bensoussan and Lions (1979, 1982), Kushner (1967, 1977), Krylov (1980), Fleming and Rishel (1975), Davis (1977).
If α is Lipschitz in y then eq. (12) is a SDE satisfying the standard Itô conditions and hence has a unique strong solution $y(t)$.
The <u>total utility</u> associated with the process $\{\alpha(t)\}_{0\le t\le t^*}$, until the time t^*, is given by

$$\rho^\alpha = \int^{t^*} U(t,x(t),\alpha(t))dt + g(x(t^*)) \tag{13}$$

where t^* is a terminal time, U is a functional of class $C_b^1(\mathbb{R}_+ \times \mathbb{R}^d \times \mathbb{R}_+ \to \mathbb{R}_-)$ measuring instantaneous utility, g is a functional of class $C_b^1(\mathbb{R}^d \to \mathbb{R}_-)$ measuring terminal utility.

The objective is to choose the function $\alpha(t) = \alpha(t,x(t))$ so as to maximize ρ^α. From the practical point of view, it is reasonable to require that the values of the control process $\alpha(t)$ at time t are to be chosen on the basis of observations of the controlled process $\{x(s)\}$ until time t. In other words, $\alpha(t)$ has to be a function of the trajectory $x_{[0,t]} = \{(s,x(s)): 0 \le s \le t\}$, also $\alpha(t) = \alpha(t,x_{[0,t]})$.

Corresponding to the strategy $\alpha(t)$ the expected utility for the process x(t) with initial condition ξ_o is given by

$$v^\alpha(\xi_o) = \mathbb{M}\left[\int_0^{t^*} U(t,x(t),\alpha(t))dt + g(x(t^*))\right] . \tag{14}$$

This gives rise to the problem of finding a function v and an optimal strategy $\alpha^o = \{\alpha^o(t,x_{[0,t]})\}$ such that for fixed ξ

$$v^{\alpha^o}(t,\xi) = v(t,\xi) = \sup_{\alpha\in A} \mathbb{M}_{(t,\xi)}\left[\int_t^{t^*} U(s,x(s),\alpha(s))ds + g(x(t^*))\right] \tag{15}$$

where the supremum is taken over all control functions restricted to the interval $[t,t^*]$, the subscript (t,ξ) indicated that the process x(s) starts at $x(t) = \xi$ and $\mathbb{M}_{(t,\xi)}$ stands for the expectation operator conditioned at (t,ξ).

The process $v(t,\xi)$ is called a <u>value function</u> and determines which strategies or controls shall hold our attention. In the case where there exists no strategy α^o we may wish to construct for each $\epsilon > 0$ a strategy $\alpha^\epsilon = \{\alpha^\epsilon(t,x_{[0,t]})\}$ such that $v(t,\xi) - \epsilon \le v^{\alpha^\epsilon}(t,\xi)$. The strategy α^o (resp. α^ϵ) is said to be optimal (ϵ-optimal) for the starting point $x(t) = \xi$. Let v be a solution of eq. (15) and let us assume v is sufficiently smooth. Then formal application of Itô's Lemma and Bellman's principle of optimality show that v satisfies $\mathbb{P}_{t,x}$-a.s. the following <u>Bellman equation</u>

$$v_t + \frac{1}{2}\sum_{i,j=1}^{d} (\sigma\sigma^*)^{ij}\, v_{\xi^{(i)}\xi^{(j)}} + \sup_{\alpha\in A}\left[\nabla_\xi v\, f(t,x,\alpha) + U(t,x,\alpha)\right] = 0, \tag{16-a}$$

$$(t,\xi) \in [0,t^*) \times \mathbb{R}^d$$

$$v(t^*,\xi) = g(\xi), \qquad \xi \in \mathbb{R}^d \tag{16-b}$$

where v_t, $v_{\xi^{(i)}}$, etc., stand for $\frac{\partial v}{\partial t}$, $\frac{\partial v}{\partial \xi^{(i)}}$, etc., and are evaluated at (t,ξ) in eq. (16-a).

Similarly we can show, using Itô's Lemma, that if a function w satisfies Bellman's equation, than it coincides with the value function $v(t,\xi)$ given by eq. (15) and

we can also see how to construct optimal and ϵ-optimal control with the aid of Bellman's equation.
Therefore Bellman's equation supplies a method for finding the value function $v(t,\xi)$ as well as optimal and ϵ-optimal strategies.
We are going to follow the so-called martingale approach to stochastic control and make heavy use of Davis (1979, 1982), Elliott (1979, 1982) and Hausmann (1981, 1982). This method recasts Bellman's principle of optimality as a supermartingale inequality and then uses Meyer's supermartingale decomposition to obtain local conditions of optimality.
Let $v(t,x)$ be a function of class $C_p^{1,2}(\mathbb{R}_+ \times \mathbb{R}^d \to \mathbb{R})$ holding a polynomial growth property which satisfies the eqs. (15), (16).
Let M_t^α be defined as

$$M_t^\alpha = \int_0^t U(s,x(s),\alpha(s)ds + v(t,x(t)) \tag{17}$$

for any admissible control $\alpha(t)$ and its corresponding trajectory.
Expanding the function $v(t,x(t))$ by the Itô formula of change of variables, eq. (8) gives

$$M_t^\alpha = v(0,\xi_0) + \int_0^t \left[v_t + \frac{1}{2}\sum_{ij=1}^{d} (\sigma\sigma^*)^{(ij)} v_{\xi^{(i)}\xi^{(j)}} + \nabla_\xi v\, f^\alpha + U\right] ds + \int_0^t \nabla_\xi v\, dB \tag{18}$$

where $f^\alpha(t,x) = f(t,x,\alpha(t,x))$.

From the assumption that v solves the eq. (16) follows straightforward that the second term of eq. (18) is a decreasing process, since then its integrand is always non-positive. Moreover, if the control α is optimal then the integrand is identically zero and if one assumes further that v is such that the last term is a martingale, one obtains a <u>supermartingale inequality</u> version of the <u>principle of optimality</u>.

P1. <u>Optimality principle</u>

For any admissible α, M_t^α is a supermartingale and α is optimal if and only if M_t^α is a martingale.

This martingale optimality principle means that the failure to switch at time s to the optimal control entails a utility loss of

$$M_s^\alpha - \mathbb{M}\,[M_t^\alpha \mid \underline{\underline{F}}_s^*]$$

if one persists in using a non-optimal control over the time interval $[s,t]$.

A striking feature of the martingale approach is that optimal controls are constructed by maximizing the Hamiltonian

$$H(t,x,\nabla_\xi v,\alpha) = \nabla_\xi v\ f(t,x,\alpha) + U(t,x,\alpha) \tag{19}$$

and an adjoint process can be obtained under some additional smoothness assumptions on the controls which, fortunately enough, are fulfilled in our problem. See Hausmann (1979, 1981) and Kushner (1972).

Summing up, in order to solve our control problem within the martingale framework, we have to undertake the following steps:

a) Define the value function v and a conditional optimal utility process M_t^α as in eqs. (15), (17).

b) Show that the optimality principle P1 holds.

c) Construct an optimal strategy by means of the Hamiltonian given by eq. (19) and and adjoint process $\{p(t)\}$ based on the representation of martingales as stochastic integrals and ideas developed by Davis (1980) and Hausmann (1981) such that $p(t,x) = \nabla_\xi v(t,x)$.

Coming back to eq. (12), let us stress the fact that control $\alpha \in A$ determines a (strong Markov) process $\{x^\alpha(t)\}$ given by eq. (12) and this in turn induces a measure, say $\tilde{\mathbb{P}}_\alpha$, on the sample space $\tilde{\Omega}$ which enables us to evaluate the utility v^α, eq. (14), corresponding to the control α. It turns out that each $\tilde{\mathbb{P}}_\alpha$ is absolutely continuous with respect to the measure $\tilde{\mathbb{P}}$ induced by $x^\alpha(t)$ with $f \equiv U \equiv 0$. This leads us to the problem of absolutely continuous changes of measures associated with changes of control, which we outline briefly next.

3.2. Absolutely Continuous Change of Measure

Let $(\Omega,\underline{\underline{F}},\mathbb{P})$ be a complete probability space endowed with a righ-continuous increasing family $\underline{F} = (\underline{\underline{F}}_t)_{0\le t\le t^*}$ of sub-σ-fields of $\underline{\underline{F}}$, each of them containing all $\mathbb{P}$-null sets of $\underline{\underline{F}}$ such that $\underline{\underline{F}}_{t^*} = \underline{\underline{F}}$ which possess the following property:

P2. <u>Extension property of a consistent family of absolutely continuous measures</u>

If μ_t is an absolutely continuous probability measure on $(\Omega,\underline{\underline{F}}_t)$ with respect to $\mathbb{P}$ such that μ_t restricted on $\underline{\underline{F}}_s$ coincides with μ_s for any $t > s \ge 0$, then there exists a probability measure μ on $(\Omega,\underline{\underline{F}})$ such that μ restricted on $\underline{\underline{F}}_t$ coincides with μ_t for every $t \ge 0$.

Let $M_2^{c,loc}$ denote the set of all locally square integrable $\underline{\underline{F}}_t$-martingales on $(\Omega,\underline{\underline{F}},P)$ with a.s. continuous paths. Let $X = (X_t)_{t\ge0}$ be an element of $M_2^{c,loc}$ and $X_0 = 0$ a.s.

For $X \in M_2^{c,loc}$, we set

$$M_t = \exp(X_t - \frac{1}{2} \langle X \rangle t) \tag{20}$$

where $\langle X \rangle$ denotes the quadratic variational process corresponding to X. For simplicity we assume that M_t is a martingale. Now, we define a measure $\hat{\mathbb{P}}_t$ on $(\Omega, \underline{\underline{F}}_t)$, for each $t \geq 0$ and $A \in \underline{\underline{F}}_t$, by the formula

$$\mathbb{P}_t(A) = \mathbb{M}[M_t : A]. \tag{21}$$

Then it can be proved easily that for any $t > s \geq 0$

$$\hat{\mathbb{P}}_t \big|_{\underline{\underline{F}}_s} = \hat{\mathbb{P}}_s .$$

Further, under the assumption given by P2, there exists a probability measure $\hat{\mathbb{P}}$ on $(\Omega, \underline{\underline{F}})$ such that

$$\hat{\mathbb{P}} \big|_{\underline{\underline{F}}_t} = \hat{\mathbb{P}}_t .$$

$\hat{\mathbb{P}}$ is called the probability measure which has density M with respect to $\mathbb{P}$. We denote $\hat{\mathbb{P}}$ as

$$\hat{\mathbb{P}} = M \cdot \mathbb{P}$$

Let us recall the following theorem obtained by Girsanov in the case when $X = (X_t)_{t \geq 0}$ is a Brownian motion.

T1. Theorem (Girsanov): (i) Let $Y \in M_2^{c,loc}$. If we define $\tilde{Y}$ by

$$\tilde{Y}_t = Y_t - \langle Y, X \rangle_t \tag{22}$$

then $\tilde{Y} \in M_2^{c,loc}$.

$\langle Y, X \rangle$ denotes the quadratic variational process corresponding to Y and X or, loosely speaking, the cross-variation process.

(ii) Let $Y^1, Y^2 \in M_2^{c,loc}$ and define $\tilde{Y}^1$ and $\tilde{Y}^2$ by eq. (22), then

$$\langle Y^1, Y^2 \rangle = \langle \tilde{Y}^1, \tilde{Y}^2 \rangle. \tag{23}$$

The transformation of measures $\mathbb{P} \to \hat{\mathbb{P}}$ is called the Girsanov transformation or transformation of drift, since it induces a drift $\langle Y, X \rangle$ for every local martingale Y.

Precisely, the Girsanov Theorem implies that every continuous local martingale Y with respect to $\mathbb{P}$ is transformed under the probability $\hat{\mathbb{P}}$ to Y = a continuous local martingale + $\langle Y,X\rangle$ as a result of the transformation of probability measures $\mathbb{P} \to \hat{\mathbb{P}} = M \cdot \mathbb{P}$.

Before closing this section, we like to point out informally how are the links between the problems of control and change of measure.

As before let $\tilde{\mathbb{P}}_\alpha$ be the induced measure associated with the control $\alpha \in A$. Let us assume further that $f \equiv U \equiv 0$, then the expected utility v^α, eq. (14), becomes

$$v^\alpha(\xi_o) = \int_{\tilde{\Omega}} g(\omega(t^*))\tilde{\mathbb{P}}_\alpha(d\omega) := \tilde{\mathbb{M}}^\alpha g(\omega(t^*))$$

where $\tilde{\Omega}$ denotes the sample space, i.e. $C([0,t^*] \to \mathbb{R}^d)$, of $\omega(t)$ coordinate functions in $\tilde{\Omega}$, and $\tilde{\mathbb{M}}^\alpha$ expectation with respect to $\tilde{\mathbb{P}}_\alpha$.
As we mentioned already, under these simplifying assumptions holds that $\tilde{\mathbb{P}}_\alpha \ll \mathbb{P}$.
This enables us to define $m_{t^*}(\alpha)$ as

$$m_{t^*}(\alpha) = \frac{d\tilde{\mathbb{P}}_\alpha}{d\mathbb{P}} \tag{24}$$

and

$$m_t(\alpha) = \mathbb{M}[m_{t^*}(\alpha)|\underline{\underline{F}}_t]. \tag{25}$$

Hence $m_t(\alpha)$ is a positive martingale, $\mathbb{M}m_t(\alpha) = 1$ and $m_o(\alpha) = 1$ a.s. Since $\underline{\underline{F}}_o$ is the completion of the trivial σ-field $\{\sigma,\Omega\}$.

Moreover, the martingale $(m_t(\alpha))_{t\geq 0}$ permits a right continuous modification with left hand limits, i.e. $m_{t-}(\alpha) = \lim_{s\uparrow t} m_s(\alpha)$ a.s. See Ch. 3 of Liptser and Shiryayev (1977).

Let us recall Lemma 6.1 of the just quoted authors.

<u>L1. Lemma</u>: Let the process $m(\alpha) = (m_t(\alpha),\underline{\underline{F}}_t)_{t\leq t^*}$ satisfy

$$\mathbb{P}\left(\int_0^{t^*} m_s(\alpha)ds < \infty\right) = 1$$

and let $m_t(\alpha) \geq 0$ a.s., $0 \leq t \leq t^*$. Then the random process $m(\alpha) = (m_t(\alpha),\underline{\underline{F}}_t)$ is a (non-negative) supermartingale. In particular, $\mathbb{M}m_t(\alpha) \leq 1$ and $(m_t(\alpha))$ is a martingale if and only if $\mathbb{M}m_t = 1$.
We like to point out that about $(m_t(\alpha))$ we only know in advance that it is a local martingale.
Therefore, using results obtained originally by Doléans-Dade, one can prove that

$(m_t(\alpha))$ is given uniquely by

$$m_t(\alpha) = \exp(m_t(\alpha) - \frac{1}{2} \langle m^c(\alpha), m^c(\alpha)\rangle_t) \prod_{s\le t} (1+\Delta m_s(\alpha))\, e^{-\Delta m_s(\alpha)}$$

where $m^c(\alpha)$ is the continuous part of local martingale $(m_t(\alpha))$, the countable product is a.s. absolutely convergent and (X_t) is a local martingale such that $(m_t(\alpha))$ holds

$$m_t(\alpha) = 1 + \int_0^t m_{s-}(\alpha) dX_s \ .$$

It turns out that $(m_t(\alpha))$ is a non-negative local martingale, provided that $\Delta m_s(\alpha) \ge -1$ for all ω and s, and according to the Lemma above is a martingale if and only if $\mathbb{M}_{t*}(\alpha) = 1$.

This result, in connection with the change of measure, is of great utility as the following Lemma due to von Schuppen and Wong points out, see Elliott (1982), ch. 13.

L2. Lemma: Suppose $\mathbb{M}_t(\alpha) = 1$ and define a measure on $(\tilde{\Omega}, \underline{F}_{t*})$ by (24). Let X be a local martingale such that the cross-variation process $\langle X, m(\alpha)\rangle$ exists. Then $\tilde{X}_t := X_t - \langle X, m(\alpha)\rangle_t$ is a $\tilde{\mathbb{P}}_\alpha$-local martingale.

From the general formula connecting Radon-Nikodym derivatives and conditional expectation one obtains

$$\tilde{\mathbb{M}}^\alpha[X_t | \underline{F}_s] = \mathbb{M}[m_t(\alpha)\tilde{X}_t | \underline{F}_s] m_s^{-1}(\alpha).$$

Hence, $\tilde{X}_t$ is a $\tilde{\mathbb{P}}_\alpha$-local martingale if and only if $m_t(\alpha)\tilde{X}_t$ is a $\mathbb{P}$-local martingale.

With respect to our control problem this means that

$$\tilde{\mathbb{M}}^\alpha g = \mathbb{M}[m_t(\alpha) g] \ . \tag{26}$$

Therefore, a look at eq. (22) suggests that our optimal control problem is equivalent to choose $\alpha \in A$ such that eq. (26) maximizes. Afterwards, a drift transformation switches us back to the original problem, i.e. one with $f \neq 0$ and $U \neq 0$, see eqs. (12), (14).

3.3. Optimal Control of Completely Observable Diffusions

Let us make the following additional assumptions on the drift f and diffusion term σ:

P3. Assumptions on f and σ

(a) $\sigma^{(ij)}(\cdot,\cdot)$ is $\underline{\underline{F}}_t$-predictable

(b) $|\sigma^{(ij)}(t,x) - \sigma^{(ij)}(t,y))| \leq \gamma \sup_{0\leq s\leq t} |x(s) - y(s)|$

(c) (t,x) is non-singular for each (t,x) and $(\sigma^{-1}(t,x))^{(ij)} \leq \gamma$

where γ is a fixed constant independent of t,i,j.

Then, there is a unique strong solution to the SDE

$$dx(t) = \sigma(t,x)\, dB(t), \qquad \xi_o \in \mathbb{R}^d \text{ given.} \tag{27}$$

Now, let $A \subseteq \mathbb{R}^d$ be compact and the set of control $\mathbb{A}$ be the set of $\underline{\underline{F}}_t$-predictable A-valued processes $\alpha = \{\alpha(t)\}$. Further:

P4. Assumptions on f and σ relative to α

(d) $f(t,x,\cdot)$ is continuous in $\alpha \in \mathbb{A}$, for each (t,x)

(e) f is $\underline{\underline{F}}_t$-predictable in (t,x) for each α

(f) $f(t,x,\alpha) \leq \gamma(1 + \sup_{s\leq t}|x_s|)$.

Now, let us define for $\alpha \in \mathbb{A}$

$$m_t(\alpha) = \exp\Big(\int (\sigma^{-1}(s,x)f(s,x,\alpha(s)))^* dB(s) - \frac{1}{2}\int_0^t |\sigma^{-1}f|^2 ds\Big). \tag{28}$$

The boundedness of σ^{-1} and the growth condition on f imply the Novikov condition, i.e. $\mathbb{M}(\frac{1}{2}\int_o^t|\sigma^{-1}f|^2ds) \leq \infty$ and this in turn guarantees $\mathbb{M}m_t(\alpha) = 1$.

Therefore, the main steps in the foregoing sections are justified, we can thus define a measure $\tilde{\mathbb{P}}_\alpha$ on $(\tilde{\Omega},\underline{\underline{F}}_{t^*})$ by eq. (24) and state the following theorem.

T2. Theorem: Under $\tilde{\mathbb{P}}_\alpha$ the process $\{x(t)\}$ satisfies

$$dx(t) = f(t,x,\alpha(t))dt + \sigma(t,x)dw^\alpha(t) \tag{29}$$

with initial condition $x(0) = \xi_o \in \mathbb{R}^d$, where $\{w^\alpha(t)\}$ is the $\tilde{\mathbb{P}}_\alpha$ - Brownian motion given by

$$dw^\alpha(t) = dB(t) - \sigma^{-1}(t,x)\, f(t,x,\alpha(t))dt\,. \tag{30}$$

Proof: This theorem follows by showing that $\{w^\alpha(t)\}$ is a Brownian motion and from the fact that stochastic integrals under $\tilde{\mathbb{P}}$ or $\mathbb{P}$ give the same process. ∎

Suppose the instantaneous utility is determined by a real, bounded measurable functional U, as in Section 3.1., satisfying the same conditions as f and that the terminal utility is given by a real, bounded measurable functional g as in 3.1. However, according to the above remark we shall take expectation with respect to the measure $\tilde{\mathbb{P}}_\alpha$. Thus, if control $\alpha \in \mathbb{A}$ is used, the total expected utility is now given by

$$v^\alpha(\xi_o) = \tilde{\mathbb{M}}^\alpha\left[\int_0^{t^*} U(t,x,(t),\alpha(t))dt + g(x(t^*))\right. . \tag{31}$$

By analogy, the terminal utility $v^\alpha(t,x,)$, for fixed $t \in [0,t^*]$ and $x(t) =$ becomes ξ

$$v^\alpha(t,\xi) = \mathbb{M}^\alpha_{(t,\xi)}\left[\int_t^{t^*} U^\alpha(s,x)ds + g(x(t^*))\right] . \tag{32}$$

Fron now on, we shall write $U(s,x(s),\alpha(s))$ as $U^\alpha(s,x)$ and similarly for f . As a conditional expectation, v^α is defined only almost surely. From eq. (32) follows that $v^\alpha(t,x)$ only depends on α restricted to $[t,t^*]$ and since all the measures $\{\tilde{\mathbb{P}}_\alpha , \alpha \in \mathbb{A}\}$ are equivalent, the null sets up to which $v^\alpha(t,x)$ is defined are also control-independent. Because U and g are bounded, $v^\alpha(t,x)$ is a well-defined element of $L^1(\tilde{\Omega},\underline{\underline{F}}_t,\mathbb{P})$ for each $\alpha \in \mathbb{A}$.

Therefore, the supremum $v(t,\xi)$ given by

$$v^{\alpha^o}(t,\xi) := v(t,\xi) = \sup_{\alpha \in \mathbb{A}} v^\alpha(t,\xi) \tag{33}$$

exists and is $\underline{\underline{F}}_t$-measurable. This result is due to the fact that the value function $v(t,x(t))$ evaluated along any trajectory corresponding to a control feasible for its initial state is a nonincreasing function of time and that $L^1(\tilde{\Omega},\underline{\underline{F}}_t,\mathbb{P})$ is a complete lattice.

To save notation, let us write $v(t)$, $\mathbb{P}_\alpha$ and $\mathbb{M}^\alpha_t$ for $v(t,x(t))$, $\tilde{\mathbb{P}}_\alpha$ and $\tilde{\mathbb{M}}^\alpha_{t,x(t)}$. Further, $\mathbb{A}^t_s$ stands for the set of controls $\alpha, \alpha \in \mathbb{A}$, restricted to the interval $[s,t] \subseteq [0,t^*]$.

Let us recall the following lemma due to Rishel (1970).

<u>L3. Lemma</u>: For each fixed $\alpha \in \mathbb{A}$ and $0 \le r \le t \le t^*$ the value function v satisfies the following principle of optimality:

$$v(r) \ge \mathbb{M}^\alpha_r\left[\int_r^t U^\alpha(s,x)ds + v(t)\right] \quad \text{a.s.} \tag{34}$$

<u>Sketch of the proof</u>: Apply the characterization of the value function v given by eq. (33) restricted to controls $\alpha \in A^{t^*}_r$ and consider the subset $\hat{A}^t_r$, $\hat{A}^t_r \subseteq A^t_r$, of admissible controls $\hat{\alpha}$ which are equal to α when restricted to $(r,t]$.

The lemma follows, then, from the facts that the family of random variables $\{v^{\alpha}(t),\ \alpha \in \mathbb{A}_r^{t^*}\} \subseteq L^1\ (\Omega, \underline{F}, \mathbb{P})$ has the ϵ-lattice property and the supremum lattice and conditional expectation operations commute. The family $\{v^{\alpha}(t),\ \alpha \in \mathbb{A}$ has the ϵ-lattice property means that for any $\alpha \in 0$ and given $\alpha^1, \alpha^2 \in \mathbb{A}$ there exists $\alpha^3 \in \mathbb{A}$ such that

$$v^{\alpha^3}(t) \geq v^{\alpha^1}(t) \ \bigvee \ v^{\alpha^2} - \epsilon \quad \text{a.s.}$$

where $\bigvee$ stands for the supremum lattice operation, see eq. (33).

Now, let us rewrite eq. (17) as

$$M_t^{\alpha} = \int_0^t U^{\alpha}(s,x)ds + v(t) \tag{35}$$

and note that for $t = 0$ and $\alpha \in \mathbb{A}$, we have

$$M_0^{\alpha} = v(0) = \bigvee_{\hat{\alpha} \in \mathbb{A}_0^{t^*}} v^{\alpha}(\xi_0) \tag{36}$$

since ξ_0 is assumed to be a fixed constant. See eq. (31). $v(0)$ represents the "<u>maximum expected utility</u>". Further, for any $\alpha \in \mathbb{A}$ and $t = t^*$ we get

$$M_{t^*}^{\alpha} = \int_0^{t^*} U^{\alpha}(s,x)ds + g(x(t^*)) = \rho^{\alpha} \tag{37}$$

the "<u>sample utility</u>" associated with the control α, see eq. (13). ∎

The principle of optimality given by P1 can now be restated as follows.

<u>T3. Theorem</u>: $\{M_t^{\alpha}\}$ is a $\mathbb{P}_{\alpha}$ -supermartingale for any admissible $\alpha \in \mathbb{A}$. The control $\alpha \in \mathbb{A}$ is optimal if and only if $\{M_t^{\alpha}\}$ is $\mathbb{P}_{\alpha}$ -martingale, i.e. if and only if α gives the maximum expected utility.

<u>Proof</u>: Since $\int_0^r U^{\alpha}(s,x)ds$ is $\underline{F}_r$ - measurable adding this expression to both sides of eq. (34), one gets

$$M_r^{\alpha} \geq \mathbb{M}_r^{\alpha}[M_t^{\alpha} | \underline{F}_r]$$

for $0 \leq r \leq t \leq t^*$. That is, $\{M_t^{\alpha}\}$ is a $\mathbb{P}_{\alpha}$ - supermartingale.

Now, if $\{M_t^{\alpha}\}$ is a $\mathbb{P}_{\alpha}$ -martingale then from eqs. (35) and (36) follows

$$\mathbb{M}^{\alpha}[M_{t^*}^{\alpha}] = \mathbb{M}^{\alpha}[M_0^{\alpha}] = v(0)$$

and α is optimal.

Conversely, if α is optimal then for any t we obtain

$$v(0) = \mathbb{M}^{\alpha}\left[\int_0^t U^{\alpha}(s,x)ds + v^{\alpha}(t)\right] \tag{38}$$

and from eq. (34) follows

$$v(0) \geq \mathbb{M}^{\alpha}\left[\int_0^t U^{\alpha}(s,x)ds + v(t)\right] . \tag{39}$$

Hence combining eqs. (38) and (39) one has

$$0 \geq \mathbb{M}^{\alpha}\left[v(t) - v^{\alpha}(t)\right]$$

which together with eq. (33) gives

$$v(t) = v^{\alpha}(t) \quad \text{a.s.} \tag{40}$$

Now, adding $\int_0^t U^{\alpha}(s,x)ds$ to both sides of eq. (40) and using eqs. (32), (35) and (37) follows

$$M_t^{\alpha} = \mathbb{M}_t^{\alpha}[M_{t^*}^{\alpha}]$$

and the theorem is proved. ∎

Let us close this section by calling attention to the fact that under the conditions we have stated in Section 1 and Section 3.2, i.e. "les conditions habituelles", the function $t \rightarrow \mathbb{M}[M_t^{\alpha}]$, for a fixed $\alpha \in \mathbb{A}$, is right-continuous. Hence the $\mathbb{P}_{\alpha}$-supermartingale $\{M_t^{\alpha}\}$ has a right-continuous modification and this is càdlàg, i.e. continuous on the right and has limits on the left. See Dellacherie et Meyer (1980), Liptser and Shiryayev (1979). Elliott (1982) gives complete proofs of the results we just mentioned which he has carefully tailored after the martingale approach at issue and concludes that the value function v has also a càdlàg version. Finally, since U^{α} and g are bounded, see Section 3.1, the $\mathbb{P}_{\alpha}$-martingale $\{M_t^{\alpha}\}$ is of class D and has therefore a Doob-Meyer decomposition. This result is crucial as we shall see in the next section.

4. The Stochastic Maximum Principle

4.1 The Doob-Meyer Decomposition and Optimality Conditions

Solving the control problem (10) by means of martingale techniques leads us to the construction of an optimal control $\{\alpha^0(t)\}$ by maximizing the Hamiltonian given by eq. (19), i.e. step c) in Section 3.1. Unfortunately, this entails, at least implicitly, solving the Bellman eq. (16) and that is precisely what we would like to avoid.

However, eq. (18) and the comments following it, suggest an alternative via the

Doob-Meyer decomposition for the family of conditional optimal utilities $\{M_t^\alpha\}$ and the representation of its martingale term as a stochastic integral. So we expect to get a process $\{p(t)\}$, the "adjoint process" in the terminology of the control theory, which is defined independently of the existence of any optimal control. This process shall play the role of the gradient of the value function, i.e. $p = \nabla_\xi v$.

Now, the above mentioned step c) of the working program resulting from the martingale approach becomes clear, and results due to Benes (1970) and Duncan and Varaiya (1971) make sure that indeed a predictable control process $\{\alpha^o(t)\}$ exists.

With regard to the second part of the step c) one proceeds as follows. First, the Doob-Meyer decomposition guarantees for any $\alpha \in A$ existence of a unique predictable decreasing process $\{A_t^\alpha\}$, with $A_o^\alpha = 0$, and a uniformly integrable $\mathbb{P}_\alpha$ -martingale $\{N_t^\alpha\}$, with $N_o^\alpha = 0$, such that

$$M_t^\alpha = v(0) + A_t^\alpha + N_t^\alpha \quad . \tag{41}$$

Then, one looks for a representation of the martingale $\{N_t^\alpha\}$ as stochastic integral with respect to the $\mathbb{P}_\alpha$ -Brownian motion $\{w^\alpha(t)\}$, see eq. (30). Let us make a few comments on this. That representation would follow from standard results, see Lipster and Shiryayev (1977), Kunita and Watanabe (1967), if the filtration $\underline{F}_t$ was generated by $\{w^\alpha(t)\}$ and $\{N_t^\alpha\}$ was square integrable.

However, one can prove that all square integrable $\underline{\underline{F}}_t$-martingales are representable as stochastic integrals of $\{w^\alpha(t)\}$, see Fujisaki, Kallianpur and Kunita (1972), Davis and Varaiya (1973), and Liptser and Shiryayev (1977).
It is worth noting that from eq. (27) and the Lipschitz property of $\sigma(t,x)$ follows that the filtration $\underline{F}_t$ is generated by $\{x(s):s \le t\}$ or, equivalently, by $\{B(s):s \le t\}$, i.e. $\underline{F}_t^B = \underline{F}_t^x = \underline{F}_t$. And further that $\{x(t)\}$ as given by eq. (29) is only a weak solution. Therefore, $\{w^\alpha(t)\}$ given by eq. (30) does not necessarily generate $\underline{F}_t$, i.e. $\underline{\underline{F}}_t \subseteq \underline{\underline{F}}_t^W$ and the inverse inclusion may not be valid. Let us finally state the following lemma.

<u>L4. Lemma</u>: Let $\{N_t^\alpha\}$, with $N_o^\alpha = 0$, be the uniformly integrable $\mathbb{P}_\alpha$ -martingale which occurs in the <u>Doob-Meyer decomposition</u>, eq. (41).
Then, there is an $\underline{\underline{F}}_t$-predictable process $\{p(t)\}$ for which there is an increasing sequence of stopping times τ_n with $\lim \tau_n = t$ a.s. and $\mathbb{M}[\int_0^\tau |p(s)|ds] < \infty$ such that

$$N_t^\alpha = \int p(s)\sigma(s,x)dw^\alpha(s) \quad \text{a.s.} \tag{42}$$

<u>Proof</u>: See Elliott (1982), corollary 16.23.

■

Now, that we have the wanted representation, eq. (42), and the adjoint process $\{p(t)\}$, let us come back to the step c) and look for conditions of optimality. From eqs. (35), (41) and (42) we have for any $\alpha \in \mathbb{A}$

$$v(t) = v(0) + A_t^\alpha + \int_0^t p(s)\sigma(s,x)dw^\alpha(s) - \int_0^t U^\alpha(s,x)ds. \tag{43}$$

Now, for $\bar{\alpha} \in \mathbb{A}$, any other admissible control, the supermartingale $\{M_t^{\bar{\alpha}}\}$ is given, according to eq. (35), by

$$M_t^{\bar{\alpha}} = \int_0^t U^{\bar{\alpha}}(s,x)ds + v(t) \tag{44}$$

which together with eqs. (30) and (43) becomes

$$M_t^{\bar{\alpha}} = v(0) + \left[A_t^{\bar{\alpha}} + \int_0^t (H_s(\bar{\alpha}_s) - H_s(\alpha_s))ds\right] + \int_0^t p(s)\sigma(s,x)dw^{\bar{\alpha}}(s) \tag{45}$$

where

$$H_s(\alpha_s) = p(s)f^\alpha(s,x) + U^\alpha(s,x). \tag{46}$$

Eq. (45) or, equivalently, eqs. (47) and (48) below are very helpful examining efficiency of alternative controls. Since a unique Doob-Meyer decomposition exists for any admissible control, it turns out, comparing eqs. (41) and (45), that for $\bar{\alpha} \in \mathbb{A}$

$$A_t^{\bar{\alpha}} = A_t^\alpha + \int_0^t (H_s(\bar{\alpha}_s) - H_s(\alpha_s))ds \tag{47}$$

$$N_t^{\bar{\alpha}} = \int_0^t p(s)\sigma(s,x)dw^{\bar{\alpha}}(s) . \tag{48}$$

At this point we like to stress the fact that the Brownian motion $\{B(t)\}$ and the adjoint process $\{p(t)\}$ as well are control independent. This result is a crucial achievement of the martingale approach.
Now, we can state necessary and sufficient conditions for a control $\alpha \in \mathbb{A}$ to be optimal.

<u>T4. Theorem:</u> a) <u>A necessary condition.</u> If $\alpha = \alpha^0 \in \mathbb{A}$ is optimal, then it maximizes (a.s. $d\mathbb{P} \times$ Lebesgue) the Hamiltonian H_s of eq. (46).

b) <u>A sufficient condition.</u> For a control $\alpha^0 \in \mathbb{A}$ consider the $\mathbb{P}_{\alpha^0}$-martingale $\{\rho_t^\alpha\}$

$$\rho_t^{\alpha^0} = \mathbb{M}_t^{\alpha^0}\left[\rho^{\alpha^0}\right] = \mathbb{M}^{\alpha^0}\left[M_{t^*}^{\alpha^0}\right] = \int_0^t U^{\alpha^0}(s,x)ds + v^{\alpha^0}(t). \tag{49}$$

Then α^o is optimal if for any other $\alpha \in \mathbb{A}$ the process $\{I_t^\alpha\}$ given by

$$I_t^\alpha = \int_0^t U^\alpha(s,x)ds + v^{\alpha^o}(t) \tag{50}$$

is a $\mathbb{P}_\alpha$-supermartingale.

<u>Proof:</u> a) If α^o is optimal, then from T3 $\{M_t^{\alpha^o}\}$ is a $\mathbb{P}_{\alpha^o}$-martingale and $A_t^{\alpha^o} = 0$. Hence the decreasing process $\{A^{\bar\alpha}\}$, eq. (47), reduces to

$$A_t^{\bar\alpha} = \int_0^t (H_s(\bar\alpha_s) - H_s(\alpha_s^o))ds$$

and the integrand has to be

$$H_s(\bar\alpha_s) \le H_s(\alpha_s^o) \quad \text{a.s.} \quad (d\mathbb{P}\, xdt) \tag{51}$$

for any other admissible $\bar\alpha \in \mathbb{A}$.

b) Suppose now that the process $\{I_t^\alpha\}$ is a $\mathbb{P}_\alpha$-supermartingale. Combining eqs. (49) and (50) one gets

$$I_t^\alpha = \rho_t^{\alpha^o} + \int_0^t (U^\alpha(s,x) - U^{\alpha^o}(s,x))ds .$$

Then the result follows, since then we have

$$v^{\alpha^o}(0) = I_0^{\alpha^o} = \rho_0^{\alpha^o} = \mathbb{M}^\alpha I_0^\alpha \ge \mathbb{M}^\alpha I_{t^*}^\alpha = v^\alpha(0) .$$

That is, α^o maximizes the total expected utility and is hence optimal. ∎

4.2. The Adjoint Process p.

We have made some progess characterizing the unique adjoint process $p = \{p(t)\}$, see L4 and eq. (42).
However, from the point of view of applications these results are still unsatisfactory. For, in order to construct an optimal process α^o, see eq. (46) and step c) in Section 3.1, we should obtain an "explicit" representation of p, i.e. in terms of the coefficients f and σ, see eq. (29), and of a functional of the diffusion x, the "<u>sample utility</u>" ρ^α given by eq. (13).
This is in general hard to obtain. Most of the existing results in this regard are due to Hausmann (1978, 1979, 1981), and rest upon the representation of functionals of Itô processes as stochastic integrals, see Liptser and Shiryayev (1977). In this section, we outline heuristically, following Davis (1980), the main ideas that shall lead us to the needed representation. To begin with, we assume that for any $\alpha \in \mathbb{A}$ the sample utility ρ^α is a smooth functional of the diffusion $x = \{x(t)\}$,

eq. (29), and that its coefficients fulfill the following additional requirements:

P5. Further Assumptions on f and σ.

(a) for all (t,ξ)

$$\sigma(t,\xi)\cdot\sigma^*(t,\xi) \geq x \quad I_d \geq 0$$

where x is a constant independent of (t,ξ) and I_d stands for the d-dimensional identity matrix.

(b) $\mathbb{P}\left\{\int_0^{t^*} |f^\alpha|^2 \, ds < \infty\right\} = 1$

(c) the functions $(\sigma(t,\xi))^{(ij)}$, $i,j = 1,\ldots,d$,

are Hölder continuous in ξ, uniformly in $(t,\xi) \in [0,t^*] \times \mathbb{R}^d$.

P6. Fréchet Differentiability of M_t^α

Further, suppose $\alpha^o \in \mathbb{A}$ is optimal and that the (random variable) functional $M_{t^*}^{\alpha^o} : \Omega \to \mathbb{R}_-$, see eq. (37), given by

$$M_{t^*}^{\alpha^o} = \int_0^{t^*} U^{\alpha^o}(s,x,\alpha\,(s,x))ds + g(x(t^*)) \tag{52}$$

is Fréchet differentiable in x, where x is given by eq. (29) and as in Section 2 and 3 is an element of the sample space Ω, the set of continuous functions from $[0,t^*]$ to $\mathbb{R}^d$. Then, there exists a map $\Pi:\Omega \to \Omega^*$ such that for $x,y \in \Omega$

$$M_{t^*}^{\alpha^o}(x+y) = M_{t^*}^{\alpha^o}(x) + \Pi(x)(y) + O(\|y\|) . \tag{53}$$

Besides for each $x \in \Omega$ there is, by the Riesz representation theorem, an $\mathbb{R}^d$-valued Radon measure $\mu_x(s)$ for $y \in \Omega$

$$\Pi(x)(y) = \int_{[0,t^*]} y(x)\ \mu_x(ds) \tag{54}$$

and $\|\Pi(x)\|_{\Omega^*} = TV(\mu_x)$ (the total variation of μ_x).

Davis (1980) denotes by $\mu_x(t)$ the right-continuous bounded variation function $\mu_x([0,t])$ corresponding to the Fréchet derivative of $M_{t^*}^\alpha$ at x.

On the other hand, since α^o is optimal, we have from eqs. (41) and (42)

$$M_t^{\alpha^o} = v(0) + \int p(s)\sigma(s,x)dw^{\alpha^o}(s) \tag{55}$$

where v, the value function, satisfies the following parabolic differential equation evaluated at (t,ξ)

$$\frac{\partial v}{\partial t} + \frac{1}{2} \sum_{i,j=1}^{d} (\sigma\sigma^*)^{(ij)} v_{\xi^{(i)}} v_{\xi^{(j)}} + \sum_{i=1}^{d} (f^{\alpha^o})^{(i)} v_{\xi^{(i)}} + U^{\alpha^o} = 0 \tag{56-a}$$

$$(t,\xi) \in [0,t^*) \times \mathbb{R}^d$$

$$v(t^*,\xi) = g(x(t^*)), \quad \xi \in \mathbb{R}^d . \tag{56-b}$$

Now, expanding $v(s,x(s))$ by the Itô rule and making use of the foregoing parabolic PDE yields

$$v(t,\xi) = \mathbb{M}^{\alpha^o}_{t\xi} \left[\int_t^{t^*} U^{\alpha^o} ds + g(x(t^*)) \right] . \tag{57}$$

Applying Itô's change of variable formula to $v(s,x(s))$ from t till t^*, see eq. (8), one gets

$$v(t^*,x(t^*)) - \int_t^{t^*} (\tfrac{\partial}{\partial s} + \mathbb{L}^{\sigma,f}) v ds = v(t,\xi) + \int_t^{t^*} \nabla_\xi v \, \sigma dw^{\alpha^o}(s)$$

which together with eq. (56) gives

$$g(x(t^*)) + \int_t^{t^*} U^{\alpha^o} ds = v(t,\xi) + \int_t^{t^*} \nabla_\xi \, v \, \sigma \, dw^{\alpha^o}(s) .$$

A look at the left-hand side of the foregoing equation and at eq. (52) with $t = 0$ gives the following representation for $M^{\alpha^o}_{t^*}$

$$M^{\alpha^o}_{t^*} = v(0) + \int_0^{t^*} \nabla_\xi \, v \, \sigma \, dw^{\sigma^o}(s) . \tag{58-a}$$

On the other hand, using eq. (57) with $t = 0$ and eq. (44) with $t = t^*$ delivers the following enlightening representation for $M^{\alpha^o}_{t^*}$

$$M^{\alpha^o}_{t^*} = \mathbb{M}^{\alpha^o} \left[M^{\alpha^o}_{t^*} \right] + \int_0^{t^*} \nabla_\xi v \, \sigma \, dw^{\alpha^o}(s) . \tag{58-b}$$

Therefore, a comparison of eq. (58-a) with eq. (55) setting $t = t^*$ suggests formally a crucial relation between p and $\nabla_\xi v$. Furthermore, taking into account the Fréchet differentiability of $M^{\alpha^o}_{t^*}$ and eq. (58-b) completes the relationship between p, $\nabla_\xi v$ and y, see eqs. (53) and (54).

In order to get additional insights into the structure of the integrands in eqs. (54) and (58) and the relationship just mentioned, we need first to find an expression for $d(\Phi\, \nabla_\xi v)$, where $\Phi(s,t)$, with $s,t \in [0,t]$, is precisely the fundamental matrix solution of

$$d\eta = \nabla_\xi f(s,x(s),\alpha^o(t,x(s)))\eta ds + \nabla_\xi \sigma^{(k)}(s,x(s))\eta dw^{(k)\alpha^o}(s) \tag{59}$$

with $\eta(s) = \Phi(s,t)$, $0 \le t \le s \le t^*$. Repeated indices imply summation from 1 to d.

P7. Smoothness Assumptions on σ, f, U and g

(d) σ is of class $C^{1,2}(\bar{Q})$

(e) f and U are of class $C^{1,1}(\bar{Q} \times A)$

(f) g is of class $C^2(G)$

where $Q = (0,t^*) \times G$, $G \subseteq \mathbb{R}^d$ an open bounded set with boundary ∂G of class C^2, so that t^* is the first exit time of the process $\{t,x(t):t \ge 0\}$ from the open set $(0,T) \times G$, $x(0) = x_o \in G$; see Gõmez (1984).

Under these additional assumptions v becomes of class $C^{1,2}(Q) \cap C^{0,1}(\bar{Q})$, see Fleming and Rishel (1975).

Before we look for an expression for $d\nabla_\xi v$ let us mention that it was Kushner (1972) who first considered the process η as a solution of eq. (29) linearized about x with the purpose of deriving a stochastic maximum principle relying on mathematical programming and the variational theory of Neustadt. Kushner's innovative ideas provide a fundamental framework upon which the representation of the adjoint process p rests which in turn constitutes in a certain way the heart of the martingale approach.

Differentiating eq. (56) with respect to ξ Hausmann (1981) obtains for $z^{(k)} = v_{\xi(k)}$ as a Schwartz distribution, $k = 1,2,\ldots,d$ the equation

$$z_s + \sum_{i=1}^{d} \left(\frac{1}{2} \sum_{j=1}^{d} (\sigma\sigma^*)^{(ij)} z_{\xi(j)} \right)_{\xi(i)} - \frac{1}{2} \sum_{i,j=1}^{d} (\sigma\sigma^*)^{(ij)}_{\xi(i)} z_{\xi(j)} + Z = 0$$

$$(s,x(s)) \in [0,t^*) \times \mathbb{R}^d \tag{60-a}$$

$$z(t^*,\xi) = v_{\xi(k)}(t^*,\xi) = g_{\xi(k)}(x(t^*)) \tag{60-b}$$

where

$$Z(s,x(s)) = \frac{1}{2} \sum_{i,j=1}^{d} (\sigma\sigma^*)^{(ij)}_{\xi(k)} v_{\xi(i)\xi(j)} + \sum_{i=1}^{d} f^{(i)\alpha^o} v_{\xi(i)\xi(k)} + f^{\alpha^o}_{\xi(k)} v_{\xi(k)} + U^{\alpha^o}_{\xi(k)}$$

that in the scalar case reduces to

$$z_s + \mathbb{L}^{\sigma,f}z + z_\xi \sigma\sigma_\xi + zf_\xi^{\alpha^o} + U_\xi^{\alpha^o} = 0. \tag{61-a}$$

$$z(t^*,\xi) = v_\xi(t^*,\xi) = g(x(t^*)) . \tag{61-b}$$

So we obtain an expression for dv_ξ, i.e. $(v_{\xi t} + \mathbb{L}^{\sigma,f}v_\xi)ds + v_{\xi\xi}\sigma dw^{\sigma^o}(s)$. Thus, we get

$$-dz^* = \left[\nabla_\xi f z^*(s,x) - \nabla_\xi \sigma^{(k)}(s,x)v_{\xi\xi}(s,x)\sigma^{(k)}(s,x) + (\nabla_\xi U^{\alpha^o}(s,x))^*\right]ds + - v_{\xi\xi}(s,x)\sigma(s,x)dw^{\alpha^o}(s) \tag{62-a}$$

$$z^*(t^*) = \nabla_\xi g(x(t^*)) \tag{62-b}$$

where $v_{\xi\xi}$ is the Hessian of the value function v. Now, applying Itô's lemma to the product $\eta(s)z(s,x(s))^*$ yields

$$d\eta z^* = (\nabla_\xi f\eta ds + \nabla_\xi \sigma^{(k)}\eta dw^{(k)\alpha^o}(s))z^* + \eta\left[(\frac{\partial z^*}{\partial s} + \mathbb{L}^{\sigma,f}z^*)ds + + \nabla_\xi z^* \sigma^{(k)}dw^{(k)\alpha^o}(s)\right] + d\eta\circ dz \tag{63}$$

and using eqs. (60) and (62) gives

$$\eta(t^*)\nabla_\xi g(x(t^*)) - \eta(t)z^*(t,\xi) = -\int_t^{t^*} \eta\nabla_\xi U^{\alpha^o} ds + \int_t^{t^*} \eta(\nabla_\xi \sigma^{(k)} z^* + \nabla_\xi z^* \sigma^{(k)})dw^{(k)\alpha^o}(s).$$

Now, taking $\eta(t) = I_d$, one obtains

$$\nabla_\xi v(t,\xi) = z(t,\xi) = \mathbb{M}_{t\xi}^{\alpha}\left[\eta(t^*)\nabla_\xi g(x(t^*)) + \int_t^{t^*} \eta\nabla_\xi U^{\alpha^o} ds)\right] \tag{64}$$

which is the representation we look for.

We like to capture the results of this subsection in the following two theorems.

<u>T5. Theorem:</u> Suppose the additional requirements (d), (e) and (f) are satisfied. then

$$p(t,\xi) = \nabla_\xi v(t,\xi)$$
$$= \mathbb{M}_{t\xi}^{\alpha^o}\left[\int_t^{t^*} \nabla_\xi U(s,x(s),\alpha^o(s,x(s)))\Phi(s,t)ds + \nabla_\xi g(x(t^*))\Phi(t^*,t)\right] . \tag{65}$$

Davis (1980) and Hausmann (1981) give rigorous proof of this result.

P8. Assumptions on ρ^{α^o}

We make the following assumptions on the functional ρ^{α^o}

(g) There exist positive integers n_1, n_2 such that

$$|\rho^{\alpha^o}(x)| + TV(\mu_x) \le n_1(1 + \|x\|)^{n_2} .$$

(h) μ_x is continuous in x in the weak topology.

T6. Theorem: Suppose assumptions (a) and (g) are satisfied. Then ρ^{α^o} has a representation of the form

$$\rho^{\alpha^o}(x) = \mathbb{M}\rho^{\alpha^o} + \int_0^{t^*} e(s)dx(s) \qquad \mathbb{P}_{\alpha^o} \text{ - a.s.} \tag{66}$$

where the integrand $e(s)$ satisfies $\mathbb{M}\int_0^{t^*} |e(s)|^2 ds < \infty$ and is given by

$$e(t) = \mathbb{M}^{\alpha^o}_{t\xi}\left[\int_{(t,t^*]} \mu_x(ds)\Phi(s,t)\right]\sigma(t,x(t)). \tag{67}$$

Here $\Phi(s,t)$ is the matrix-valued process defined for $0 \le t \le s \le t^*$ by

$$d\Phi(s,t) = F(s,x(s))\Phi(s,t)ds + E^{(k)}(s,x(s))\Phi(s,t)dw^{(k)\alpha^o}(s) \tag{68}$$

where

$$\begin{aligned} F(s,x(s)))^{(ij)} &= (f_{\xi^{(j)}}(s,\xi))^{(i)} \\ (E^{(k)}(s,x(s))^{(ij)} &= (\sigma^{(k)}_{\xi^j}(s,\xi))^{(ik)} . \end{aligned} \tag{69}$$

Proof: See Davis (1980) and Hausmann (1981) for a rigorous proof of this theorem.

Recall that $\rho^{\alpha^o} = M^{\alpha^o}_{t^*}$. The Fréchet derivative of $M^{\alpha^o}_{t^*}$ is then

$$\mu_\xi(ds) = \left(\int_t^{t^*} \nabla_\xi U^{\alpha^o} ds + \nabla_\xi g(x(t^*))\right)\delta_1(ds) \tag{70}$$

where δ_t is the Dirac measure at t.

Further, one proves that $y(x)$ in eq. (54) is also given by eq. (67), which clearly establishes the suggested relation between the integrands in eqs. (54), (55) and (58). Moreover, we obtain the following crucial result

L5. Lemma: Assume (d) - (f) as in P7. Then Hamiltonian H, eq. (46), is differentiable $\mathbb{P}_{\alpha^o}$ - almost everywhere and for any $k = 1,2,\dots,d$.

$$\frac{\partial}{\partial \xi^{(k)}} H(t,x,\nabla_\xi v) = \sum_{i=1}^{d} v_{\xi^{(i)}\xi^{(k)}}(t,x) f^{(i)\alpha^o}(t,x)$$

$$+ \nabla_\xi v f^{\alpha^o}_{\xi^{(k)}}(t,x) + U^{\alpha^o}_{\xi^{(k)}}(t,x)$$

$$\mathbb{P}_{\alpha^o} - \text{a.e.} \tag{71}$$

Proof: See Hausmann (1981), lemma 5.1. ∎

The relevance of the lemma comes from the fact that no derivatives with respect to α are needed, since we are using precisely the optimal control α^o.
Under the notation of Theorem T5, eq. (62) is a well-known equation satisfied by the adjoint variable. If σ is independent of ξ, then the drift term alone gives the deterministic adjoint equation

$$-\dot{p}^* = \nabla_\xi f^{\alpha^o} p^* + (\nabla_\xi U^{\alpha^o})^*. \tag{72}$$

Finally, if $\sigma = 0$, then eq. (62) reduces to the deterministic adjoint equation.

Now we have all the results needed to cope with our economic problem formulated in Section 2.
On a final balance of results and efforts we like to say a few words. First of all, we should point out that the semi-martingale version of the stochastic maximum principle put us in a position to deal with the optimization problem (10) in quite the same way as we did in the deterministic case by means of Pontryagin's principle, see Gómez (1984). This is a great achievement of the martingale approach to optimal control which merits to be stressed because the logic of the procedure and main line of argumentation remain transparent, lucid and elegant in spite of the increasing degree of complexity of the stochastic analysis and techniques involved. We have preferred to argue at a basic level and proceed in an intuitive and heuristic manner because we are primarily interested in applications and confidently we are able, relying on the original works, to give rigorous and formal proofs of our results.

5. The Controlled Labour-Surplus Economy

5.1. Economics of the Intertemporal Employment Policy

Planning is together with markets and rationing a major method of coordination of resources and activities. It calls for clear delineation of objectives, for working out a pattern of coordination of resource allocation over time. Such matters take us into questions of optimal utilization of resources and optimal economic

growth. The growth pattern for an economy includes the movement of the whole system of prices, quantities and flows throughout time.
Fundamentally, growth means here increases in resource utilization and technological knowledge. However, the course of capital accumulation enters as both cause and effect as a look at eq. (3-a) - (3-g) shows.
Indeed, taking income, i.e. production, as received one determines the path of capital, from the selected consumption-investment mix, i.e. $Y_2 = C_2 + I_2$, by means of the equation $\dot{K}_2 = I_2 - \delta_2 K_2$ and the initial capital endowment ${}^oK_2$.
Consequently, the higher the fraction of income invested, the higher the expansion rate of capital, production, investment, consumption, etc., ... However, this sort of policy is likely to be opposed due to the fact that people usually prefer present to future consumption.
Here is where controversies, that go beyond the boundaries of orthodox economics, come into play, i.e. social value and choice, time preference and social rates of discount, planning objectives and institutional constraints, attitudes and motivation, etc., ... At issue is the elaboration of an economic policy that leads to the elimination over the long-run of the labour-surplus, i.e. one which directs the labour force from less productive or idleness to more highly productive employment.

Since the government cannot rely solely on the market to bring about the desired level of (secondary) employment, it is natural to ask how one gets an optimal employment policy and further how the economy might evolve, if one succeeds influencing employment (or alternatively investment) decisions and social attitudes by means of the social policy forwarded by the goverment and articulated in terms of the utility of aggregate consumption U and the terminal pay-off g, see Section 2.

The main problem in the labour-surplus economy is that employment (i.e. present consumption) and wages conflict with investment (future consumption) and growth. In a neoclassic world capitalists would settle at $\underline{L}_2$, the investment-maximizing employment level. Therefore, in order to increase L_2 beyond $\underline{L}_2$ one needs governmental invervention which we embody in the following set of assumptions.

<u>Assumption</u> (J_3)

(a) At the beginning of every time period the control board sets a minimum wage rate $\bar{W}_2$ in the secondary sector which remains constant until the end of the period.

(b) With every $\bar{W}_2$ one associates the amount of labour ${}^oL_2$ available for employment in the secondary sector. The supply of labour (to the secondary sector) is infinitely elastic up to the point of full employment.

(c) The fractions $(1 - s_{K_2})$ and $(1 - s_{L_2})$ of profit and wage income are consumed.

The government is able to oblige capitalists to save any proportion of profits that it desires them to save, provided it does not attempt to force capitalists to save more than $\bar{s}_K$ units of each unit of profits.

Under unemployment the foregoing assumptions put certain constraints on the choice variables s_{K_2} and W_2, i.e. $W_2 \geq \bar{W}_2$ and $0 < s_K \leq \bar{s}_K$. This means the choice of the consumption-investment mix cannot be made independently of the choice of the level of employment. Therefore, the only degree of freedom remaining to determine the output level is the choice of the level of employment.

To begin with, we like to introduce the concept of accounting price of capital (investment), which we denote ${}_*P_K$, by means of the following marginal rate of transformation (MRT)

$$ {}_*P_K = -\left(\frac{dC_2}{dI_2}\right)_{J_3} = -\left(\frac{C_{L_2}}{I_{L_2}}\right)_{J_3} $$

$$ = \frac{(s_{K_2} - s_{L_2})W_2 + (1 - s_{K_2})Y_{L_2}}{(s_{K_2} - s_{L_2})W_2 - s_{K_2} Y_{L_2}} \tag{73} $$

where C_{L_2}, I_{L_2} and Y_{L_2} stand for the partial derivatives of consumption C_2, investment I_2 and production Y_2 with respect to labour L_2. J_3 hints at the fact that under the assumption (J_3) the only choice variable available is L_2.
${}_*P_K$ indicates how much consumption the economy has to make available, i.e. sacrifice, in order to get one additional unit of investment, i.e. ${}_*P_K$ defines a social supply price of investment in terms of consumption.
${}_*P_K$ is invariably higher than one. This contrasts sharply with the conventional nominal price of investment P_K, $P_K(t) = 1$ at any $t \in R$, what means that the physical substitution of consumption for investment takes place at any time on a one-for-one basis. ${}_*P_K$ approaches 1 only as the marginal productivity of labour Y_{L_2} goes to zero and employment becomes technological rather than institutional.
The fact that ${}_*P_K \geq 1$ makes investment at the margin more valuable than consumption and provides the key to the proper valuation in the labour-surplus economy. For investment has not a value on its own but for the future consumption (employment) it provides. Otherwise one cannot reasonably explain giving up the output foregone by allowing $({}^oL_2 - \underline{L}_2)$ workers to be idle.
Based on ${}_*P_K$ one calculates accounting prices, i.e. social values of profits ${}_*P_\pi$, wage and rental rates, ${}_*W_2$ and ${}_*R_2$ respectively, see Marglin (1976). The relation between ${}_*W_2$ and ${}_*R_2$ and the corresponding nominal wage and rental rates turns out to be

$$ {}_*W_2 \leq W_2 , \quad {}_*R_2 > R_2 \tag{74} $$

and this may lead to greater utilization of labour, provided the extra profits due to the premium attached to investments by means of $_*P_K$ and $_*R_2$ are directed to new investments.

In order to say more, we need the marginal rate of substitution (MRS) which shall follow from the government social policy formulated by means of U and g. The MRS reveals how much consumption the decision-maker is willing to give up for a marginal unit of investment given its technology and institutional constraints and defines thus a social demand price of investment in terms of consumption. In consequence, the consumption-investment mix which equates the social supply and demand price solves partially our optimization problem.
Indeed, the MRT and MRS, under the assumptions in Sections 2 and 5.1, determine unequivocally the loci of admissible consumption-investment mixes given technology, institutions and social preferences. Therefore, the optimal combination of consumption and investment can to some extent be described in terms of the well-known tangency condition. However, the characterization of the optimum is not simple at all. In the next sub-sections we shall work out the desired economic policy by means of the stochastic techniques presented in Sections 3 and 4. To our knowledge, problems of this kind have not been treated yet in the literature. Gômez (1984) deals with a deterministic version of it; related questions are considered in Gômez (1983, 1984) and (1985). Marglin (1976) has been the main source of inspiration. Lucid and stimulating control theoretical presentations of economic growth theory can be found in Burmeister and Dobell (1970), Burmeister (1980), Cass and Shell (1976), Tintner and Sengupta (1972), Pitchford (1974), Arrow and Kurz (1970), Aoki (1976) and many others. In order to facilitate the arguments we suppress the explicit time dependence in the production and utility functionals, i.e. $Y_2 = F(K_2(t),L_2(t))$ and consider $U(k_2(t),\alpha(t))$ instead of eqs. (3-a) and (10) respectively. However, the explicit consideration of time t would not affect our results.

5.2. The Pontryagin Path of Labour Allocation

We have purposely employed the same notation in the formulation of the economic problem as well as in the presentation of the needed mathematical tools. Therefore, we shall make use of the results, functionals and variables of Sections 3 and 4 without further economic interpretation, unless it is not apparent.
To save notation we suppress the subindex 2, referring to the secondary sector, unless any confusion may arise.
Let us consider the control $\alpha \in \mathbb{A}$. Recall $\alpha(t) = \alpha(s_K,L)$. Taking into account eqs. (3-b) and (3-c), which define I and C as functionals of $\{s_K(t),L(t)\}$, and the SDE (2'), which describes the labour supply L, one obtains, with the help of Itô's lemma, the following stochastic version of the accounting price of capital $_*P_K$, we call it $^b_*P_K$,

$$ {}_*^b P_K(t) = $$

$$ = -\frac{\left\{\left[(s_K - s_L)W + (1-s_K)Y_L\right]a + (1-s_K)Y_{LL}b^2L\right\}dt + \left[(s_K - s_L)W + (1-s_K)Y_L\right]bLdW^\alpha(t)}{\left\{\left[(s_K - s_L)W - s_K Y_L\right]a + s_K Y_{LL}b^2L\right\}dt + \left[(s_K - s_L)W - s_K Y_L\right]bLdW^\alpha(t)} \tag{75} $$

where Y_{LL} stands for $\frac{\partial^2 Y}{\partial L^2}$. For simplicity we shall refer to the right-side of eq. (75) as SAPK, short-hand for stochastic accounting price of capital.
As one easily sees, ${}_*^b P_K$ coincides with ${}_*P_K$ in the deterministic case, i.e. if $b = 0$.
Unfortunately, SAPK is a little cumbersome and does not enable us to trace out clearly the effects on ${}_*^b P_K$ of changes of control as P_K does. This is, of course, due to the presence of the Wiener process $W^\alpha(s)$ and it is the price we pay for the randomness we consider.

Let us now characterize an optimal economic policy. We go back to theorem T4 a) and assume $\alpha^0 \in \mathbb{A}$ is optimal. This means, α^0 maximizes (a.s. $d\mathbb{P}\times$ Lebesgue) the Hamiltonian $H_s(\alpha_s^0)$ given by eq. (46). Hence, we have

$$ H_s(\alpha_s^0) = p(s)\, f^{\alpha^0}(s,k(s)) + U^{\alpha^0}(s,k(s)) \tag{76} $$

at any $s \in [0,t^*]$. f^{α^0} and U^{α^0} are given by eqs. (9) and (10), respectively. According to the remark following eq. (10'), we rewrite eq. (76) as

$$ H_s(I^0,C^0) = p(s)\, f^{\alpha^0}(I^0(s)) + U(I^0,C^0) \tag{76'} $$

where $(I^0(s),C^0(s))$ denotes time path of the investment-consumption mix associated with the optimal control $\alpha^0 = \alpha(s_K^0,L^0)$. Further we like to stress the fact that the first term of $f^\alpha(s,k(s))$, the drift vector in eq. (9), is $s_K(s)y(s)$ which is equivalent to $(I^0(s)/L^0(s))$ at any time s. Since $H_s(I^0,C^0(s))$ has a maximum at $I^0 = I(s)$, the control set A is all of $\mathbb{R}$, and since $U(C^0(I^0(s))$ is differentiable in I^0, we must have

$$ 0 = \frac{\partial}{\partial I^0} H_s(I^0(s),C^0(s)) = p(s) - \left[\frac{\partial}{\partial C^0} U(C^0(s))\right]\left(-\frac{dC^0}{dI^0}\right) $$

and further using eq. (73) and $U_{C^0}(s) = \frac{\partial}{\partial C^0} U(C^0(s))$ we get

$$ p(s) = U_{C^0}(s)\; {}_*^b P_K(s) \qquad \text{a.s. } d\mathbb{P} \times \text{Lebesgue.} \tag{77} $$

Hence eq. (77) holds for all $s \in [0,t^*]$ with possible exceptions on $d\mathbb{P}_{\alpha^o} \times ds$ - null sets. It is for that reason a moment-to-moment relation known in dynamic economics as the dynamic efficiency condition and means that at a.s. any time s the social utility derived from the decision to invest according to I^o should equate the consumption utility loss associated with the consumption the economy has to sacrifice in order to further investments as the control α^o requires.
The dynamic efficiency condition given by eq. (77) amounts to the already mentioned tangency condition between the consumption-investment transformation functional and the consumption-investment utility substitution functional articulated by means of the family of Hamiltonians (isoquants) $\{H_s(I,C)\}$. In analogy with a functional occurring in classical mechanics, the Hamiltonian, eq. (76), measures the total utility (energy) of current output: The utility of consumption $U(C(s))$ corresponds to the potential energy and $p(s)f^{\alpha}(s,k)$, the instrumental utility of investment, to the kinetic energy. The control decision α diverts consumption from the current consumption-investment mix associated with the status quo to investment building thereby an alternative mix.
Equivalently, α frees potential energy which changes over to kinetic energy by means of the drift $f^{\alpha}(s,k)$ of the dynamics, eq. (9). This fact deserves to be noted because it relates social utility of consumption to capital accumulation and offers us, by that means, a connection between economic development and the learning process attached to it, see Gómez (1984b).

Let us further characterize the optimum. By analogy with the optimization problem given by eq. (51), and taking into account eqs. (76), (3) and (11), T4 a) has the following equivalent formulation

$$H_s(\alpha^o) = \max_{s_K(s),L(s)} \{H_s(I(s_K(s),L(s)),C(s_K(s),L(s)))\} . \tag{78}$$

Hence, the static first-order conditions which one obtains from eq. (78) and the differentiability of H_s, fully describe the following three phases, the economy undergoes in every time period before entering the neoclassic era.

Phase I

$$L^o(s) < {}^o\bar{L} \qquad {}^b_*P_K = SAPK(\alpha^o) \quad d\mathbb{P}_{\alpha^o} \times ds \text{ - a.s.}$$
$$s^o_K(s) = \bar{s}_K$$

Phase II

$$L^o(s) = {}^o\bar{L} \qquad 1 \le {}^b_*P_K \le SAPK(\alpha^o) \quad d\mathbb{P}_{\alpha^o} \times ds \text{ - a.s.}$$
$$s^o_K(s) = \bar{s}_K$$

Phase III

$L^o(s) = {}^o L$

$$ {}^{b}_{*}p_K = 1 \qquad\qquad d\,\mathbb{P}_{\alpha^o} \times ds \text{ - a.s.} $$

$s^o_K(s) < s_K$

The Hamiltonian together with the initial constraints determines whether the economy finds itself in phase I, II or III. A priori one may say that the economy optimally develops by moving from phase I to II and III for enough low initial capital intensity. The phases are to be interpreted as follows: optimality dictates employment, $L^o(s) = {}^o\bar{L}$, or unemployment, $L^o(s) < {}^o\bar{L}$, and binding savings constraints, $s_K(s) = \bar{s}_K$, or not, $s_K < \bar{s}_K$, then the relative social desirability of the alternative mix of investment and consumption has to be measured by its effect on the Hamiltonian, i.e. to evaluate the alternative (I_2,C_2) mix one uses the value of ${}^{b}_{*}p_K$, that has been assigned to the phase by the first-order conditions, see right-hand side of the corresponding phase.
The α^o in SAPK(α^o) stresses the dependence of the Wiener process on α^o, i.e. $w^{\alpha^o}(s)$. The relations characterizing the three phases hold $d\mathbb{P}_{\alpha^o} \times ds$ a.s.
Now, let us look at the optimality of the conditions imposed on the adjoint process $\{p(s)\}$.
According to T5 and T6 p satisfies eqs. (65) and (68) as well as the instantaneous intertemporal consistency requirement given by eq. (72) which can be written in the following familiar form

$$ -\dot{p} = \frac{\partial H}{\partial k} \, . \tag{79} $$

Eq. (79) provides the relationship between various measures of capital productivity and rates of discount. Marglin (1976) shows that $-\dot{p}/p$, i.e. the percentage rate at which the marginal utility of investment decays over time, is the appropriate rate for discounting future investment.

The transversality condition is given by

$$ g(k(t^*)) = -p(t^*)(k(t^*) - k^*)^- \tag{79} $$

where k^* is set at the level which allows full employment to be sustained from t^* on. Here z^- means the negative part of z, i.e. $z^- = -\min(0,z)$.

Inversely, a stochastic Pontryagin path of employment and savings, i.e. one that satisfies eqs. (77), (78), (79), (80) and (11), solves the stochastic control problem (10).

Goméz (1986) studies matters concerning the attainability and reversibility of the different phases. One would like to know also how long does it take to go through each phase and the smallest size of the capital impulse required. We shall deal with these questions somewhere else.

Second Author's Acknowledgements

I should like to express my gratitude to the organizers of BiBoS for inviting me to join the Research Center Bielefeld-Bochum-Stochastics and to the Stiftung Volkswagenwerk for financial support.

My special thanks go to S. Albeverio and Ph. Blanchard for many suggestions, discussions and encouragements which made this work possible and my stay at BiBoS so instructive and enjoyable.

I am indebted to several participants in the Seminar of BiBoS for making valuable comments. Last not least I gratefully acknowledge the expert secretarial assistance of Mrs. Jegerlehner.

References

Aoki, M. (1976). Optimal Control and System Theory in Dynamic Economic Analysis. North-Holland, Amsterdam.

Arnold, L. (1974). Stochastic Differential Equations and its Applications. Wiley, New York.

Arrow, K. and Kurz, M. (1970). Public Investment, the Rate of Return and Optimal Fiscal Policy. The John Hopkins University Press, Baltimore.

Beneš, V.E. (1970). Existence of optimal strategies bases on specified information, for a class of stochastic decision problems, SIAM Journal on control, 9, pp. 354-371.

Bensoussan, A. and Lions, J. (1978). Applications des Inêqualités Varionnelles en Contrôle Stochastique. Dunod, Paris.

Bensoussan, A. (1982). Stochastic Control by Functional Analysis Methods. North-Holland, Amsterdam.

Bourguignon, F. (1974). A particular class of continuous-time stochastic growth-models. Journal Economic Theory, 9, pp. 141-158.

Burmeister, E. and Dobell, R. (1970). Mathematical Theories of Economic Growth, Macmillan, London.

Cass, D. and K. Shell (1976). The Hamiltonian Approach to Dynamic Economics, Academic Press, New York.

Davis, M. (1977). Linear Estimation and Stochastic Control, Chapman and Hall, London.

Davis, M. (1979). Martingale methods in stochastic control. Lect. Notes in Control and Information Sciences, Vol. 16, Springer Verlag, Berlin.

Davis, M. (1980). Functionals of diffusion processes of stochastic integrals. Mathematical Proceedings of the Cambridge Philosophical Society, 87, pp. 157-166.

Davis, M. (1982). Stochastic control with tracking of exogeneous parameters. Lecture Notes in Control and Information Sciences, Vol. 43. Springer Verlag, Berlin.

Davis, M. and P. Varaiya (1973). Dynamic Programming conditions for partially observable stochastic systems, SIAM Journal on Control and Optimization, 11, pp. 226-261.

Dellacherie, C. and P. Meyer (1980). Probabilité et Potentiel. Théorie des Martingales. Hermann, Paris.

Duncan, T. and P. Varaiya (1971). On solutions of a stochastic control system. SIAM Journal on Control, 9, pp. 354-371.

Elliott, R. 1979). The Martingale Calculus and its Applications. Lecture Notes in Control, Vol. 16, Springer Verlag, Berlin.

Elliott, R. (1982). Stochastic Calculus and Applications. Springer Verlag, Berlin.

Fleming, W. and R. Rishel (1975). Deterministic and Stochastic Optimal Control. Springer Verlag, N.Y.

Fujisaki, M.; G. Kallianpur and H. Kunita (1972). Stochastic differential equations for the nonlinear filtering problem. Osaka Journal of Mathematics, 9, pp. 19-40.

Gihman, I.I. and A.V. Skorohod (1972). Stochastic Differential Equations. Springer Verlag, Berlin.

Gihman, I.I. and A.V. Skorohod (1979). The Theory of Stochastic Processes III. Springer Verlag, Berlin.

Gómez, G. (1981). Controlling a dual economy. Part I. Div. of Appl. Math. Brown University.

Gõmez, G. (1983). The intertemporal labour allocation inherent in the optimal stopping of the dual economy: the static case. European Meeting of the Econometric Society. Madrid, 3-7 September, 1984.

Gómez, G. (1984a). Modelling the economic development by means of impulsive control techniques. Mathematical Modelling in Sciences and Technology, pp. 802-806, X.J. Avual and R.E. Kalman (eds.), Pergamon Press, N.Y.

Gômez, G. (1984b). On the Markov Stopping Rule Associated with the Problem of Controlling a Dual Economy. Dynamic Modelling and Control of National Economies, 1983, pp. 197-204. T. Basar and L.F. Pau (eds.), Pergamon Press, N.Y.

Gômez, G. (1985a). A mathematical dynamic model of the dual economy emphasizing unemployment, migration and structural change. Dept. of Mathematics, University of Erlangen-Nürnberg, Erlangen, FRG.

Gômez, G. (1985b) Optimal stopping times in the economic development planning. Dept. of Mathematics, University of Erlangen-Nürnberg, Erlangen, FRG.

Gómez, G. (1985c). Controlling a dual economy. Part II, III. Work in progress.

Goméz, G. (1986). Attainability and Reversibility of a Golden Age for the Labour Surplus Economy: A Stochastic Variational Approach. Research Centre Bielefeld-Bochum-Stochastics, University of Bielefeld, Bielefeld, FRG.

Hausmann, U. (1978). On the stochastic maximum principle. SIAM Journal on Control. 16, pp. 236-251.

Hausmann, U. (1979). On the integral representation of functionals of Itô process. Stochastics, Vol. 3, pp. 17-27.

Hausmann, U. (1980). Existence of partially observable optimal stochastic controls. Lecture Notes in Control, 36, Springer Verlag, Berlin.

Hausmann, U. (1981). On the adjoint process for optimal control of diffusion processes. SIAM Journal of Control and Optimization, Vol. 19, pp. 221-243.

Hausmann, U. (1982). Extremal control for completely observable diffusions. Lecture Notes in Control, Vol. 42, Springer Verlag, Berlin.

Ikeda, N. and S. Watanabe (1981). Stochastic Differential Equations and Diffusion Processes. North-Holland, Amsterdam.

Jöreskog, K.G. and Wold, H. (1982). Systems Under Direct Observation, North-Holland, Amsterdam.

Krylov, N.V. (1980). Controlled Diffusion Processes, Springer Verlag, N.Y.

Kushner, H.J. (1967). Stochastic Stability and Control. Academic Press, N.Y.

Kushner, H.J. (1977). Probability Methods for Approximations in Stochastic Control and for Elliptic Equations. Academic Press, N.Y.

Kunita, H. and S. Watanabe (1967). On square integrable martingales, Nagoya Mathematical Journal, 30, pp. 209-245.

Liptser, R.S. and A.N. Shiryayev (1977). Stochastics of Random Processes I. Springer Verlag, Berlin.

Malliaris, A. (1981). Stochastic Methods in Economics and Finance, North-Holland, Amsterdam.

Marglin, S.A. (1976). Value and Price in the Labour-Surplus Economy. Oxford University Press, London.

Merton, R. (1975). An asymptotic theory of growth under uncertainty, Review of Economic Studies, 42, pp. 375-393.

Pitchford, D. (1974). Population in Economic Growth, North-Holland, Amsterdam.

Rishel, R. (1970). Necessary and sufficient conditions for continuous-time stochastic optimal control, SIAM Journal on Control, Vol. 8, pp. 559-571.

Schuss, Z. (1980). Theory and Applications of Stochastic Differential Equations. Wiley, N.Y.

Tintner, G. and Sengupta, J. (1972). Stochastic Economics. Academic Press, N.Y.

A CENTRAL LIMIT THEOREM FOR THE LAPLACIAN IN REGIONS WITH MANY SMALL HOLES

R. FIGARI, S. TETA
Research Center Bielefeld-Bochum-Stochastics
and
Dipartimento di Fisica, Università di Napoli

E. ORLANDI
Dipartimento di Matematica, Università di Roma
"La Sapienza"

Introduction

In classical field theory, the presence of macroscopic media is generally taken into account through tensor-valued functions of space coordinates, time, the field itself and possibly of some external parameter. These quantities, which enter the field equations as coefficients, are meant to describe the local response of the medium to the physical field. Thermal conductivity, refraction index, and dielectric tensor are typical examples.

In some cases, the field equations are trivial inside the region occupied by particular macroscopic media, which are then described purely in terms of boundary conditions on their surfaces. As examples of the above situation, we mention the case of conductors in electrostatic problems and the case of hard core scatterers in the Schrödinger equation for a quantum particle.

Very often in the applications only a partial knowledge of the local structure of the medium is available, e.g. of statistical type. This leads quite naturally to study partial differential equations with random coefficients and/or with boundary conditions on random surfaces. The main goal is to analyze the conditions under which relevant physical quantities show a behavior weakly dependent on the specific microscopic realisation of the medium (deterministic behavior of a random medium).

The model we shall discuss in this paper concerns precisely a situation of the above type, inasmuch it studies a partial differential equation with boundary conditions on randomly placed surfaces. This model presents in a suitable limit a deterministic behavior which can be described analytically. Moreover, for this model we also exhibit explicitly the behavior of the fluctuations around the asymptotic limit.

The Model

Let Ω be an open region of $\mathbb{R}^3$ with smooth boundary and let $\underline{w}^{(m)} \equiv \{w_1,\dots,w_m\} \in (\Omega)^m$ be any choice of m points w_i, $i = 1,\dots,m$, $w_i \in \Omega$.

Let B be any simply connected open neighborhood of the origin. For any fixed $\underline{w}^{(m)}$ and any positive real number ν, let us define the family of open neighborhoods of the w_i : $B_i^\nu \equiv \{x \in \Omega | m^\nu(x - w_i) \in B\}$ and the region $\Omega^{(m)} \equiv \Omega \setminus \bigcup_{i=1}^{m} \bar{B}_i^\nu$. $\Omega^{(m)}$ consists then of the region Ω with a "hole" of linear dimension proportional to $m^{-\nu}$ around each of the w_i.

Δ_m will indicate the Laplacian in $\Omega^{(m)}$ with Dirichlet boundary conditions on $\partial\Omega^{(m)}$. With G_m^λ we will denote the operator $(-\Delta_m + \lambda)^{-1}$, for $\lambda > 0$, and with $G_m^\lambda(x,y)$ the corresponding integral kernel. Since the very beginning we are dropping the dependence of Δ_m and G_m^λ by $\underline{w}^{(m)}$. In the following we will adopt this convention whenever there will be no risk of confusion.

Δ_o will denote the Laplacian on Ω with Dirichlet boundary conditions on $\partial\Omega$. $(-\Delta_o + \lambda)^{-1}$ and its integral kernel will be indicated by G_o^λ and $G_o^\lambda(x,y)$ respectively.

We want to analyze the case in which the points w_i are independent, identically distributed in Ω. We make the assumption that their common distribution admits a continuous density $V(x)$, $x \in \Omega$.

The problem is to characterize the asymptotic behavior of Δ_m as m goes to infinity. Relevant questions about Δ_m we want to answer are:

- is there any deterministic limit operator to which Δ_m tends with "high" probability?
- is it possible to characterize the fluctuations of Δ_m around the limit operator?

A complete answer to the first question was given by many authors in the last years (see [1] - [11] and references quoted there). Concerning the model described above, we can summarize the essential result as follows:

Let $u_m(x;\underline{w}^{(m)})$ be the solution of the problem

$$\begin{cases} -\Delta_m u_m(x;\underline{w}^{(m)}) + \lambda u_m(x;\underline{w}^{(m)}) = f(x) & x \in \Omega^{(m)} \\ u_m(x;w^{(m)}) = 0 & x \in \partial\Omega^{(m)} \end{cases} \tag{1}$$

for $f \in L^2(\Omega^{(m)})$, then

(i) if $\nu > 1$ $\quad \|u_m - \bar{u}\|_{L^2(\Omega^{(m)})} \xrightarrow[m\to\infty]{} 0$

where $\bar{u}$ is the solution of

$$\begin{cases} (-\Delta_0 + \lambda)\bar{u}(x) = f(x) & x \in \Omega \\ \bar{u}(x) = 0 & x \in \partial\Omega \end{cases}$$

(ii) if $\nu < 1$ $\quad \|u_m\|_{L^2(\Omega^{(m)})} \xrightarrow[m\uparrow\infty]{} 0$

(iii) if $\nu = 1$ $\quad \|u_m - u\|_{L^2(\Omega^{(m)})} \xrightarrow[m\uparrow\infty]{} 0$

where u is the solution of

$$\begin{cases} A^\lambda u \equiv (-\Delta_0 + \alpha V + \lambda)\, u(x) = f(x) & x \in \Omega \\ u(x) = 0 & x \in \partial\Omega \end{cases}$$

where α is the electrostatic capacity of the set B:

$$\alpha = \int_{\partial B} \frac{\partial\varphi}{\partial\hat{n}}\,(y)\, d\,S(y)$$

with $\hat{n}$ representing the inner normal to ∂B and φ (the capacitory potential) being the solution of

$$\begin{cases} (\Delta\varphi)(x) = 0 & x \in \mathbb{R}^3\backslash B \\ \varphi(x) = 1 & x \in \partial B \\ \lim_{|x|\mapsto\infty} \varphi(x) = 0 \end{cases}$$

Using the Green's identity to express the solution u_m of problem (1), we get

$$u_m(x;\underline{w}^{(m)}) = (G_0^\lambda f)(x) + \sum_{i=1}^{m} \int_{\partial B_i^\nu} \frac{\partial u_m}{\partial\hat{n}}\,(y;\underline{w}^{(m)}) G_0^\lambda(x,y) dS(y)\ . \tag{2}$$

Apart from the non-random term $(G_0^\lambda f)(x)$ the right hand side of (2), for any fixed x, is the sum of the identically (not independently) distributed random contributions coming from each surface. The result stated above is then a law of large numbers for this sum.

In the following, we present an approach to this problem introduced by Ozawa in [10] which allows to characterize easily the limit operator and to prove a corresponding central limit theorem. We will only consider the $\nu = 1$ case.

The Approximation Procedure

Due to the boundary conditions there is no explicit way to express the fundamental solution G_m^λ of problem (1) in terms of G_o^λ. This fact makes it difficult to analyze directly the asymptotic behavior of G_m^λ.

We then want first to find an explicit approximation of G_m^λ. In particular, we want to show that the operator H_m^λ defined by the integral kernel:

$$H_m^\lambda(x,y;\underline{w}^{(m)}) \equiv G_o^\lambda(x,y) + \sum_{i=1}^{m} q_i(x)\, G_o^\lambda(w_i,y) \tag{3}$$

tends to coincide, for suitably chosen $q_i(x)$, with G_m^λ, when m becomes large. This means that the effect of the boundary conditions on each ∂B_i can be approximated by putting suitably chosen image charges q_i on each w_i.

We will choose the q_i in such a way that the average value of $H_m^\lambda(x,\cdot)$ on each ∂B_i is equal to 0 up to terms of order m^{-1}. This amounts to fix the q_i in such a way that they satisfy the linear system

$$G_o^\lambda(x,w_J) + \sum_{\substack{i=1\\ i\neq J}}^{m} q_i(x;\underline{w}^{(m)})G_o^\lambda(w_i,w_J) + \frac{q_J(x;\underline{w}^{(m)})}{\alpha/m} = 0 \tag{4}$$

for each $J = 1,\ldots,m$.

In fact, notice that for $\lambda = 0$ $G_o^0(w_i,w_J)$ (resp. $G_o^0(x,w_J)$) is exactly the average value of $G_o^0(w_i,y)$ (resp. of $G_o^0(x,y)$) on ∂B_J as far as w_i (resp. x) is outside B_J (which is more and more likely as m grows). Moreover, $q_J/\alpha/m$ is, for $\lambda = 0$, the potential of the "conductor" B_J (whose capacity is just α/m) when the total charge on it is q_J. One easily checks that $\lambda \neq 0$ introduces in this picture only terms of order m^{-1}.

We introduce the short notation

$$\bar{q}(x) \equiv \{q_1(x),\ldots,q_m(x)\} \qquad \bar{G}_o^\lambda(x) \equiv \{G_o^\lambda(x,w_1),\ldots,G_o^\lambda(x,w_m)\}$$

$$\bar{G}_o^\lambda f \equiv \{(G_o^\lambda f)(w_1),\ldots,(G_o^\lambda f)(w_m)\}$$

$$\{\bar{\bar{G}}_o^\lambda\}_{iJ} \equiv \begin{cases} G_o^\lambda(w_i,w_J) & i \neq J \\ 0 & i = J \end{cases}.$$

Formally, (4) is solved by

$$\bar{q}(x) = -\bar{G}_o^\lambda(x)\left(\bar{\bar{G}}_o^\lambda + \frac{m}{\alpha}\mathbb{1}\right)^{-1} \tag{5}$$

where $\mathbb{1}$ is the unit matrix on $\mathbb{R}^m$. Our Definition (3) would then become

$$H_m^\lambda(x,y) \equiv G_0^\lambda(x,y) - \frac{\alpha}{m}\bar{G}_0^\lambda(x)\left(\frac{\lambda}{m}\bar{\bar{G}}_0^\lambda + \mathbb{1}\right)^{-1}\bar{G}_0^\lambda(y)\ . \tag{6}$$

That (6) can be taken as a definition requires only the invertibility of the matrix $(\frac{\alpha}{m}\bar{\bar{G}}_0^\lambda + \mathbb{1})$. Indicating with $|||\cdot|||$ the norm of a matrix $\mathbb{R}^m$, we have

$$|||\frac{\alpha}{m}\bar{\bar{G}}_0^\lambda||| \le \alpha\left\{\frac{1}{m^2}\sum_{\substack{i,j=1\\ i\neq j}}^{m}\left[G_0^\lambda(w_i,w_j)\right]^2\right\}^{1/2}.$$

By the law of large numbers

$$\alpha\left\{\frac{1}{m^2}\sum_{\substack{i,j=1\\ i\neq j}}^{m}\left[G_0^\lambda(w_i,w_j)\right]^2\right\}^{1/2} \xrightarrow[\text{a.s.}]{} \alpha\left\{\int_{\Omega\times\Omega} V(x)(G_0^\lambda(x,y))^2 V(y)dxdy\right\}^{1/2}$$

which is finite and less than 1 for λ sufficiently large. $(\frac{\alpha}{m}\bar{\bar{G}}_0^\lambda + \mathbb{1})$ is then invertible justifying (6) as definition of H_m^λ.

The analysis of the asymptotic behavior of G_m^λ is worked out in two steps. The first one consists in checking the effectiveness of the approximation of G_m^λ by H_m^λ. The second one consists in the analysis of the asymptotic behavior of H_m^λ.

An estimate of the difference between G_m^λ and H_m^λ was given in [10] and can be stated as:

Theorem 1: Let $H^1(\Omega)$ be the standard Sobolev space with norm $\|\cdot\|_{H^1}$. Then for any $f,g \in H^1(\Omega)$ and for any sequence of $\{w^{(m)}\}$ belonging to a set of configurations of probability increasing to 1 when m goes to infinity

$$m^{1/2}(g,(G_m^\lambda - H_m^\lambda)f) \le D(\underline{w}^{(m)})\,\|g\|_{H^1}\|f\|_{H^1}$$

with $\lim_{m\to\infty} D(\underline{w}^{(m)}) = 0$. ($(\cdot,\cdot)$ indicates here the inner product in $L^2(\Omega)$).

When B is a sphere, one can prove Theorem 1 in the stronger form:

Let B be a sphere. Then for some constant c

$$\|G_m^\lambda - H_m^\lambda\|_{L^2(\Omega^{(m)})} < c\, m^{-\beta}$$

for any $\beta < 2/3$, uniformly on a set of configurations of points of probability going to 1 as m goes to infinity.

The reader can find the details of the proof of Theorem 1 in [12] where the same notation introduced above was used.

One can get an intuitive idea of the proof noticing that, by the maximum principle, it is enough to bound the boundary values of the difference of the two kernels. For $y \in \partial B_J$ we have

$$(G_m^\lambda - H_m^\lambda)(x,y) = -H_m(x,y)$$

$$= \left[G_o^\lambda(x,w_J) - G_o^\lambda(x,y)\right] + \sum_{i \neq J} q_i(x)\left[G_o^\lambda(w_i,w_J) - G_o^\lambda(w_i,y)\right] + q_J(x)\left[\frac{m}{\alpha} - G_o^\lambda(w_J,y)\right] \quad (7)$$

where the null term (4) was subtracted in the last line.

Let us consider for simplicity the case in which B is a sphere of radius r. In this case $\alpha = 4\pi r$ and

$$G_o^\lambda(w_J,y) - \frac{m}{\alpha} \sim \frac{m}{\alpha}\,(e^{-\sqrt{\lambda}\alpha/4\pi m} - 1)$$

for y on ∂B_J.

Taking into account the Definition (4) for the charges q_i, one can easily verify that each term in (7) is going to 0 as m goes to infinity. In fact, it would be trivial, starting from (7), to get the result for any $\beta < 1/2$. This would be enough to guarantee that the limits attained by G_m^λ and H_m^λ are the same. However, the strong form of the result (involving also $1/2 \leq \beta < 2/3$) is needed to identify the fluctuations of H_m^λ and G_m^λ around the limit operator.

Asymptotic Behavior and Fluctuations

The convergence result for the solutions of Problem 1 mentioned at the end of the second Section states that G_m^λ converges to $(A^\lambda)^{-1} = (-\Delta + \alpha V + \lambda)^{-1}$ when m goes to infinity. Looking at the explicit form of H_m^λ, one is now able to get easily an intuitive idea of the possible steps of a proof.

Notice that by the law of large numbers

$$\frac{\alpha}{m}\sum_{J=1}^{m} G_o^\lambda(\cdot,w_J)f(w_J) \xrightarrow[m\uparrow\infty]{} \alpha(G_o^\lambda V f)(\cdot) \quad (8)$$

for any continuous $f \in L^2(\Omega^{(m)})$, where V is intended to be the multiplication operator by the function $V(x)$. (The precise notion of convergence will be given in Theorem 2 below.)

If we can prove, as a kind of generalization of (8), that for any $s > 1$, and f continuous in $L^2(\Omega^{(m)})$

$$\frac{\alpha}{m} G_o^\lambda(\cdot)\,(\frac{\alpha}{m}\bar{\bar{G}}_o^\lambda)^s(\bar{G}_o f) \to \alpha\left[G_o^\lambda V\,(\alpha G_o^\lambda V)^s\, G_o^\lambda f\right](\cdot) \quad (9)$$

then the convergence of H_m^λ (and hence of G_m^λ) to $(A^\lambda)^{-1}$ would follow. In fact, (9) would imply that

$$\frac{\alpha}{m}\bar{G}_o^\lambda(x)\left(\frac{\alpha}{m}\bar{\bar{G}}_o^\lambda + \mathbb{1}\right)^{-1}\bar{G}_o^\lambda(y) = \frac{\alpha}{m}\sum_{s=0}^{\infty}(-)^s\,\bar{G}_o^\lambda(x)\left(\frac{\alpha}{m}\bar{\bar{G}}_o^\lambda\right)^s\bar{G}_o^\lambda(y)\xrightarrow[m\uparrow\infty]{}\alpha\sum_{s=0}^{\infty}(-)^s$$

$$\times\left[G_o^\lambda V(\alpha G_o^\lambda V)^s\,G_o^\lambda\right](x,y) = (\alpha G_o^\lambda V + 1)^{-1}\,(x,y)$$

which gives immediately

$$H_m^\lambda \xrightarrow[m\uparrow\infty]{} (-\Delta + \alpha V + \lambda)^{-1}\,.$$

Notice that in the development of the s-th power of the matrix $\frac{\alpha}{m}\bar{\bar{G}}_o^\lambda$ there are "many" terms involving $(s+1)$ distinct w_i. Each of these terms has an average value coinciding with the right hand side of (9). The law of large numbers guarantees then the convergence of the sum of all these terms to the right hand side of (9).

This suggests that what one has to prove is that the contribution of terms with repeated w_i in $(\frac{\alpha}{m}\bar{\bar{G}}_o^\lambda)^s$ becomes smaller and smaller, on a set of configuration of large probability, when m goes to infinity. It turns out that this is true in a very strong sense. This fact enables us to disregard terms with repeated w_i even in the analysis of the fluctuations of H_m^λ around the limit operator.

The final result we can get on the asymptotic behavior of G_m^λ is contained in

<u>Theorem 2</u>: (a) For all the $\underline{w}^{(m)}$ belonging to a set of measure going to 1 as m goes to infinity $\|G_m^\lambda(\underline{w}^m)\chi_m(\underline{w}^{(m)}) - A^\lambda\|\xrightarrow[m\uparrow\infty]{}0$.

(b) For any $f,g \in H^1(\Omega)$ the random field

$$\xi_g^\lambda(f;\underline{w}^{(m)}) \equiv m^{1/2}\left(f,\left[G_m^\lambda(\underline{w}^{(m)})\chi_m(\underline{w}^{(m)}) - A^\lambda\right]g\right) \tag{10}$$

converges in distribution to the gaussian random field $\bar{\xi}_g^\lambda(f)$ of mean 0 and covariance:

$$E\left(\bar{\xi}_g^\lambda(f)\,\bar{\xi}_g^\lambda(f')\right) = \alpha^2\left[(A^\lambda f A^\lambda g,\ A^\lambda f' A^\lambda g)_{L^2_V} - (A^\lambda f, A^\lambda g)_{L^2_V}(A^\lambda f', A^\lambda g)\right]_{L^2_V} \tag{11}$$

(here $(\cdot,\cdot)_{L^2_V} = (\cdot, V\,\cdot)$, χ_m is the characteristic function of $\Omega^{(m)}$ and $G_m^\lambda\chi_m$ is extended to all Ω setting its value equal to 0 on $\bigcup_{J=1}^{m}\bar{B}_J$).

As a consequence of Theorem 2 a complete analysis of the asymptotic behavior of the eigenvalues of Δ_m, when Ω is a bounded region, can be worked out. For the details see [12].

Sketch of the proof: as a consequence of Theorem 1 it will be sufficient to prove the statements (a) and (b) with G^λ_m substituted with H^λ_m .

According to the intuitive picture we presented above, we introduce the following definitions

$$\left\{\bar{N}^\lambda(s)\right\}_{iJ} = \sum_{\substack{i_1,i_2,\dots,i_{s-1}=1 \\ i_k \neq i_\ell \quad k\neq\ell \\ i_k\neq i \quad \forall k \\ i_k \neq J \quad \forall k \\ i\neq J}}^{m} G^\lambda_o(w_i,w_{i_1})G^\lambda_o(w_{i_1},w_{i_2})\dots G^\lambda_o(w_{i_{s-2}},w_{i_{s-1}})G^\lambda_o(w_{i_{s-1}},w_J)$$

$$\left\{\bar{\bar{I}}^\lambda(s)\right\}_{iJ} = \left\{(\bar{\bar{G}}^\lambda_o)^s\right\}_{iJ} - \left\{\bar{\bar{N}}^\lambda(s)\right\}_{iJ}$$

$$N^\lambda_m(x,y) = G^\lambda_o(x,y) - \frac{\alpha}{m}\,\bar{G}^\lambda_o(x)\left[\sum_{s=0}^{\infty}(-)^s\,\bar{\bar{N}}^\lambda(s)\,(\tfrac{\alpha}{m})^s\right]\bar{G}^\lambda_o(y)$$

$$I^\lambda_m(x,y) = -\frac{\alpha}{m}\,\bar{G}^\lambda_o(x)\left[\sum_{s=0}^{\infty}(-)^s\,\bar{\bar{I}}^\lambda(s)\,(\tfrac{\alpha}{m})^s\right]\bar{G}^\lambda_o(y) \tag{12}$$

in such a way that $N^\lambda_m(x,y) + I^\lambda_m(x,y) = H^\lambda_m(x,y)$.

With N^λ_m and I^λ_m we will indicate the operator in $L^2(\Omega)$ corresponding respectively to the kernels $N^\lambda_m(x,y)$ and $I^\lambda_m(x,y)$.

To any term $G^\lambda_o(i_1,i_2)\dots G^\lambda_o(i_s,i_{s+1})$ appearing in $(\bar{\bar{G}}^\lambda_o)^s$ we associate the oriented path of s steps $\{i_1,i_2,\dots,i_{s+1}\}$. By definition $\bar{\bar{N}}^\lambda(s)$ is the sum of terms in $(\bar{\bar{G}}^\lambda_o)^s$ corresponding to non-selfintersecting graphs while $\bar{\bar{I}}^\lambda(s)$ is the sum of terms corresponding to graphs with at least one repeated point.

We want to prove

Lemma: for λ large enough, on a set of configurations of probability going to 1 as m goes to infinity

$$\lim_{m\to\infty} m^{1/2}\,\|I^\lambda_m\|_{L^2(\Omega)} = 0 \ .$$

Proof: for any $f,g \in L^2(\Omega)$ we have from the definition (12)

$$|(g,m^{1/2}\,I^\lambda_m f)| \le \sum_{s=2}^{\infty} \frac{\alpha^{s+1}}{m^{s+1/2}}\,|\bar{G}^\lambda_g\,\bar{\bar{I}}^\lambda(s)\,\bar{G}^\lambda_f| \ .$$

Let us consider the set of all terms in $\bar{\bar{I}}^\lambda(s)$ whose graph has the first repeated point after n steps and comes back for the last time to this same point after a loop of ℓ steps $(2 \le \ell \le s-n)$. Their total contribution to $\frac{\alpha^{s+1}}{m^{s+1/2}} |\bar{G}^\lambda_g\, I^\lambda(s)\, \bar{G}^\lambda_f|$ will be indicated by $A^{(m)}_{\ell,n}(g,f)$. Its explicit form is:

$$A^{(m)}_{\ell,n}(g,f) = \frac{s+1}{m^{s+1/2}} \sum_{k_{\pm 1},\dots,k=1}^{m} \left\{\bar{G}^\lambda_g\right\}_{k_1} \left\{\bar{\bar{N}}^\lambda(\ell-1)\right\}_{k_1,k_2} \left\{\bar{\bar{G}}^\lambda_o\right\}_{k_2k_3}$$

$$\left\{\bar{\bar{G}}^\lambda_o\right\}_{k_3k_4} \left\{(\bar{\bar{G}}^\lambda_o)^{n-1}\right\}_{k_4k_5} \left\{\bar{\bar{G}}^\lambda_o\right\}_{k_5k_3} \left\{\bar{\bar{G}}^\lambda_o\right\}_{k_3k_6}$$

$$\left\{(\bar{\bar{G}}_o)^{s-\ell-n-1}\right\}_{k_6k_7} \left\{\bar{G}_g\right\}_{k_7} .$$

Using Schwartz inequality one can separate the contribution coming from the "vertex" (consisting of the four steps leaving or reaching the first repeated point) from the rest to give the bound

$$A^{(m)}_{\ell,n}(g,f) \le \text{cost.}\ \alpha^{s+1} \|f\|_{L^2(\Omega)} \|g\|_{L^2(\Omega)} \left|\left|\left|\frac{\bar{\bar{G}}^\lambda_o}{m}\right|\right|\right|^{s-4} m^{-7/2} \sum_{p,k,r} \left\{\bar{\bar{G}}^\lambda_o\right\}^2_{p,k} \left\{\bar{\bar{G}}_o\right\}^2_{kr}$$

The vertex part is bounded as follows

$$\eta(\underline{w}^{(m)}) \equiv m^{-7/2} \sum_{\substack{p,k,r\\ p\neq k\\ p\neq r\\ k\neq r}} (G^\lambda_o(w_p,w_k))^2 (G^\lambda_o(w_k,w_r))^2 + m^{-7/2} \sum_{\substack{p,k\\ p\neq k}} (G^\lambda_o(w_p,w_k))^4$$

$$\le m^{-7/2} \sum_{\substack{p,k,r\\ p\neq k\\ p\neq r\\ k\neq r}} \frac{1}{|w_p-w_k|^2} \frac{1}{|w_k-w_r|^2} + m^{-7/2} \sum_{\substack{k,p\\ k\neq p}} \frac{1}{|w_p-w_k|^4} .$$

The first term is a positive quantity whose average value is bounded by $m^{-1/2} E[|w_p-w_k|^{-2} |w_k-w_r|^{-2}]$. In particular, it is going to 0 in probability as m goes to infinity.

For the second term we have:

$$m^{-7/2} \sum_{\substack{k,p\\ k\neq p}} \frac{1}{|w_p-w_k|^4} \le \frac{m^{-3/2}}{\min\limits_{p\neq k} |w_p-w_k|^{1+\varepsilon}} \frac{1}{m^2} \sum_{\substack{k,p\\ k\neq p}} \frac{1}{|w_p-w_k|^{3-\varepsilon}}$$

for any $\varepsilon > 0$. By the law of large numbers $m^{-2} \sum_{k\neq p}^{m} |w_p-w_k|^{3-\varepsilon}$ is bounded on a set of probability increasing to 1 as m goes to infinity. Moreover, the probability that $m^{-3/2} \min\limits_{p\neq k} |w_p-w_k|^{1+\varepsilon} > \delta$ is, for any fixed $\delta > 0$, going rapidly to 0 as one can immediately realize using our assumption of smoothness of the distribution density $V(x)$. We have then shown that, in probability, $\lim_{m\uparrow\infty} \eta(\underline{w}^{(m)}) = 0$.

On the other hand, from

$$\frac{e^{-\sqrt{\lambda}|w_i - w_j|}}{|w_i - w_j|} \le \frac{\lambda^{-\beta/2}}{|w_i - w_j|^{1+\beta}}$$

we have:

$$|||\bar{\bar{G}}_0^\lambda||| \le \frac{1}{m}\left\{\sum_{\substack{i,j\\ i\neq j}} \frac{e^{-2\sqrt{\lambda}|w_i - w_j|}}{16\pi^2|w_i - w_j|^2}\right\}^{1/2} \le \lambda^{-\beta/2}\left\{\frac{1}{m^2}\sum_{\substack{i,\sigma\\ i\neq j}} \frac{1}{|w_i - w_j|^{2+2\beta}}\right\}^{1/2} .$$

Taking $\beta < 1/2$ the sum in curled brackets is bounded uniformly on a set of $\underline{w}^{(m)}$ of measure going to 1 when m goes to ∞ and we have finally

$$A_{\ell,n}^{(m)}(f,g) \le \|f\|_{L^2(\Omega)} \|g\|_{L^2(\Omega)} \alpha^{s+1} (\text{cost.}\ \lambda^{-\beta/2})^{s-4} \eta(\underline{w}^{(m)}) . \tag{13}$$

The estimate (13) is independent of ℓ and m. Moreover, the possible choices of of the couple ℓ,n are obviously bounded by s^2. We have then

$$|(g, m^{1/2} I_m^\lambda f)| \le \|f\|_{L^2(\Omega)} \|g\|_{L^2(\Omega)} \eta(\underline{w}^{(m)}) \sum_{s=2}^{\infty} \alpha^{s+1} (\text{cost.}\lambda^{-\beta/2})^s s^2$$

which converges to 0 in probability for λ sufficiently large. The lemma is then proved.

To prove Theorem 2 we are now left to prove statements (a) and (b) with N_m^λ replacing G_m^λ. In particular, instead of the random field $\xi_g^\lambda(f)$ defined in (10) we can analyze the large m behavior of the random field

$$\begin{aligned}\Theta_g^\lambda(f;\underline{w}^{(m)}) &\equiv m^{1/2}(f,[N_m^\lambda(\underline{w}^{(m)}) - A^\lambda]g) \\ &= \sum_{s=1}^{\infty} (-\alpha)^s \left[m^{-s+1/2}\, \bar{G}_f^\lambda\, \bar{\bar{N}}^\lambda(s-1)\bar{G}_g^\lambda - m^{1/2}(f, G_0^\lambda(VG_0^\lambda)^s g)\right] .\end{aligned}$$

We want to stress that $\bar{G}_f^\lambda\, \bar{\bar{N}}^\lambda(s)\bar{G}_g^\lambda$ does not contain terms whose average value is infinite (like e.g. $(G_0^\lambda f)(w_i)(G_0^\lambda(w_i,w_j))^s(G_0^\lambda g)(w_j)$). In particular we have $\lim_{m\uparrow\infty} E[\Theta_g^\lambda(f)] = 0$.

Analogously, any product like $\bar{G}_f^\lambda\, \bar{\bar{N}}^\lambda(s)\bar{G}_g^\lambda\, \bar{G}_f^\lambda\, \bar{\bar{N}}^\lambda(s')\bar{G}_g^\lambda$ does not contain terms with infinite average value. The computation of the covariance of $\Theta_g^\lambda(f)$ becomes mainly a combinatorial exercise.

We refer to [12] for the details of the computation which gives

(i) $E(\Theta_g^\lambda(f)\ \Theta_g(f'))$

$$=\alpha^2\left[(A^\lambda f\ A^\lambda g, A^\lambda f'\ A^\lambda g)_{L_V^2(\Omega)} - (A^\lambda f, A^\lambda g)_{L_V^2(\Omega)}\ (A^\lambda f', A^\lambda g)_{L_V^2(\Omega)}\right]$$
$$+\ O(\tfrac{1}{m})\ .$$

An immediate consequence of this result is that

$$m^\nu \|N_m^\lambda - A^\lambda\|_{L^2(\Omega)} \xrightarrow[m\to\infty]{} \quad \forall \nu < \tfrac{1}{2}$$

thus proving the first statement of the theorem.

(ii) Up to term of order $\frac{1}{m}$ the covariance of the random field $\Theta_g^\lambda(f)$ coincides with the covariance of

$$\bar{\Theta}_g^\lambda(f,\underline{w}^{(m)}) = m^{-1/2} \sum_{i=1}^{m} \left[K_g^\lambda(f;w_i) - E(K_g(f))\right]$$

where

$$K_g^\lambda(f;w_i) = \sum_{s=1}^{\infty} \sum_{n=0}^{s-1} (G_0^\lambda(VG_0^\lambda)^n f)(w_i)(G_0^\lambda(VG_0^\lambda)^{s-n-1} g)(w_i)\ .$$

Notice that $\bar{\Theta}_g^\lambda(f)$ is explicitly expressed as the sum of identically distributed, independent random variable with mean 0 and covariance given by the right hand side of (11) up to term of order $\frac{1}{m}$. The central limit theorem implies then that $\bar{\Theta}_g^\lambda(f)$ converges in distribution to the gaussian random field $\xi_g^\lambda(f)$ defined in (11).

Only some more combinatorics is needed at this point to prove that

$$\lim_{m\to\infty} E\left[(\Theta_g^\lambda(f) - \bar{\Theta}_g^\lambda(f))^2\right] = 0$$

showing that the limit attained by $\Theta_g^\lambda(f)$ and the one obtained by $\bar{\Theta}_g^\lambda(f)$ are the same and concluding the proof of Theorem 2.

We want to conclude by mentioning some extensions of the results presented before.

Only minor changes are requested to prove analogous results for any second order, strictly elliptic, differential operator in divergence form $\sum_{i,J} \frac{\partial}{\partial x_i}(a_{iJ}(x)\frac{\partial}{\partial x_J})$ with smooth a_{iJ}. What was relevant in the proofs was in fact only the singular behavior of the Green's function around the diagonal $x=y$. In particular, the results remain true for the Laplacian on a smooth 3-dimensional manifold.

The treatment of the 2-dimensional case is formally identical with the previous one if the linear size of the obstacles is chosen proportional to $e^{-m/2\pi\alpha}$.

On the contrary, our proofs are not easily generalized to the n-dimensional case with $n \geq 4$. In fact, we used extensively the integrability of the square of the Green function for the Laplacian around the singularity. As it is well-known, this fails to be true in dimensions larger or equal to 4 .

REFERENCES

[1] HRUSLOV, E.Ja., The method of orthogonal projections and the Dirichlet problem in domains with a fine-grained boundary. Math. USSR Sb 17 (1972) 37-59

[2] HRUSLOV, E.Ja., MARCHENKO, V.A., Boundary value problems in regions with fine-grained boundaries. Naukova Dumka, Kiev, 1974

[3] HRUSLOV, E.Ja., The first boundary value problem in domains with a complicated boundary for higher order equations. Math. USSR Sb. 32 (1977) 535-549

[4] KAC, M., Probabilistic methods in some problems of scattering theory. Rocky Mountain J. Math. 4 (1974) 511-538

[5] RAUCH, J., a) The mathematical theory of crushed ice.
b) Scattering by many tiny obstacles. In: "Partial Differential Equations and Related Topics." Lect. Notes in Math. 446, J. Goldstein (ed.), Springer (1975), resp. 370-379 and 380-389

[6] RAUCH, J., TAYLOR, M., Potential and scattering theory on wildly perturbed domains. J. Funct. Anal. 18 (1975) 27-59

[7] RAUCH, J., TAYLOR, M., Electrostatic Screening. J. Math. Phys. 16 (1975) 284-288

[8] PAPANICOLAU, G., VARADHAN, S.R.S., Diffusions in regions with many small holes. In: Lect. Notes in Control and Information 75, Springer (1980) 190-206

[9] SIMON, B., Functional Integration and Quantum Physics. Academic Press (1979) Pag.[231-245]

[10] OZAWA, S., On an elaboration of M. Kac's theorem concerning eigenvalues of the Laplacian in a region with randomly distributed small obstacles. Comm. Math. Phys. 91 (1983) 473-487

[11] CIORANESCU, D., MURAT, F., Un terme étrange venu d'ailleurs. In: "Nonlinear Partial Differential Equations and their Applications, Collège de France Séminaire, H. Brezis, J. Lions, eds., Vol. II R.N.M. 60 Pitman (1982) 98-138; Vol. III R.N.M. 70 Pitman (1982) 154-178

[12] FIGARI, R., ORLANDI, E., TETA, S., The Laplacian in Regions with many small obstacles, Fluctuations Around the Limit Operator. To appear in Journal of Statistical Physics

ON DIRICHLET FORMS WITH RANDOM DATA
--RECURRENCE AND HOMOGENIZATION

M.Fukushima, S.Nakao and M.Takeda

§1. Introduction

Consider a differential operator L^{ω} of the type

$$(L^{\omega}u)(x) = \frac{1}{m^{\omega}(x)} \sum_{i,j=1}^{d} \frac{\partial}{\partial x_i}\left(a_{ij}^{\omega}(x) \frac{\partial u(x)}{\partial x_j} \right), \quad x \in R^d,$$

on R^d with coefficients m^{ω}, a_{ij}^{ω} depending on a random parameter $\omega \in \Omega$. We assume that $a_{ij} = a_{ji}$, and for any compact $K \subset R^d$, $\lambda_K |\xi|^2 \leq \sum_{i,j=1}^{d} a_{ij}(x)\xi_i\xi_j \leq \Lambda_K |\xi|^2$, $x \in K$, $\gamma_K \leq m(x) \leq \Gamma_K$, $x \in K$, $\lambda_K, \Lambda_K, \gamma_K, \Gamma_K$ being positive constants.

L^{ω} of the above type represents an infinitesimal generator of a symmetrizable diffusion process "in random media". Since coefficients are supposed to be random fields on R^d, it is not very natural to assume their differentiability in $x \in R^d$. Accordingly the SDE theory is very much helpful neither in constructing the diffusion nor in performing the relevant computations. Two alternative theories available are the Dirichlet space theory ([2]) and PDE theory ([11]). We like to know how the first theory should work intrinsically and to what extent the second one should be called for in addition. Thus we look at the associated Dirichlet form

$$E^{\omega}(u,v) = \sum_{i,j=1}^{d} \int_{R^d} u_{x_i}(x)\, v_{x_j}(x)\, a_{ij}^{\omega}(x)\, dx$$

on $L^2(R^d; m^{\omega}(x)dx)$ --the Dirichlet form with random data a_{ij} and m.

In §2 we examine the recurrence problem in the special case that

$$a_{ij}^{\omega}(x) = \delta_{ij}\, e^{-X^{\omega}(x)}, \quad m^{\omega}(x) = e^{-X^{\omega}(x)}$$

for some random field X on R^d. L^{ω} is then reduced to

$$L^{\omega}u = \Delta u - \nabla X^{\omega}(x)\cdot\nabla u$$

the Laplacian with random drift. When $d = 1$ and $X^{\omega}(x)$ is the one-dimensional Brownian motion, the associated diffusion is known to be recurrent for a.e. $\omega \in \Omega$ (Brox [1]). In higher dimensions, we treat the case of centered Gaussian random field $X^{\omega}(x)$ with slowly increasing variance and the case that $X^{\omega}(x) = B^{\omega}(|x|)$

for a one-dimensional parameter Brownian motion $B^{\omega}(t)$. The situation will be seen to be the same as the case of zero drift in the former case, while the recurrence takes place regardless of the dimension d in the latter case. But we do not know about the case that $X^{\omega}(x)$ is the Lévy's Brownian motion with multidimensional parameter x.

In §4, $a^{\omega}_{ij}(x)$ are assumed to be stationary ergodic random fields and further we assume that $m^{\omega}(x)$ equals 1 identically. The associated homogenization problem concerns scaled limits of L^{ω} and of the corresponding diffusion, which has been investigated by Kozlov et al. [10], Papanicolaou-Varadhan [14] and Osada [12]. Osada[12] proved the weak convergence of $\varepsilon X^{\omega}_{t/\varepsilon^2}$ as $\varepsilon \downarrow 0$ for a.e. fixed $\omega \in \Omega$ where X^{ω}_t is the diffusion associated with a non-symmetric random operator L^{ω} more general than the present one. He utilized a PDE method by extending the results of Nash concerning the transition density function.

However a kind of smoothness of coefficients of L^{ω} was required in [12] in the step of using Ito's formula to calculate the quadratic variation of a martingale generated by a weakly harmonic function. In this paper, we show that such a smoothness condition can be dispensed with in the present symmetric case by reducing the above computation of the martingale to the Dirichlet space theory in the manner formulated in §3. We mention that Osada has recently succeeded ([13]) to remove the smoothness condition from his general setting of [12] by using an associated non-symmetric Dirichlet form $\mathcal{E}^{\omega}$.

§2 Recurrence of random energy forms

Let $\mathcal{E}$ be a Dirichet form on an L^2-space $L^2(X;m)$ and $\{T_t, t > 0\}$ be an associated semigroup on $L^2(X;m)$. Then T_t extends to a contractive, positive linear operator on $L^1(X;m)$ and the Green function $Gf(x)$ ($\leq +\infty$) makes sense m-a.e. for any non-negative $f \in L^1(X;m)$. We say that $\mathcal{E}$ is transient if $Gf(x)$ is finite m-a.e. for some strictly positive (or, equivalently, for any non-negative) $f \in L^1(X;m)$. $\mathcal{E}$ is said to be irreducible if $\{T_t, t > 0\}$ admits no proper invariant set : $T_t(I_B u) = I_B T_t u$, $u \in L^2$, $t > 0$, implies either $m(B) = 0$ or $m(X - B) = 0$. $\mathcal{E}$ is called recurrent if $\mathcal{E}$ is irreducible and non-transient.

We consider a random field $X(x)= X^{\omega}(x)$, $x \in R^d$, defined on a probabaility space $(\Omega, \mathcal{B}, P)$. Thus $X^{\omega}(x)$ is, for a fixed $x \in R^d$, a random variable on Ω and, for P-a.e. fixed $\omega \in \Omega$, a locally bounded measurable function in x.
Then

(2.1) $$\mathcal{E}^{\omega}(u,v) = \sum_{i=1}^{d} \int_{R^d} u_{x_i}(x) v_{x_i}(x)\, e^{-X^{\omega}(x)}\, dx, \quad u, v \in C_o^{\infty}(R^d),$$

is, for P-a.e.fixed $\omega \in \Omega$, closable on $L^2(R^d; e^{-X^{\omega}(x)}dx)$ and the closure is a Dirichlet form-- a so called energy form ([3]). In this section we study the recurrence of the random energy form $\mathcal{E}^{\omega}$ for special random fields $X^{\omega}(x)$.

The energy form (2.1) is always irreducible ([4]). According to Ichihara's test [8], $\mathcal{E}^{\omega}$ is recurrent if

(2.2) $$\int_1^{\infty} r^{1-d} \left(\int_{S^{d-1}} e^{-X^{\omega}(r\sigma)}\, d\sigma \right)^{-1} dr = \infty$$

and it is transient if

(2.3) $$\int_1^{\infty} r^{1-d} e^{X^{\omega}(r\sigma)}\, dr < \infty$$

for $\sigma \in S^{d-1}$ with positive surface measure.
A simpler proof of Ichihara's test will be given in [6]. When $X(x)$ is deterministic and constant, then we see from (2.2) and (2.3) the well known fact that $\mathcal{E}$ is recurrent or transient according as $d \leq 2$ or $d \geq 3$.

We can guess that, if $X^{\omega}(x)$ does not vary too much at infinity, then the recurrence situation would be the same as the constant case. In fact we have the following :

<u>Theorem 2.1</u> Suppose that $X(x)$, $x \in R^d$, is a centered Gaussian random field. Let $R(x) = E(X(x)^2) = \int_{\Omega} X^{\omega}(x)^2 P(d\omega)$.

(i) If $d \geq 3$ and $R(x) \leq \alpha \log |x|$, $|x| > A$, for some A and $0 < \alpha < 2d-4$, then $\mathcal{E}^{\omega}$ is transient for P-a.e. ω.

(ii) If $d = 2$ and $R(x) \leq 2 \log \log |x|$, $|x| > A$ for some A, then $\mathcal{E}^{\omega}$ is recurrent for P-a.e. ω.

<u>Proof</u> (i)

$$E\left(\int_{S^{d-1}} \int_A^{\infty} r^{1-d} e^{X^{\omega}(r\sigma)}\, dr d\sigma \right) = \int_{S^{d-1}} \int_A^{\infty} r^{1-d} e^{\frac{1}{2}R(r\sigma)}\, dr d\sigma$$

$$\leq \int_{S^{d-1}} \int_A^{\infty} r^{1-d} r^{\alpha/2}\, dr\, d\sigma < \infty .$$

Hence we get (2.3) for a.e. $\sigma \in S^{d-1}$ for P-a.e. $\omega \in \Omega$.

(ii) By Schwarz, we have, for any $\alpha > 1$,

$$\int_A^{\infty} r^{-1} \left(\int_0^{2\pi} e^{-X(r\sigma)}\, d\sigma \right)^{-1} dr \cdot \int_A^{\infty} r^{-1} (\log r)^{-2\alpha} \left(\int_0^{2\pi} e^{-X(r\sigma)}\, d\sigma \right) dr$$

$$\geq \left(\int_A^{\infty} r^{-1} (\log r)^{-\alpha}\, dr \right)^2 = \frac{1}{(\alpha - 1)^2} \frac{1}{(\log A)^{2\alpha-2}} .$$ Hence,

$$E[\{\int_A^\infty r^{-1}(\int_0^{2\pi} e^{X(r\sigma)}\,d\sigma)^{-1}\,dr\}^{-1}]$$

$$\leq (\alpha-1)^2 (\log A)^{2\alpha-2}\int_A^\infty r^{-1}(\log r)^{-2\alpha} e^{\frac{1}{2}R(r\sigma)}\,dr = \frac{(\alpha-1)^2}{2\alpha-2} = \frac{\alpha-1}{2}$$

which decreases to zero as $\alpha \downarrow 1$. Therefore (2.2) holds for P-a.e. $\omega \in \Omega$. q.e.d.

When $R(x)$ grows at infinity more rapidly, then the situation would change radically. This will be illustrated by the following theorem.

Theorem 2.2 Let $B^\omega(t)$, $t > 0$, be the one dimensional Brownian motion and let $X^\omega(x) = B^\omega(|x|)$, $x \in R^d$. Then the associated energy form $\mathcal{E}^\omega$ is recurrent for a.e. ω regardless of the dimension d.

Proof In this case, (2.2) reads

$$(2.4) \qquad \int_1^\infty e^{B^\omega(r) - (d-1)\log r}\,dr = \infty$$

and consequently, it suffices to show

$$(2.5) \qquad |\{ r \geq 0 : B(r) \geq (d-1)\log r\}| = \infty \quad \text{P-a.e.}$$

where $|\ |$ denotes the Lebesgue measure. But we can assert more than (2.5) as follows :

Lemma 2.3 Let $h(t)$ be a real function on $[0,\infty)$ such that

(i) $h(t) \uparrow \infty$, $r \uparrow \infty$

(ii) $h(t)$ is concave on (A,∞) for some $A > 0$.

(iii) h belongs to the lower class $\mathcal{L}$ for the Brownian motion $B(t)$.

then

$$|\{t \geq 0 : B(t) \geq h(t)\}| = \infty \quad \text{P-a.e.}$$

Proof Let us fix $\alpha > 0$ and define stopping times τ_n, η_n, $n = 1,2,\cdots$, by

$$\tau_1 = \inf\{t > A : B(t) = h(t) + \alpha\}, \quad \eta_1 = \inf\{t > \tau_1 : B(t) = h(t)\},$$

$$\tau_n = \inf\{t > \eta_{n-1} : B(t) = h(t) + \alpha\}, \quad \eta_n = \inf\{t > \tau_n : B(t) = h(t)\}.$$

By assumption (iii), those stopping times are finite a.s.

On the other hand, the assumption (ii) implies

$$h(\tau_n + t) \leq (h(\tau_n) - h(\tau_{n-1})) + h(\tau_{n-1} + t) \quad \text{and}$$

$$\eta_n - \tau_n = \inf\{ t > 0 : B(\tau_n + t) = h(\tau_n + t)\}$$

$$\geq \inf\{t > 0 : B(\tau_n + t) = (h(\tau_n) - h(\tau_{n-1})) + h(\tau_{n-1} + t)\}.$$

Since $B(\tau_n + \cdot) - (h(\tau_n) - h(\tau_{n-1}))$ has the same law as $B(\tau_{n-1} + \cdot)$, we get

$E[(\eta_n - \tau_n)\wedge 1] \geqq E[(\eta_{n-1} - \tau_{n-1})\wedge 1]$.

Now we have for $X_n = \frac{(\eta_n - \tau_n)\wedge 1}{E[(\eta_n - \tau_n)\wedge 1]}$, the bound $E(X_n^2) \leqq \frac{1}{(E[(\eta_1 - \tau_1)\wedge 1])^2}$.

Hence $|\{ t \geqq 0 : B(t) \geqq h(t)\}| \geqq \sum_{n=1}^{N} (\eta_n - \tau_n) \geqq E[(\eta_1 - \tau_1)\wedge 1] \sum_{n=1}^{N} X_n$ which diverges a.s. as $N \to \infty$ by the law of the large number. q.e.d.

In view of Lemma 2.3, we know that theorem 2.2 holds for a more general $X^{\omega}(x)$. For instance suppose that $X^{\omega}(x)$ is of the form

$$X^{\omega}(x) = B^{\omega}(|x|) + Y^{\omega}(x)$$

where Y^{ω} is a random field such that the function

$$(d-1)\log r + \log \int_{S^{d-1}} e^{-Y^{\omega}(r\sigma)}\, d\sigma$$

is dominated by a function $h(r)$ appearing in Lemma 2.3 for a.e. ω. then $\mathcal{E}^{\omega}$ is recurrent a.s. this is the case when $Y^{\omega}(x) = C\, B^{\omega}(x;d)$ with $0 \leqq C < 1/\sqrt{d}$, where $B^{\omega}(x;d)$ is the d-dimensional parameter Lévy's Brownian motion independent of $B^{\omega}(t)$.

We do not know if Theorem 2.2 remains to be true when $X^{\omega}(x) = B^{\omega}(x;d)$. We know in this case that the associated semigroup T_t is conservative $(T_t 1 = 1)$ for a.e.ω by virtue of Ichihara's explosion test [9].

§3 $\mathcal{E}$-harmonic functions and martingales

Let m be a positive Radon measure on R^d with $\mathrm{Supp}[m] = R^d$. Consider a C_0^{∞}-regular Dirichlet space $(\mathcal{F},\mathcal{E})$ on $L^2(R^d;m)$ possessing the local property and an associated diffusion process $M = (X_t, P_x)$ on R^d. For simplicity, we assume that M is conservative : its transition function $p_t(x,dy)$ satisfies $p_t(x,R^d) = 1$, $t \quad 0$, $x \quad R^d$. See Ichihara [9] for the explosion (or, conservativeness) test applicable to the special Dirichlet spaces in §2 and §4.

A function $u \in \mathcal{F}_{loc}$ is called $\mathcal{E}$-harmonic if

(3.1) $\quad \mathcal{E}(u,v) = 0 \quad$ for any $v \in C_0^{\infty}(D)$.

Taking a quasi-contionuous version $\tilde{u}$ of u, we have then by the Dirichlet space theory [2 ; pp164] that the process $\tilde{u}(X_t) - \tilde{u}(X_0)$ is a continuous local martingale with respect to P_x for q.e. $x \in R^d$ with the quadratic variation being the positive continuous additive functional corresponding to the energy measure $\mu_{<u>}$ of u defined by

$$(3.2) \quad \int_{R^d} v(x)\mu_{<u>}(dx) = 2\,\mathcal{E}(uv, v) - \mathcal{E}(u^2, v), \quad v \in C_0^\infty(R^d).$$

Within the framework of the Dirichlet space theory only, we can not control the location of the exceptional set of x of zero capacity in the above statement. But in many cases such as the homogenization problem treated in the next section, we would like to know if the statement holds P_{x_0}-a.s. for a preassigned point $x_0 \in R^d$. The question is as follows ; what kind of additional information is necessary to achieve this ? We give an answer by the next theorem.

Theorem 3.1 Fix a point $x_0 \in R^d$ and assume the following conditions for the process $\underline{\underline{M}} = (X_t, P_x)$ and for a function u.

(i) The transition function p_t of $\underline{\underline{M}}$ satisfies $p_t(x_0, A) = 0$ for any t if $\mathrm{Cap}(A) = 0$.

(ii) $u \in \mathcal{F}_{loc}$. u is continuous and $\mathcal{E}$-harmonic.

(iii) $\mu_{<u>} \ll m$ and the density function f satisfies

$$E_{x_0}\left(\int_0^t f(X_s)ds\right) < \infty, \quad t > 0.$$

Then $M_t = u(X_t) - u(X_0)$ is a P_{x_0}-square integrable martingale with

$$(3.3) \quad \langle M\rangle_t = \int_0^t f(X_s)ds, \quad t > 0, \quad P_{x_0}\text{-a.s.}$$

Proof Denote the right hand side of (3.3) by ϕ_t. Let $\tau_n = \inf\{\, t > 0 : |X_t| > n\}$. According to the general theory [2], there exist a sequence of square integrable martingale additive functionals $M_t^{(n)}$ with a common properly exceptional set $N \subset R^d$ such that

$$(3.4) \quad \begin{cases} P_x(\, M_t = M_t^{(n)},\ t \leqq \tau_n) = 1 & x \in R^d - N \\ E_x((\, M_{t\wedge\tau_n}^{(n)})^2) = E_x(\,\phi_{t\wedge\tau_n}) & x \in R^d - N. \end{cases}$$

On the other hand, we have from

$$\int_{R^d} p_s(x_0, dy)\, E_y(\phi_t) = E_{x_0}\left(\int_s^{s+t} f(X_v)dv\right) < \infty$$

that $E_y(\phi_t)$ is finite for $p_s(x_o,dy)$-a.e. $y \in R^d$. By (3.4), we see then, for $p_s(x_o,dy)$-a.e. $y \in R^d - N$, $E_y(M_{t\wedge\tau_n}^2) = E_y(\phi_{t\wedge\tau_n}) \leq E_y(\phi_t) < \infty$.

Accordingly, $\{M_{t\wedge\tau_n}\}_n$ is an $L^2(P_y)$-bounded P_y-martingale for such y. Since $p_s(x_o, N) = 0$ by our hypothesis, we can conclude that, for each $s, t > 0$,

(3.5) $$E_y(M_t^2) = E_y(\phi_t) < \infty, \quad E_y(M_t) = 0, \quad \text{for } p_s(x_o,dy)\text{-a.e. } y \in R^d.$$

(3.5) implies, for $0 < s < t$,

$$E_{x_o}((u(X_t) - u(X_s))^2) = \int_{R^d} p_s(x_o,dy)\, E_y(M_{t-s}^2) \leq \int_{R^d} p_s(x_o,dy) E_y(\phi_t)$$

$$\leq E_{x_o}(\phi_{s+t}).$$

By Fatou's lemma, we have $E_{x_o}(M_t^2) \leq E_{x_o}(\phi_t) < \infty$ the P_{x_o}-square integrability of M_t. Since both M_t and ϕ_t are additive functionals of $\underline{\underline{M}}$, it also follows from (3.5) that, for $0 < s < t$, $E_{x_o}(M_t | \mathcal{B}_s) = M_s + E_{X_s}(M_{t-s}) = M_s$ P_{x_o}-a.s. and

$$E_{x_o}(M_t^2 - \phi_t | \mathcal{B}_s) = M_s^2 - \phi_s + 2M_s E_{X_s}(M_{t-s}) + E_{X_s}(M_{t-s}^2 - \phi_{t-s}) = M_s^2 - \phi_s \quad P_{x_o}\text{-a.s.}$$

q.e.d.

§4 Homogenization in random media

We consider a probability space $(\Omega, \mathcal{B}, \mu)$ and an ergodic flow $\{T_x : x \in R^d\}$ defined on it :

(4.1) T_x is a μ-measure preserving transformation of Ω

(4.2) T_0 = identity. $T_{x+y} = T_x \circ T_y$.

(4.3) The map $(x,\omega) \rightarrow T_x\omega$ from $R^d \times \Omega$ into Ω is measurable.

(4.4) Any random variable (namely, real valued $\mathcal{B}$-measurable function on Ω) which is T_x-invariant μ-a.e. for every $x \in R^d$ is constant μ-a.e.

We then consider random variables $a_{ij}(\omega)$, $1 \leq i,j \leq d$, such that $a_{ij} = a_{ji}$ and

(4.5) $$\lambda_o|\xi|^2 \leq \sum_{i,j=1}^{d} a_{ij}(\omega)\xi_i\xi_j \leq \Lambda_o|\xi|^2, \quad \omega \in \Omega, \quad \xi \in R^d,$$

for some constants $0 < \lambda_o \leq \Lambda_o$. The associated Dirichlet space $(\mathcal{F}^\omega, \mathcal{E}^\omega)$ on $L^2(R^d,dx)$ can then be defined by

(4.6) $$\mathcal{F}^\omega = H^1(R^d), \quad \mathcal{E}^\omega(u, v) = \sum_{i,j=1}^{d} \int_{R^d} u_{x_i}(x) v_{x_j}(x) a_{ij}(T_x\omega)dx, \quad u, v \in \mathcal{F}^\omega.$$

The general theory of Dirichlet spaces admits an associated diffusion process $\underline{\underline{M}}^{\omega} = (X_t^{\omega}, P_x^{\omega})$ on R^d such that the transition function p_t^{ω} of $\underline{\underline{M}}^{\omega}$ is a realization of the L^2-semigroup T_t^{ω} generated by the Dirichlet space (4.6). However, for the present purpose, we have to adopt a more specific realization of T_t^{ω} due to Nash [11] ; $p_t^{\omega}(x,dy)$ has a density $p_t^{\omega}(x,y)$ such that

$$(4.7)\qquad C_1 t^{-d/2}\exp\left(-\frac{C_2|x-y|^2}{t}\right) \leq p_t^{\omega}(x,y) \leq C_3 t^{-d/2}\exp\left(-\frac{C_4|x-y|^2}{t}\right), \quad x, y \in R^d.$$

For $\varepsilon > 0$, let us introduce a Dirichlet space $(F^{\omega,\varepsilon}, E^{\omega,\varepsilon})$ by (4.6) with T_x being replaced by $T_{x/\varepsilon}$. The associated diffusion in the above sense is designated by $\underline{\underline{M}}^{\omega,\varepsilon} = (X_t^{\omega,\varepsilon}, P_x^{\omega,\varepsilon})$. Since $\underline{\underline{M}}^{\omega}$, $\underline{\underline{M}}^{\omega,\varepsilon}$ are conservative, we may assume that P_x^{ω}, $P_x^{\omega,\varepsilon}$ are probability measures on the space $C([0,\infty)\rightsquigarrow R^d)$.

<u>Theorem 4.1</u> For μ-a.e. $\omega \in \Omega$, $P_0^{\omega,\varepsilon}$ converges as $\varepsilon \downarrow 0$ weakly to a probability measure Q on $C([0,\infty)\rightsquigarrow R^d)$. Here Q corresponds to a Gaussian diffusion on R^d starting at 0 with generator

$$\hat{A}u = \sum_{i,j=1}^{d} \hat{a}_{ij}\, u_{x_i x_j}$$

where $\{\hat{a}_{ij}\}$ is a constant matrix defined by

$$(4.8)\qquad \hat{a}_{ij} = \int_{\Omega} \sum_{k,h=1}^{d} (1-\psi_k^i)\, a_{kh}\, (1-\psi_h^j)(\omega)\, d\mu(\omega)$$

for random variables $\psi_k^i \in L^2(\Omega,\mu)$ specified in Lemma 4.2.

This theorem was proven by Osada [12] for a more general random operator L^{ω} than the present $L^{\omega} = \sum \frac{\partial}{\partial x_i}\left(a_{ij}(T_x\omega)\frac{\partial}{\partial x_j}\right)$ but by assuming a smoothness condition in x for the coefficients so that Ito's formula was applicable. Recently Osada [13] has also succeeded to remove the smoothness condition and our present theorem itself will be included in his new one. The novelty of our present approach is that we are able to reduce a part of the proof in [12] to a general Dirichlet space theory formulated in the preceding section.

We refer to Kozlov et al [10] and Papanicolaou-Varadhan[14] for the proof of the next lemma.

Lemma 4.2 There exist unique functions $\chi^j(x,\omega)$, $1 \leq j \leq d$, and $\psi_i^j(\omega)$, $1 \leq i,j \leq d$, satisfying the following :

(i) $\chi^j(x) \in L^2(\Omega;\mu)$ for each $x \in R^d$ and $\psi_i^j \in L^2(\Omega;\mu)$.

(ii) For μ-a.e. fixed $\omega \in \Omega$, $\chi^j(\cdot,\omega)$ is continuous, belongs to $H^1_{loc}(R^d)$ and the function $\beta^j(x,\omega) = x^j - \chi^j(x,\omega)$ is $\mathcal{E}^\omega$-harmonic. Moreover

$$\frac{\partial}{\partial x^i} \chi^j(x,\omega) = \psi_i^j(T_x\omega) \quad \text{and} \quad \chi^j(0,\omega) = 0.$$

The following lemma is taken from Osada [12 ; Proposition 2.1]. See also Papanicolaou-Varadhan [15].

Lemma 4.3 For a fixed $\omega \in \Omega$, we define a stochastic motion on Ω by

$$\underline{\underline{X}}_t^\omega(\tilde{\omega}) = T_{X_t^\omega(\tilde{\omega})}\omega, \qquad t \geq 0,$$

where $\tilde{\omega}$ denotes the point of the sample space $\tilde{\Omega}$ of the diffusion $\underline{\underline{M}}^\omega$. Then the probability measure P_0^ω on Ω induces by $\underline{\underline{X}}^\omega$ a probability measure Q_ω on the space of all trajectories on Ω. $(\underline{\underline{X}}_t^\omega, Q_\omega)$ is then a Markov process with state space Ω starting at ω possessing μ as its invariant measure. Moreover $\underline{\underline{X}}_t^\omega$ is ergodic under $\int_\Omega \mu(d\omega)Q_\omega(\cdot)$.

Proof of Theorem 4.1 The tightness of $P_0^{\omega,\varepsilon}$ in ε is immediate from the estimate (4.7). As for the convergence of finite dimensional distributions, we proceed as follows. Since $\varepsilon X_{t/\varepsilon^2}^\omega$ under P_0^ω has the same law as $P_0^{\omega,\varepsilon}$, it suffices to prove that any finite dimensional distribution of $\varepsilon X_{t/\varepsilon^2}^\omega$ under P_0^ω converges as $\varepsilon \downarrow 0$ to that of Q.

Using functions appearing in Lemma 4.2, we have $x^j = \beta^j + \chi^j$, $1 \leq j \leq d$, which shall be written as $x = \beta + \chi$ in the vector notation. Then

$$(4.9) \qquad \varepsilon X_{t/\varepsilon^2}^\omega = \varepsilon\beta(X_{t/\varepsilon^2}^\omega, \omega) + \varepsilon\chi(X_{t/\varepsilon^2}^\omega, \omega).$$

We now apply Theorem 3.1 to the function $u(x,\omega) = \lambda\cdot\beta(x,\omega) = \sum_{j=1}^d \lambda_j \beta^j(x,\omega)$ for a fixed $\lambda = (\lambda_1,\cdots,\lambda_d)$. We take x_o to be the origin 0. Condition (i) of Theorem 3.1 is satisfied due to the present specific choice of $\underline{\underline{M}}^\omega$. The function u satisfies condition (ii) of Theorem 3.1 for μ-a.e. $\omega \in \Omega$ by virtue of Lemma 4.2. As for condition (iii), we have from Lemma 4.2 that

$$f(x,\omega) = 2\sum_{i,j=1}^d u_{x_i}(x,\omega)a_{ij}(T_x\omega)u_{x_j}(x,\omega) = 2\lambda\cdot q(T_x\omega)\cdot\lambda^*$$

where q is a matrix with component

$$q^{ij}(\omega) = \sum_{k,h=1}^{d} (1 - \psi_k^i(\omega))a_{kh}(\omega)(1 - \psi_h^j(\omega)).$$

Hence, in view of Lemma 4.3,

$$\int \mu(d\omega)\, E_0^{\omega}(\int_0^t f(X_s^{\omega},\omega)ds) = 2\int \mu(d\omega)\, E^{Q_\omega}(\int_0^t \lambda\cdot q(\underline{\underline{X}}_s^{\omega})\cdot\lambda^*\, ds) = 2t\int \lambda\cdot q(\omega)\cdot\lambda^*\mu(d\omega),$$

which is finite because $\psi_k^i \in L^2(\Omega,\mu)$ by Lemma 4.2. Consequently $E_0^{\omega}(\int_0^t f(X_s^{\omega},\omega)ds)$ is finite for μ-a.e. $\omega \in \Omega$ and the condition (iii) of Theorem 3.1 is fulfilled for the present function u for μ-a.e $\omega \in \Omega$.

Therefore we see by Theorem 3.1 that, for μ-a.e.$\omega \in \Omega$, the process $M_t^{\omega,\varepsilon} = \varepsilon\beta(X_{t/\varepsilon^2}^{\omega},\omega)$ is R^d-valued square integrable P_0^{ω}-martingale with co-variation matrix given by

$$(4.10) \qquad \langle M^{\omega,\varepsilon}\rangle_t = 2\varepsilon^2\int_0^{t/\varepsilon^2} q(\underline{\underline{X}}_s^{\omega})ds.$$

Denote by $\hat{a}$ the matrix with component $\hat{a}_{ij}$ of (4.8). Apply Lemma 4.3 and the individual ergodic theorem to (4.10) to get, for each $t > 0$,

$$(4.11) \qquad \int_{\Omega} P_0^{\omega}(\lim_{\varepsilon\downarrow 0} \langle M^{\omega,\varepsilon}\rangle_t = 2\hat{a}\cdot t\,)\mu(d\omega) = 1.$$

In particular $\langle M^{\omega,\varepsilon}\rangle_t$ converges, as $\varepsilon \downarrow 0$, to $2\hat{a}t$ P_0^{ω}-a.s. for μ-a.e. fixed $\omega \in \Omega$, and we can conclude for such ω that the finite dimensional distribution of the process $M_t^{\omega,\varepsilon}$ converges as $\varepsilon \downarrow 0$ to that of Brownian motion with covariance matrix $2\hat{a}\cdot t$ P_0^{ω}-a.s.

As for the second term $\varepsilon\chi(X_{t/\varepsilon^2}^{\omega},\omega)$ of the right hand side of (4.9), Osada[12] has proven its convergence to the trivial process 0 in the sense of finite dimensional distribution by using the strong estimate (4.7). Thus we see the desired convergence of $\varepsilon X_{t/\varepsilon^2}^{\omega}$. q.e.d.

References

[1] T. Brox, A one-dimensional diffusion process in a Wiener media, to appear

[2] M. Fukushima, Dirichlet forms and Markov processes, Kodansha and North-Holland, 1980.

[3] M. Fukushima, A generalized stochastic calculus in homogenization, Quantum fields--algebras, Processes, L.Streit (ed.) Soringer-Verlag, Wien/New York, 1980.

[4] M. Fukushima, Markov processes and functional analysis, Proc. International Math. Cinf. , Chen, Ng, Wick (eds.) North-Holland, 1982.

[5] M. Fukushima, Energy forms and diffusion processes, Mathematics + Physics I, L.Streit (ed.) World Scientific, Singapore, Philadlphia, 1985.

[6] M. Fukushima, On recurrence criteria in the Dirichlet space theory, Proc. Symp. University Warwick, D.Elworthy (ed.), to appear

[7] M. Fukushima and S. Nakao, On spectra of the Schrödinger operator with a white Gaussian noise potential, Z.Wahrscheinlichkeitstheorie verw.Gebiete 37 (1977), 267-274.

[8] K. Ichihara, Some global properties of symmetric diffusion processes, Publ. RIMS, Kyoto Univ. 14(1978), 441-486.

[9] K. Ichihara, Explosion problems for symmetric diffusion processes,Proc.Japan Acad. Vol.60, No 7 ser A, (1984), 243-245.

[10] S.M. Kozlov, O.A. Olenik, Kha T'en Ngoan and V.V. Zhikov, Averaging and G-convergence of differential operators, Russian Math. Surveys, 34:5(1979), 69-147.

[11] J. Nash, Continuity of solutions of parabolic and elliptic equations, Amer. J. Math., 80(1958), 931-953.

[12] H. Osada, Homogenization of diffusion processes with random stationary coefficients, Proc. 4-th Japan-USSR Symp. on Prob. Th., Lecture Notes in Math. 1021, Springer-Verlag, 1983.

[13] H. Osada, Homogenization problem for diffusion processes associated with a generalized divergence form, to appear

[14] G. Papanicolaou and S.R.S. Varadhan, Boundary value problems with rapidly oscillating random coefficients, Coll. Math. Soc. Janos Bolyai 27, North-Holland, 1979.

[15] G. Papanicolaou and S.R.S. Varadhan, Diffusions with random coefficients, Essay in honor of C.R.Rao (G.Kallianpur, P.R.Krishnaiah, J.K.Ghosh eds), North-Holland, 1982.

Masatoshi Fukushima
Department of Mathematics, College of General Education
Osaka University

Shintaro Nakao
Department of Mathematics, Osaka University

Masayoshi Takeda
Department of mathematics, Osaka university

Toyonaka, Osaka, Japan

A NICOLAI MAP FOR SUPERSYMMETRIC QUANTUM MECHANICS ON RIEMANNIAN MANIFOLDS

R. Graham and D. Roekaerts
Universität-GHS Essen
Fachbereich Physik
D-4300 Essen, West-Germany

Abstract : It is shown that a Stratonovich stochastic differential equation plays the role of a Nicolai map in the zero-fermion sector of supersymmetric quantummechanics on Riemannian manifolds.

Recently we have shown[1] that the Stratonovich stochastic differential equation for the Markov process $q(t) = (q^1(t),\dots,q^n(t))$ with diffusion matrix $g^{\mu\nu}(q), \mu,\nu=1,\dots,n$ and drift vector $h^\mu(q) = g^{\mu\nu}V_{;\nu}(q)$ is a Nicolai map in the zero-fermion sector of supersymmetric quantum mechanics on the Riemannian manifold with local coordinates $q = (q^1,\dots,q^n)$ and metric tensor $g^{\mu\nu}(q)$, $V(q)$ being the superpotential. Here we shall present our (heuristic) proof of this result in some more detail.

Let us first recall the meaning of the terms Nicolai map and supersymmetric quantum mechanics. (See also the contribution by Girardello to this volume) Consider a system described by ordinary canonical variables $B = (B^1,\dots,B^n)$ and Grassmann variables $F = (F^1,\dots,F^m)$, i.e. $[B^i,B^j] = \{F^i,F^j\} = [B^i,F^j] = 0$, and with Lagrangian [2]

$$L(B,\dot{B},F,\dot{F}) = L_1(B,\dot{B}) + L_2(B,\dot{B},F,\dot{F}) \tag{1}$$

(i.e. all terms independent of F are contained in L_1).(For the general idea the restriction to a finite number of degrees of freedom is unessential). In the path integral approach the transition amplitudes of the quantum theory for Euclidean time $\tau = it$ are given by [3]

$$I = \int \mathcal{D}B(\tau)\ \mathcal{D}F(\tau)\ \exp\{-\int_0^T L(B,\dot{B},F,\dot{F})\ d\tau\} \tag{2}$$

(Precise definition to be given below). By integrating out the F-variables one obtains

$$I = \int \mathcal{D}B(\tau)\ J[B(\tau)]\ \exp\{-\int_0^T L_1(B,\dot{B})\ d\tau\} \qquad (3)$$

A Nicolai map is a, in general nonlinear and nonlocal, transformation of variables $B = B(\bar{B})$ such that a) $L_1(B,\dot{B}) = L_o(\bar{B},\dot{\bar{B}})$ + total derivative, with L_o quadratic, and b) the Jacobian of the transformation cancels the functional $J[B(\tau)]$. Hence, by this map I is reduced to a Gaussian functional integral, subject however to a nonlinear boundary condition. The existence of a Nicolai map expresses a supersymmetry of the system.[4,5] Of special interest is the case in which the Nicolai map is local since then it can be considered as a stochastic differential equation relating the Gaussian stochastic process $\bar{B}(\tau)$ to the stochastic process $B(\tau)$.[6,7] Until recently all systems for which a local Nicolai map is known had L_2 quadratic in the Grassmann variables. Supersymmetric quantum mechanics on a Riemannian manifold M provides the simplest example of a model with Lagrangian containing a term quartic in these variables.[8] One has $L = L_1 + L_2$ with

$$L_1 = \tfrac{1}{2}g_{\mu\nu}(q)\dot{q}^\mu\dot{q}^\nu + \tfrac{1}{2}g^{\mu\nu}(q)V_{;\mu}(q)V_{;\nu}(q) \qquad (4)$$

$$L_2 = \frac{1}{2}\left(\psi^*_i\frac{D\psi^i}{D\tau} - \frac{D\psi^*_i}{D\tau}\psi^i\right) - \psi^{*i}\psi^j e_i{}^\mu(q)e_j{}^\nu(q)V_{;\mu\nu}(q)$$
$$- \frac{1}{4}\psi^{*i}\psi^{*j}\psi^k\psi^l\ R_{ijkl}(q) \qquad (5)$$

The dynamical variables are $q^\mu, \mu = 1,\ldots,n$ which are local coordinates on M and ψ^i, ψ^{i*}, $i = 1,\ldots,n$ which are pairs of complex conjugate Grassmann variables tangent to M. $g_{\mu\nu}(q)$ is the metric tensor and $e_i{}^\mu(q)$, $\mu = 1,\ldots,n$; $i = 1,\ldots,n$ is an orthogonal frame or 'Vielbein' at q with inverse $e^i{}_\mu(q)$. ($e_i{}^\mu e_i{}^\nu = g^{\mu\nu}$ and $e_i{}^\mu e^j{}_\mu = \delta_i{}^j$). R_{ijkl} is the curvature tensor in the Vielbein basis and V(q) is a Riemann scalar called superpotential. The covariant derivative of Riemann tensors (Greek indices) is denoted by a semi colon and the covariant derivative of a Euclidean vector and Riemann scalar by $D/D\tau$.[9] Explicitly one has

$$\frac{D\psi^i}{D\tau} = \frac{\partial}{\partial\tau}\psi^i + e^i{}_\nu\ e_j{}^\nu{}_{;\lambda}\dot{q}^\lambda\psi^j,\ i = 1,\ldots,n \qquad (6)$$

with

$$e_j{}^\nu{}_{;\lambda} = \frac{\partial e_j{}^\nu}{\partial q^\lambda} + \Gamma_\lambda{}^\nu{}_\alpha\ e_j{}^\alpha,\ j = 1,\ldots,n\ ,\ \Gamma_\lambda{}^\nu{}_\alpha \text{ the affine connection.}$$

The Lagrangian (4)(5) can be obtained as a simplification of the supersymmetric nonlinear sigma model by requiring the fields to be functions of time only [8] or introducing supercoordinates directly in the finite dimensional case.[1] How canonical quantization of (4)(5) leads to the exterior algebra was discussed by Witten in Ref. 8. (See also Refs. 10

and 11) The connection with Morse theory is explained in Ref. 12. A quantum mechanical state with fermion number p, represented by an antisymmetric wavefunction carrying p indices, can be identified with a differential p-form. In the absence of a superpotential the Hamiltonian operator is the Laplacian acting on forms. Fermion number p is conserved and different values of p define different 'fermion sectors'.

The Euclidean functional integral (2) is now

$$I = \int \mathcal{D}q(\tau)\mathcal{D}\psi(\tau)\mathcal{D}\psi^*(\tau)\ \exp\{-\int_0^T L(q,\dot{q},\psi,\dot{\psi},\psi^*,\dot{\psi}^*)\ d\tau\} \tag{7}$$

Which transition amplitude (7) represents is only specified by a precise definition of this formal expression, including the boundary conditions. Here we shall define (7) as the limit of a multidimensional integral.[13] All paths $q(\tau)$ are between fixed initial and final points. The boundary conditions on the integration over Grassmann variables depend on the fermion sector considered.[14-16] Therefore the integrand of the path integral over $q(\tau)$ which remains after the path integral of $\psi^*(\tau)$ and $\psi(\tau)$ is done also depends on the fermion sector. This implies that the Nicolai map that transforms this path integral into Gaussian form will be sector dependent. Here we restrict ourselves to the study of the zero-fermion sector of the theory ($p = 0$). (The $p=n$ fermion sector can be treated in the same way. In the other sectors no exact local Nicolai maps are known at present. Perturbative results on the torus are obtained in Ref. 17.) The result of the functional integral over Grassmann variables also depends on the limiting procedure by which it is defined and which fixes a correspondence rule between products of Grassmann variables and products of fermion creation and annihilation operators.[13,15,18] (Cf. Eqs. 13-15). Nicolai's procedure which employs a transformation of variables with the usual rules of calculus in the functional integral over $q(\tau)$ is compatible only with the restricted class of formally covariant definitions of this functional integral (Ref. 13, chapter 6). Once the Nicolai transformation that reduces the path integral to Gaussian form has been found, the quantum Hamiltonian in the sector considered is also known. The Schrödinger equation for Euclidean time can be obtained from the Fokker-Planck equation for the transition probability density of the stochastic process $q(\tau)$ defined by the Nicolai map.

The Stratonovich stochastic differential equation for the Markovian process with diffusion matrix $g^{\mu\nu}(q)$ and drift vector

$$h^\mu(q) = -g^{\mu\nu}V_{;\nu}(q) \tag{8}$$

is given by

$$\dot{q}^{\mu} + g^{\mu\kappa}V_{;\kappa}(q) = -k^{\mu}(q) + e_i{}^{\mu}(q)\circ\xi^{i} \tag{9}$$

with

$$k^{\mu}(q) = \frac{1}{2}g^{\mu\nu}e_i{}^{\lambda}e^{i}{}_{\nu;\lambda} = \frac{1}{2}g^{\mu\nu}e_i{}^{\lambda}\left(\frac{\partial e^{i}{}_{\nu}}{\partial q^{\lambda}} - \frac{\partial e^{i}{}_{\lambda}}{\partial q^{\nu}}\right) \tag{10}$$

and $\xi^i(t)$ Gaussian white noise.[19] It is important to remark that although the terms $-k^{\mu}(q)$ and $e_i{}^{\mu}(q)\circ\xi^i$ are not separately invariant under local orthogonal rotations of the 'Vielbein' ($\bar{e}_j{}^{\mu}(q) = \Omega_j{}^{i}(q)e_i{}^{\mu}(q)$ with $\Omega_j{}^{i}\,\Omega_i{}^{k} = \delta_j{}^{k}$), the drift $h^{\mu}(q)$ and the diffusion $g^{\mu\nu}(q)$, and hence the stochastic process, are independent of these rotations. This implies that there are in general infinitely many different Stratonovich stochastic differential equations belonging to a given Markov process. We have found [1] that a given equation (9) has the property of a Nicolai map, i.e. cancellation of functional Jacobian $\delta q^{\mu}(\tau)/\delta\xi^{i}(\tau')$ and the functional $J[q(\tau)]$, only when the covariant derivative of the 'Vielbein' (considered as a set of Riemann vectors) vanishes

$$e_i{}^{\nu}{}_{;\lambda} = 0, \quad i = 1,\dots,n \tag{11}$$

In flat space this can be imposed globally, in curved space at most along an arbitrary but fixed curve. Then for different paths different choices of the 'Vielbein' basis are required. All these choices are related by local orthogonal rotations and, as mentioned, are stochastically equivalent. We remark that the condition (11) is also an essential part of the construction of diffusion processes on manifolds by 'rolling' the manifold along a curve in Euclidean space.[20] We also remark that a different, but stochastically equivalent, Nicolai map for the model under consideration here was constructed by Claudson and Halpern.[16,17] In Ref. 1 we have compared our approach and that of Refs. 16,17.

To give the heuristic proof of our result we start with the calculation of the functional integral over Grassmann variables

$$J[q(\tau)] = \int \mathcal{D}\psi(\tau)\mathcal{D}\psi^*(\tau)\,\exp\{-\int_0^T L_2(q,\dot{q},\psi,\dot{\psi},\psi^*,\dot{\psi}^*)\,d\tau\} \tag{12}$$

with L_2 given by (5). $q(\tau)$ is an arbitrary but fixed curve between $q(0) = q_o$ and $q(T) = q_T$ and the 'Vielbein' is chosen such that (11) is satisfied along this curve. Then $D\psi^i/D\tau = \partial\psi^i/\partial\tau$ and (12) is defined by the limit ($t_m = m\varepsilon$, $m = 0,1,\dots,N+1$; $t_{N+1} = T$)

$$J[q(\tau)] = \lim_{N\to\infty} \int \prod_{m=1}^{N} d\psi(t_m)d\psi^*(t_m)$$
$$\exp\{-\varepsilon \sum_{m=1}^{N} \tfrac{1}{2}[\psi_i^*(t_m)\frac{\psi^i(t_m)-\psi^i(t_{m-1})}{\varepsilon} - \frac{\psi_i^*(t_{m+1})-\psi_i^*(t_m)}{\varepsilon}\psi^i(t_m)]\}$$
$$\exp\{\varepsilon \sum_{m=1}^{N} [e_i{}^\mu e_j{}^\nu V_{;\mu\nu}(q(t_m))\psi_i^*(t_m)\frac{\psi^i(t_m)+\psi^i(t_{m-1})}{2}$$
$$+ \tfrac{1}{4}R_{ijkl}(q(t_m))\psi^{*i}(t_m)\psi^{*j}(t_m)\frac{\psi^k(t_m)+\psi^k(t_{m-1})}{2}$$
$$\frac{\psi^l(t_m)+\psi^l(t_{m-1})}{2}]\} \qquad (13)$$

where

$$\psi^*(t_{N+1}) = \psi(0) = 0. \qquad (14)$$

That we are considering transition amplitudes in the zero-fermion sector is reflected in the boundary condition (14) and in the discretisation of the terms $(\psi_i^* \dot{\psi}^i - \dot{\psi}_i^* \psi^i)/2$ in L_2.[15] The discretisation of $\psi_i^*(\tau)\psi^i(\tau)$ as $\psi_i^*(t_m)(\psi^i(t_m)+\psi^i(t_{m-1}))/2$ is equivalent to the choice of the symmetric correspondence rule in the operator formalism.[18]
Using the usual definition of integration over Grassmann variables (13) can be calculated.[3] More elegantly one can proceed directly in the continuum notation (12) using a representation by Feynman diagrams and

$$\int \mathcal{D}\psi(\tau)\mathcal{D}\psi^*(\tau)\ \psi_k^*(\tau_1)\psi^l(\tau_2)\ \exp\{-\int_0^T \tfrac{1}{2}(\psi_i^*\dot{\psi}^i-\dot{\psi}_i^*\psi^i)d\tau\}\ \delta_k{}^l\ \theta(\tau_2-\tau_1) \qquad (15)$$

with $\theta(0) = 1/2$, in accordance with the correspondence rule. The result is

$$J[q(\tau)] = \exp\{\int_0^T (\tfrac{1}{2}g^{\mu\nu}V_{;\mu\nu} - \tfrac{1}{8}R)\ d\tau\} \qquad (16)$$

where R is the scalar curvature.

Next we calculate the Jacobian of the map (9). We use a discrete representation. (References to the mathematical literature can be found in Ref. 22).

$$\xi^i(t_m) = e^i{}_\mu(\bar{q}_m)[\tfrac{1}{\varepsilon}\Delta q_m^\mu + g^{\mu\kappa}V_{;\kappa}(\bar{q}_m) + k^\mu(\bar{q}_m)] \qquad (17)$$

where $t_m = m\varepsilon$, $m = 0,1,\ldots,N+1$; $t_{N+1} = T$, $\bar{q}_m = (q(t_m)+q(t_{m-1}))/2$ and $\Delta q_m = q(t_m)-q(t_{m-1})$. The functional Jacobian Det $\delta\xi^i(\tau)/\delta q^\nu(\tau')$, $0\le\tau\le T$, $0\le\tau'\le T$ is represented by[13]

$$\lim_{N\to\infty} \prod_{m=1}^{N+1} \det \frac{\partial\xi^i(t_m)}{\partial q^\nu(t_m)} \qquad (18)$$

One has that

$$\varepsilon\frac{\partial\xi^i(t_m)}{\partial q^\nu(t_m)} = e^i{}_\mu(\bar{q}_m)\{\delta_\nu{}^\mu + \tfrac{1}{2}e_k{}^\mu e^k{}_{\lambda,\nu}(\bar{q}_m)\Delta q_m^\lambda$$

$$+ \tfrac{1}{2}\varepsilon e_k{}^\mu[e^k{}_\lambda(g^{\lambda\kappa}V_{;\kappa}(\bar{q}_m) + k^\lambda(\bar{q}_m))]_{,\nu}\} \tag{19}$$

$$= e^i{}_\mu(q(t_m))\{\delta_\nu{}^\mu + \tfrac{1}{2}e_k{}^\mu(e^k{}_{\lambda,\nu}(\bar{q}_m) - e^k{}_{\nu,\lambda}(\bar{q}_m))\Delta q_m^\lambda$$

$$+ \tfrac{1}{2}\varepsilon e_k{}^\mu[e^k{}_\lambda(g^{\lambda\kappa}V_{;\kappa}(\bar{q}_m) + k^\lambda(\bar{q}_m))]_{,\nu} + O((\Delta q_m)^2,\ \varepsilon(\Delta q_m)) \tag{20}$$

and, using standard formulae,

$$\det\frac{\partial\xi^i(t_m)}{\partial q^\nu(t_m)} = \varepsilon^{-n}[\det g_{\mu\nu}(q(t_m))]^{1/2}\exp\{\tfrac{1}{2}e_k{}^\mu(e^k{}_{\lambda,\mu}-e^k{}_{\mu,\lambda})\Delta q_m^\lambda$$

$$+ \tfrac{1}{2}\varepsilon e_k{}^\mu[e^k{}_\lambda(g^{\lambda\kappa}V_{;\kappa} + k^\lambda)]_{;\mu} + O((\Delta q_m)^2,\ \varepsilon(\Delta q_m),\varepsilon^2)\} \tag{21}$$

Consequently, in the discrete representation the Gaussian functional integral over ξ-variables is related to the functional integral over q-variables with $q(t_o) = q_o$ and $q(t_{n+1}) = q$ fixed (the nonlinear boundary condition mentioned in the introduction) as follows :

$$\bar{I} \equiv \int\prod_{m=1}^{N+1}\left[d\xi(t_m)\left[\frac{\varepsilon}{2\pi}\right]^{n/2}\right]\exp\{-\tfrac{1}{2}\varepsilon\sum_{m=1}^{N+1}\xi_i(t_m)\xi^i(t_m)\}\ \delta(q(t_{N+1})-q_T)$$

$$= \int\prod_{m=1}^{N+1}\left[dq(t_m)\ [(2\pi\varepsilon)^{-n}\det g_{\mu\nu}(q(t_m))]^{1/2}\right]\delta(q(t_{N+1}) - q_T)$$

$$\exp\{-\varepsilon\sum_{m=1}^{N+1}\tfrac{1}{2}g_{\mu\nu}(\bar{q}_m)(\tfrac{1}{\varepsilon}\Delta q_m^\mu + g^{\mu\kappa}V_{;\kappa}(\bar{q}_m) + k^\mu(\bar{q}_m))(\tfrac{1}{\varepsilon}\Delta q_m^\nu + g^{\nu\lambda}V_{;\lambda}(\bar{q}_m) + k^\nu(\bar{q}_m))\}$$

$$\exp\{\varepsilon\sum_{m=1}^{N+1}[\tfrac{1}{2}e_k{}^\mu(e^k{}_{\lambda,\mu}-e^k{}_{\mu,\lambda})\Delta q_m^\lambda + \tfrac{1}{2}\varepsilon e_k{}^\mu(e^k{}_\lambda(g^{\lambda\kappa}V_{;\kappa}+k^\lambda))_{;\mu}$$

$$+ O((\Delta q_m)^2,\ \varepsilon(\Delta q_m),\ \varepsilon^2)]\} \tag{22}$$

Let us pay attention now to the terms of order $(\Delta q_m)^2$ or $\varepsilon(\Delta q_m)$ or ε^2 which were not calculated in (22). The terms of order $\varepsilon(\Delta q_m)$ or order ε^2 will not contribute when the limit $N\to\infty$ is taken. The terms of order $(\Delta q_m)^2$ however do contribute since effectively $(\Delta q_m)^2 \sim \varepsilon$. In Ref. 13 they were calculated and several classes of not formally covariant functional integral representations for the transition probability density of the process $q(\tau)$ were obtained. However, when a nonlinear change of variables such as (9) is made in the continuum expression of a covariantly defined functional integral the terms of order $(\Delta q_m)^2$ do not contribute in the continuum expression because the usual rules of calculus apply, by definition, in a covariant functional integral.

The method to obtain the $(\Delta q_m)^2$-terms in the discretized expression, when needed, forms however an essential part of the definition of the covariant functional integral. As a result, since we assume our path integrals to be defined in a covariant manner, we can neglect the $(\Delta q_m)^2$-terms in the search for a Nicolai map. In continuum notation, and using (8) and (10), (22) reads $(q(0) = q_o,\ q(T) = q_T)$

$$I = \int \mathcal{D}q(\tau)\ \exp\{-\int_0^T \tfrac{1}{2} g_{\mu\nu}(q)(\dot{q}^\mu - h^\mu(q) + k^\mu(q))(\dot{q}^\nu - h^\nu(q) + k^\nu(q))\,d\tau\}$$

$$\exp\{-\int_0^T [-k_\lambda \dot{q}^\lambda + k_\lambda h^\lambda - k_\lambda k^\lambda - \tfrac{1}{2} h^\lambda{}_{;\lambda} - \tfrac{1}{2} k^\lambda{}_{;\lambda}]\,d\tau\} \qquad (23)$$

This expression was obtained first in Ref. 21. It is Riemann covariant but not Euclidean invariant, since k^μ is not a Euclidean invariant. For any fixed path $q(\tau)$ however one can choose a 'Vielbein' such that (11) is satisfied. Then, along this path

$$k^\mu(q) = 0, \quad k^\lambda{}_{;\lambda} = -\tfrac{1}{4} R \qquad (24)$$

and the integrand of (23) becomes

$$\exp\{-\int_0^T [\tfrac{1}{2} g_{\mu\nu}(q)\dot{q}^\mu \dot{q}^\nu + \tfrac{1}{2} h^\mu(q) h_\mu(q) - h_\mu(q)\dot{q}^\mu]\,d\tau\}$$

$$\exp\{-\int_0^T [-\tfrac{1}{2} h^\mu{}_{;\mu}(q) + \tfrac{1}{8} R(q)]\,d\tau\} \qquad (25)$$

which equals

$$\exp\{-\int_0^T [L_1(q,\dot{q}) + \text{total derivative}]\,d\tau\}\, J[q(\tau)] \qquad (26)$$

This completes the calculation.

The probability density functional we have obtained here through the change of variables from Gaussian white noise to the stochastic process q was also obtained by Dekker in a completely different and algebraically much more tedious way, starting from the Fokker-Planck equation of the process and using Fourier analysis of the paths.[23] In Ref. 1 we have also commented on the relation between supersymmetry breaking and the appearance of a nonvanishing global circulation in the stationary state of the stochastic process. In future work also the other fermion sectors of the model will be considered.

One of us (D.R.) acknowledges a fellowship by the Alexander von Humboldt foundation.

References

1. R. Graham, D. Roekaerts, Phys. Lett. 109A (1985) 436
2. R. Casalbuoni, Nuov. Cim. 33A (1976) 115
3. L.D. Faddeev, in Methods in Field Theory, R. Balian, J. Zinn-Justin (Eds.) (North Holland, Amsterdam, 1976)
4. H. Nicolai, Phys. Lett. 89B (1980) 341
5. H. Nicolai, Nucl. Phys. B176 (1980) 419
6. G. Parisi, N. Sourlas, Nucl. Phys. B206 (1982) 321
7. S. Cecotti, L. Girardello, Ann. Phys. 145 (1983) 81
8. E. Witten, Nucl. Phys. B202 (1982) 253
9. S. Weinberg, Gravitation and Cosmology (John Wiley and Sons, New York, 1972)
10. A.C. Davis, A.J. Mac Farlane, P.C. Popat, J.W. Van Holten, J.Phys. A17 (1984) 2945
11. M. Claudson, M.B. Halpern, Nucl. Phys. B250 (1985) 689
12. E. Witten, J.Diff. Geom. 17 (1982) 661
13. F. Langouche, D. Roekaerts, E. Tirapegui, Functional Integration and Semiclassical Expansions (Reidel, Dordrecht, 1982)
14. J.W. Van Holten, Erice EPS Study Conf. (1981) 505
15. H. Ezawa, J.R. Klauder, preprint ZiF Bielefeld N°74 (1984)
16. M. Claudson, M.B. Halpern, Phys. Rev. D31 (1985) 3310
17. M. Claudson, M.B. Halpern, preprint UCB-PTH-84/28 (Berkeley, 1984)
18. T. Suzuki, J..Math. Phys. 21 (1980) 918
19. R. Graham, Z. Physik B26 (1977) 397
20. N. Ikeda, S. Watanabe, Stochastic Differential Eauations and Diffusion Processes (North Holland, Amsterdam, 1981)
21. R. Graham, Springer Tracts in Modern Physics 66 (1973) 1
22. Z. Haba, preprint BIBOS N°18 (Bielefeld 1984)
23. H. Dekker, Physica 103A (1980) 586 and Phys. Rev. A24 (1981) 3182

STOCHASTIC EQUATIONS FOR SOME EUCLIDEAN FIELDS

Z. Haba
Research Center Bielefeld-Bochum-Stochastics*
Bielefeld University, D-4800 Bielefeld 1, FRG
and
Institute of Theoretical Physics
University of Wroclaw, Poland

The Euclidean functional integral has become a powerful tool in the rigorous construction of models of quantum field theory, especially $P(\varphi)$ interactions. The usefulness of the functional integral in the quantum mechanics is well-known. However, the conventional functional approach to the quantum mechanics on a manifold, although possible (e.g. through the lattice approximation) appears less promising. This is so, because there is no natural decomposition of the functional measure on the manifold into Gaussian (free) part and a perturbation. Fields with values in a manifold appear in models aspiring to a unification and geometrization of interactions. The Yang-Mills theory, which is of geometric origin, is the most important example of the relevance of geometry to quantum physics. It appears that the infrared difficulties in gauge theories result from the use for quantization of improper geometrical objects.

The stochastic equations seem to be the proper tool for a study of the (imaginary time) quantum mechanics on a manifold [1]. We derive stochastic equations for Euclidean Markov fields with values in a manifold. We believe that a deep understanding of the geometry of these fields (including the gauge fields) can lead to a solution of the stochastic equations and to results of physical significance.

I. Quantum Mechanics on a Manifold

We illustrate in this section some aspects of the stochastic description, which are relevant to quantum field theories. Consider the stochastic process generated by the Laplace-Beltrami operator Δ_M on the Riemannian manifold M. In some coordinates ξ

*Supported by Stiftung Volkswagenwerk

$$d\xi_\mu = \frac{1}{2} g^{\alpha\beta}(\Gamma_{\mu\alpha\beta} - e_{a\beta}\,\partial_\alpha\, e_{a\mu})dt + e_{a\mu}\, db^a \tag{I.1}$$

where $e_{a\mu}$ is the vierbein $(e_{a\alpha}\, e_{a\beta} = g_{\alpha\beta})$ and b^a is the Wiener process

$$E[b^a_t\, b^{a'}_{t'}] = \delta^{aa'} \min(t,t') \tag{I.2}$$

$f db$ denotes in this paper the Stratonovitch differential [1] defined by

$$\int f\, db = \lim \sum_i f(\tfrac{1}{2}(t_{i+1} + t_i))(b(t_{i+1}) - b(t_i)) \ .$$

The stochastic equations simplify on a class of complex manifolds called Kähler manifolds. The tangent space of a complex manifold is an orthogonal sum of holomorphic $(\frac{\partial}{\partial z})$ and antiholomorphic $(\frac{\partial}{\partial \bar{z}})$ vectors

$$TM = (TM)^+ + (TM)^- \ . \tag{I.3}$$

On a Kähler manifold this decomposition of the tangent space is preserved during the parallel transport. The non-covariant term in eq. (I.1) vanishes in the complex coordinates w^α. Then

$$dw_\alpha = e_{a\alpha}(w)db^a \tag{I.4}$$

where b^a is the complex Brownian motion.

Consider now a compact Lie group G. Eq. (I.1) on the group can be expressed in the form

$$g^{-1}dg = db \tag{I.5}$$

where b is a matrix.

The solution of eq. (I.5) has the form of the time-ordered exponential

$$g_t = T(\exp \int^t db) \ . \tag{I.6}$$

As an application of the formula (I.6) consider $SU(N)$ for large N. Then

$$E[\mathrm{Tr}(g_t^+ g_{t'})] \simeq \exp \frac{1}{2} \mathrm{Tr}\, E[(\int_t^{t'} db)^2] = \exp - c\, |t-t'| . \qquad (I.7)$$

So, we get the mass gap.

Let now $M = G/H$ be a symmetric homogeneous space. Consider the bundle $\Pi: G \to M$ with the group $H \subset G$ as the fiber. We can write $g = vh$, where $h \in H$ and $v \in G/H$ (we embed G/H in G). The Lie algebra $L(G)$ of G is a direct sum

$$L(G) = L(H) + L(G/H)$$

with

$$[L(H), L(G/H)] \subset L(G/H).$$

Let P_v be the projection of $L(G)$ onto $L(G/H)$. Then, eq. (I.5) can be decomposed into two equations

$$\begin{aligned} h^{-1}dh &= (1 - P_v)db \\ v^{-1}dv &= h\, P_v db\, h^{-1} . \end{aligned} \qquad (I.8)$$

Eq. (I.8) may be considered as a Brownian motion on the bundle $\Pi: G \to G/H$ with the connection P_v.

As an important example consider $G = SU(n+1)$ and $G = Sp(n+1)$ (the symplectic group, which can be described as a unitary group with quaternionic matrix elements) with $H = SU(n)$ and $H = Sp(n)$, respectively. In this case v can be parametrized as follows [2]

$$v = \gamma^{-1} \begin{pmatrix} 1 & -w^+ \\ w & \alpha(w) \end{pmatrix} \qquad (I.9)$$

where $\gamma = (1 + w^+w)^{1/2}$ and $\alpha = \gamma(1 + w^+w)^{-1/2}$, w is a column and $w^+ = (\bar{w}_1, \ldots, \bar{w}_n)$. w's are the complex coordinates of the $CP(n)$ ($G = SU(n+1)$) and the quaternionic coordinates of the $HP(n)$ manifolds.

From eq. (I.8) we get the equation for w (h can be absorbed into the definition of w)

$$(1 + w^+w)^{-1}\, \alpha(w)\, dw = db \qquad (I.10)$$

where b is the complex (quaternionic) Brownian motion.

In the search for solutions also other descriptions can be useful, e.g. the Brownian motion on the sphere $S^2 = SU(2)/U(1)$ can be described by an equation defined on a linear space (R^3), whose solutions stay on a submanifold (S^2) [3]

$$d\vec{n} = P(n)\, d\vec{b} \tag{I.11}$$

where $P(n)$ is a matrix with matrix elements

$$P(n)_{ij} = \delta_{ij} - n_i n_j / n^2 \ , \quad n^2 = \Sigma n_i n_i \ . \tag{I.12}$$

$P(n)$ projects the vector $d\vec{b}$ onto the tangent space of S^2, so that $d\vec{n}$ is tangent to S^2, hence $\vec{n} d\vec{n} = 0$ implies that $\vec{n}^2 = \text{const.}$ The generalization of eq. (I.11) to the CP(n) model described by a Hermitian matrix φ with $\text{Tr}\,\varphi^2 = 1$ [4] reads

$$d\varphi = (\text{Tr}\varphi^2)^{-1}[\varphi,[\varphi,db]] \ . \tag{I.13}$$

All the models of Brownian motion on a symmetric space can immediately be solved using the solution of the Brownian motion on a group (I.6). It is sufficient to extract the coset G/H from the group. The solution of the CP(n) model can be expressed by elementary functions, e.g. the Brownian motion on the sphere (I.11) is solved by $\vec{n} = \vec{b}\,|\vec{b}|^{-1}$. Note that it would be difficult to solve the nonlinear equations (I.1) directly in coordinates.

II. Two-Dimensional Fields with Values in a Complex Manifold

There exists in the two-dimensional Euclidean space an analogue to the complex Brownian motion - the complex massless scalar free field. This is the Gaussian random field with the covariance

$$E[\varphi_o^a(z)\ \overline{\varphi_o^b(z')}] = -\frac{1}{4\pi}\ln|z-z'|^2\,\delta^{ab} \tag{II.1}$$

(other two-point correlation functions vanishing). The complex scalar field can be considered as a random map $\varphi_o : \mathbb{C} \to \mathbb{C}^n$. $T\mathbb{C}^n$ splits into the holomorphic $(T\mathbb{C}^n)^+$ and antiholomorphic $(T\mathbb{C})^-$ parts. The pull-back to $T^*\mathbb{C}$ of a basis $\{e^a\}$ of $(T^*\mathbb{C}^n)^+$ has the form

$$d\varphi_o^a(z) = \partial\varphi_o^a\,dz + \bar{\partial}\varphi_o^a(z)\,d\bar{z} \tag{II.2}$$

where $\bar{\partial} = \frac{\partial}{\partial \bar{z}}$. It is easy to check that $\bar{\partial}\varphi_o$ (as well as $\partial\varphi_o$) is a complex white noise, i.e.

$$E[\bar{\partial}\varphi_o^a(z) \quad \overline{\bar{\partial}\varphi_o^b(z')}] = \delta(z-z')\delta^{ab} . \tag{II.3}$$

So, $\bar{\partial}\varphi_o$ is a generalization of $\frac{d}{dt} b_t$.

Consider now a map $\varphi: \mathbb{C} \to M$, where M is a symmetric space G/H with a complex structure, i.e. the cotangent space $(T^*M)_g$ is the direct sum of holomorphic $(T^*M)_g^+$ and antiholomorphic $(T^*M)_g^-$ parts. Let L_g be the left translation on G from the unit element 1 to g and $\{e^a\}$ a basis of $(T^*M)_1^+ \sim (T^*\mathbb{C})^+$. Then, we can compare the $(T^*\mathbb{C})^-$ part of the pull-back of $\{e^a\}$ by the composition of the maps $L_g^{-1}\varphi$ to the $(T^*\mathbb{C})^-$ part of the pull-back of $\{e^a\}$ by φ_o (II.2). This leads to the equation

$$(1 + w^+w)^{-1} \quad \alpha(w) \; \bar{\partial}w = \bar{\partial}\varphi_o \tag{II.4}$$

which is a generalization of eq. (I.10) to two dimensions.

Eq. (II.4) is invariant under the Euclidean group

$$z \to e^{i\alpha}z + c . \tag{II.5}$$

Solutions of stochastic equations, which are of first order in time t , have the Markov property in the t-direction [1]. Now, the Markov property and the Euclidean invariance are sufficient for a construction of relativistic fields from Euclidean fields according to Nelson's reconstruction theorem [5]. A direct interpretation of the Euclidean field in the physical space-time may also be possible [6].

In our paper [7] we have obtained eq. (II.4) as a stochastic equation on the manifold $F(R,M)$ of maps $R \to M$. F is a Hilbert manifold modelled on $L^2(R)$ with the scalar product in $(TF)_\sigma$ (the tangent space at $\sigma \in F$) defined by

$$(v,v') = \int dx \; (v(x), v'(x))_{\sigma(x)} \tag{II.6}$$

where $(\; , \;)_\sigma$ is the Riemannian structure in $(TM)_\sigma$. The two-dimensional white noise $\bar{\partial}\varphi_o$ may be considered as a time derivative $\dot{b}$ of an $L^2(R)$-valued Wiener process b_t defined by

$$E[b_t(f)\ b_{t'}(f')] = \min(t,t')(f,f') \tag{II.7}$$

where $(f,f') = \int dx\ \bar{f}(x)\ f'(x)$.

We modify eq. (I.10) (which could also be considered as an equation for a Brownian motion on F) by an addition of a drift term β. The drift should fulfil the following requirements: i) the solutions of the modified equation should also stay on M, ii) the exponential decay of correlations (mass gap) should be preserved. The preservation of some correlation functions means that the generator of the stochastic process can be modified only by an addition of the generator of an isometry [8]. If we treat the process (I.10) as defined on F, then its generator is the Laplace-Beltrami operator defined by the metric (II.6). Hence, the drift has to be a generator of an isometry of the metric (II.6), i.e. its Killing vector. The metric (II.6) has many Killing vectors. The choice of $K = (i\partial_x w, -i\partial_x \bar{w})$ as the drift β in eq. (II.4) comes from the requirement of the Euclidean invariance. The vector K is the generator of the isometry $R_\alpha T_a R_\alpha^{-1}$, where R_α is the rotation $w \to e^{i\alpha} w$ and T_a is the translation $w(x) \to w(x+a)$.

These two, a priori different, interpretations of the random field, either as a Euclidean covariant random map $R^2 \to M$ or as a stochastic process on the manifold $F(R,M)$ of maps, admit different regularization schemes. In the first interpretation we may use the covariant regularization, whereas the second interpretation requires the preservation of the Markov property, i.e. only a regularization in space coordinates is admissible, e.g. $b_t(x) \to (-\varepsilon\partial_x^2 + 1)^{-1} b_t(x)$ or the lattice regularization in space. If the stochastic equation (II.4) is regularized only in the spatial coordinate, then w_t can be treated as a Markov process with values in a Hilbert space. In such a case the functional measure corresponding to the solution w_t of eq. (II.4) is determined by the Girsanov formula (see [1], [9]-[11])

$$d\mu = d\mu_0 \quad \exp[-\frac{1}{2}\int g_{\alpha\bar{\beta}}(w)\ \partial_x w \quad \overline{\partial_x w^\beta} \quad + Q] \tag{II.8}$$

where $d\mu_0$ is the functional measure corresponding to the stochastic process w_t without the drift $\partial_x w$ and Q is the topological charge.

The measure $d\mu_0$ can be obtained from the short-time propagator for the stochastic process (I.10). We can conclude in this way that the stochastic equation (II.4) describes a field theory with the Lagrangian

$$L_B = \frac{1}{2} g_{\alpha\bar{\beta}}(w)\ \partial_\mu w^\alpha\ \overline{\partial_\mu w^\beta} - Q\ . \qquad \text{(II.9)}$$

On the other hand, eq. (II.4) could be treated as a prescription for a non-linear transformation of the Gaussian measure corresponding to the free field φ_o. From the Jacobian of this transformation we get a fermionic contribution to the total Lagrangian L

$$L = L_B + L_F = L_B + \bar{\psi}\ \not{D}\ \psi \qquad \text{(II.10)}$$

where $\not{D} = \gamma^\mu D_\mu$ and D_μ is the covariant derivative along w. If a spatial lattice regularization is applied, then the formulas (II.8) - (II.9) and (II.10) do not contradict each other (see [11]). In fact, the Girsanov formula (II.8) (with the lattice regularization of the exponential factor) is a rigorous version of the functional transformation [12]. So, the fermionic determinant is absent (det $\not{D} = 1$). The only way to determine, whether eq. (II.4) describes the σ-model with fermions or without, is to study the removal of the ultraviolet regularization. We have shown [11] in a model with a holomorphic potential [13] that the spatial regularization cannot be removed, whereas the covariant one can. It is more difficult to resolve this problem in the model (II.4). It appears that the spatial regularization as well as the covariant one can be removed in the perturbation theory (at least in the S^2-model) with the proper renormalization of the coupling constant. It remains unclear whether det $\not{D} \neq 1$ on the support of the functional measure after a rigorous removal of the ultraviolet regularization.

We can obtain a straightforward generalization of eq. (II.4) by an addition of a random connection χ. Then,

$$(1 + w^+ w)^{-1}\ \alpha(w)\ (\bar{\partial} + \bar{\chi})w = \bar{\partial}\varphi_o \qquad \text{(II.11)}$$

where χ is a complex white noise independent of $\bar{\partial}\varphi_o$. In such a case we get the four-fermion interaction $\bar{\psi}\psi\bar{\psi}\psi$ in the Lagrangian. We are studying this model now (such models can be ultraviolet finite).

Eq. (II.4) can be embedded back in the group G. As $d\omega_o = \bar{\partial}\varphi_o\, d\bar{z}$ is an (L(G)-valued) 1-form it can be integrated along a curve γ. We are looking for a certain generalization of eq. (I.6) that could solve eq. (II.4) (see refs. [14], [15] for another generalization of the formula (I.6)). The integral $T(\exp \int_\gamma d\omega_o)$ depends on the curve γ. Hence, it cannot be a solution of eq. (II.4). $d\omega_o$ must be rotated during the integration along γ in such a way that the integral does not depend on the curve.

Other forms of eq. (II.11) can be useful in a search for solutions. So, the generalization of eq. (I.11) to two dimensions has the form

$$d\vec{n} = p(n)\ \partial_x \vec{n}\ dt + p(n) d\vec{\chi} + P(n)\ d\vec{b} \qquad \text{(II.12)}$$

where the matrix $(p(n))_{ij} = \varepsilon_{ijk}\ n_{k/n}$ is the square root of P ($p^2 = P$). The analogue of eq. (I.13) for $CP(n)$ reads

$$d\varphi = i[\varphi, \partial_x \varphi] dt + i[\varphi, d\chi] + (\mathrm{Tr}\varphi^2)^{-1}[\varphi,[\varphi, db]]. \qquad \text{(II.13)}$$

The additional Brownian motion χ in eqs. (II.12) - (II.13) leads to the four-fermion terms in the Lagrangian.

III. Gauge Theories in Four Dimensions

It appears that a generalization of the stochastic equations of Section II to four dimensions should be related to the quaternionic structure of $R^4 = \mathbb{C} \times \mathbb{C}$. A point $x \in R^4$ can be expressed in the quaternion basis $\{e_\mu\}$

$$x = \sum_{\mu=0}^{3} x_\mu\ e_\mu$$

where the quaternion algebra

$$e_i e_j = -\delta_{ij} + \varepsilon_{ijk} e_k\ ,\ e_o e_i = e_i e_o = e_i,\ e_i^2 = -e_o$$

can be realized by means of the Pauli matrices.
The action of the Euclidean group $x \to U_1\ x U_2^{-1} + a$, where $U_i \in SU(2)$ and $a \in GL(2,C)$, is a generalization of eq. (II.5). We can also define a first order differential operator (the Hamilton operator [2])

$$\mathcal{D} = \Sigma\, e_\mu \frac{\partial}{\partial x_\mu}$$

which is a generalization of $\frac{\partial}{\partial z}$. Let A_μ be a vector potential and $A = \Sigma A_\mu\ e_\mu$. Then the equation

$$\mathcal{D} A = \dot{b} \qquad \text{(III.1)}$$

where $\dot{b} = \Sigma \dot{b}_\mu e_\mu$ and $\dot{b}_\mu$ is the four-dimensional white noise, defines the electromagnetic field in the Feynman gauge. In fact, it is easy to check that

$$E[f(A(b))] = \int d\mu_o(A) f(A) \qquad \text{(III.2)}$$

where μ_o is the Gaussian measure with the covariance

$$\int d\mu_o(A)\, A_\mu(x)\, A_\nu(x') = \delta_{\mu\nu}(-\Delta)^{-1}(x,x'). \qquad \text{(III.3)}$$

Eq. (III.1) is the analogue of $\bar{\partial}\varphi_o = \dot{b}$ (Sec. II) and $\frac{d}{dt} b = \dot{b}$ (Sec. I). However, we are unable to generalize directly eqs. (III.1) - (III.3) (with the preservation of the Euclidean invariance) to the non-abelian case, because the problem of gauge degrees of freedom is more involved there. A possible way of generalization is to consider the path-dependent phase factors [16] (with $\mathcal{D} \to \Sigma\, e_\mu \frac{\delta}{\delta\xi_\mu}$) or the quaternionic σ-models [2], [17].

We pursue here a more conventional formulation in terms of the potentials, which is a generalization of our description of the random field as a Brownian motion on a manifold of maps. The formulation is non-covariant with respect to the Euclidean group, but this dose not preclude the possibility that we get covariant equations through a change of variables (cp. eqs. (II.11) and (II.12)).

First we need to introduce some notions from differential geometry. Let $\Pi: P \to M$ be a principal fiber bundle with a group G as a fiber. Let $\mathcal{B}$ be the space of (irreducible) connections ω on P and $\mathcal{G}$ an infinite dimensional Lie group of gauge transformations

$$\omega \to \omega^g = g^{-1}\omega g + g^{-1} dg\ . \qquad \text{(III.4)}$$

Consider the coset $\mathcal{M} = \mathcal{B}/\mathcal{G}$. Then, $\Pi: \mathcal{B} \to \mathcal{M}$ is a principal fiber bundle [18]. Let Λ^o be the space of $L(G)$-valued functions on M. Then, the vertical subspace V_ω of $T\mathcal{B}$ consists of functions of the form $\nabla_\omega \lambda$, where $\lambda \in \Lambda^o$ and ∇_ω denotes the covariant derivative. The horizontal subspace of $T\mathcal{B}$ can be defined as the orthogonal complement of V_ω in $\mathcal{B}$. This horizontal subspace is determined by the connection form

$$\Omega = (\nabla^*_\omega \nabla_\omega)^{-1} \nabla^*_\omega\ . \qquad \text{(III.5)}$$

The connection allows to identify H_ω with $(TM)_{\Pi(\omega)}$ and embed M as a submanifold in B. In particular, if $\bar{X},\bar{Y}$ are vectors in TB, then their horizontal parts correspond to vectors X,Y in $(TM)_{\Pi(\omega)}$ with the scalar product

$$g(X,Y) = (\bar{X},P_\omega \bar{Y}) \qquad \text{(III.6)}$$

where $(\, , \,)$ is the L^2-scalar product in B and

$$P_\omega = 1 - \nabla_\omega (\nabla^*_\omega \nabla_\omega)^{-1} \nabla^*_\omega . \qquad \text{(III.7)}$$

A stochastic process A_t, which is to describe the quantum Yang-Mills theory should take its values in M rather than in B, because $\mathrm{Tr}F^2$ does not depend on the gauge degrees of freedom. This is similar to the case of fields with values in a sphere (Secs. I-II), where the Lagrangian does not depend on the radial component of $\vec{n} \in R^3$. We would like to write down a stochastic equation in a form independent of coordinates. Such an approach is inspired by eqs. (I.11) and (II.12) (see also [3]). So, we would like to find a stochastic equation for $A \in B$, whose solutions stay on a submanifold $\bar{M}$ being an embedding of M in B.

If the curve A_t is to be the lift to B of $\Pi(A_t) \in M$ with respect to the connection Ω (III.5), then its tangent $\frac{d}{dt} A_t$ must be an element of H_A. This leads to the equation (cp. with eq. (I.11))

$$dA_t = P_A \, db_t \qquad \text{(III.8)}$$

which can also be expressed as an equation on the fiber bundle $\Pi: B \to B/G$ (cp. with eq. (I.8))

$$\begin{aligned} d\omega^g &= g^{-1} \, db \, g \\ g^{-1} dg &= \Omega db \end{aligned} \qquad \text{(III.9)}$$

where ω^g is defined in eq. (III.4) and $\Omega db = (\nabla^* \nabla)^{-1} \nabla^*_j db_j$.

Following the discussion of Sec. II we may still modify the Brownian motion on M by an addition of a Killing vector corresponding to an isometry of the metric (III.6). It is easy to see that the scalar product (III.6) is invariant under a rotation of the potential A_k. Moreover, it is invariant under the translation of the fiber (it is not invariant under the ordinary translations)

$$A(x) \rightarrow \exp(iA_k(x)\Delta x_k)\ A(x+\Delta x_k)\ \exp(-iA_k(x)\Delta x_k).$$

The sum of commutators $\Sigma[R_k, P_k]$ of the generators of the above mentioned transformations (R_k is the generator of the rotation around the k-th axis) is equal to

$$\varepsilon_{ijk}\, F_{jk} \frac{\delta}{\delta A_i} .$$

The addition of this Killing vector as a drift to eq. (III.8) leads to the stochastic equation

$$dA_t = {}^*F\, dt + P_A\, db_t$$

where (III.10)

$${}^*F_i = \frac{1}{2}\, \varepsilon_{ijk}\, F_{jk} .$$

Consider now the functional measure $d\mu(A)$ corresponding to the solution of eq. (III.10). Let $d\mu_o(A)$ be the probability measure corresponding to the solution of eq. (III.8). Then, from the Girsanov formula, we get

$$d\mu(A) = d\mu_o(A)\ \exp[-\frac{1}{2}\int P_A^{-1}\, {}^*F\ \ P_A^{-1}\, {}^*F + \int P_A^{-1}\, {}^*F\ db]$$

$$= d\mu_o(A)\ \exp[-\frac{1}{4}\int F_{jk}\, F_{jk} + Q] \qquad \text{(III.11)}$$

where Q is the topological charge.

In the derivation of eq. (III.11) the horizontality of ${}^*F$ $(P_A{}^*F = {}^*F)$ and the equality of Ito and Stratonovitch integrals in Q were used. The measure $d\mu_o$ can be obtained from the short-time propagator for the process (III.8). This problem can be treated rigorously on the lattice, where the propagator on M can be derived from the propagator on B (the Cartesian product of groups attached to bonds of the lattice). We get as a result that the functional measure (III.11) coincides (up to the topological charge) with the standard functional measure for the pure Yang-Mills theory in any spatial gauge (see [19] and a paper in preparation).

In the temporal gauge eq. (III.10) has been derived earlier by Nicolai [20] (without P_A) and interpreted as an equation for the Yang-Mills theory with fermions. Stochastic equations for gauge fields have been studied by Asorey and Mitter [21] (see also [22]). These authors write the stochastic equation in coordinates and treat the spatial part of $F_{\mu\nu} F_{\mu\nu}$ as a potential. Then, the stochastic equation has the form of the ground state equation [23].

The form of the equation (III.10) suggests that the solution for the Wilson loop $\exp i \int A_\mu d\xi^\mu$, expressed by $F_{\mu\nu}$ through the Stokes theorem, could be obtained in a form of a surface integral over the white noise (see [14], [15] for another approach to such integrals). We suggest here an elementary approach to the solution of eq. (III.10) originating from Yang's formulation of the self-duality equation [24]. In Yang's complex coordinates eq. (III.10) reads

$$F_{yz} = \eta_2 - i\eta_1$$
$$\frac{1}{2}(F_{y\bar{y}} + F_{z\bar{z}}) = -i\eta_3 \qquad \text{(III.12)}$$

where $\eta = P_A \dot{b}$. When we introduce a new variable $g \in GL(n,C)$ such that after a complex gauge transformation implemented by g $A'_y = 0$, then the non-local part of η drops out from the equation for g. So, we get a simple linear perturbation by noise of Yang's equations.

By means of the dimensional reduction [25] we get from eq. (III.10) the equation for the two-dimensional abelian Higgs model (with the φ^4 interaction) discussed in our earlier papers [7], [11] and an equation for the three-dimensional non-abelian Higgs model (with fermions resulting from the Jacobian)

$$dA_i = \varepsilon_{ij} \nabla_j \varphi \, dt - \nabla_i G \varphi \, db' + (\delta_{ij} - \nabla_i G \nabla_j^*) db_j$$
$$d\varphi = \varepsilon_{ij} F_{ij} \, dt + (1 - \varphi G \varphi) db' - \varphi G \nabla_j^* db_j \qquad \text{(III.13)}$$

where $G = (\nabla^* \nabla + \varphi\varphi)^{-1}$.

The mechanism of dimensional reduction leading to eq. (III.13) suggests that it might be possible to get in this way an interaction of gauge fields with scalar and Fermi fields in four dimensions. A stochastic equation for R^2-gravity (which has instantons) can be derived following the argument leading to eq. (III.10). It may be that some simple stochastic partial differential equations describe unified models of particle interactions.

References

[1] N. Ikeda and S. Watanabe, Stochastic Differential Equations and Diffusion Processes, North Holland, 1981

K.D. Elworthy, Stochastic Differential Equations on Manifolds, Cambridge Univ. 1981

[2] F. Gürsey and H.C. Tze, Ann. Phys. 128, 29 (1980)

[3] M. van den Berg and J.T. Lewis, Bull. Lond. Math. Soc. 17. 144 (1985)

[4] A.M. Perelomov, Physica 4D, 1 (1981)

[5] E. Nelson, Journ Funct. Anal. 12, 97 (1973)

[6] E. Nelson, these Proceedings

[7] Z. Haba, Journ. Phys. A18, L347 (1985)

[8] E. Seiler, Acta Phys. Austr., Supp. XXVI, p. 259, 1984

[9] Z. Haba, Journ. Phys. A18, 1641 (1985)

[10] G. Jona-Lasinio, these Proceedings

[11] Z. Haba, BiBoS preprint Nr. 18 , 1985

[12] L. Gross, Tran. Amer. Math. Soc. 94, 404 (1960)

[13] G. Parisi and N. Sourlas, Nucl. Phys. B206, 321 (1982)

S. Cecotti and L. Girardello, Ann. Phys. 145, 81 (1983)

[14] S. Albeverio and R. Høegh-Krohn, in Stochastic Analysis and Applications, M. Pinsky, Ed., p. 1, 1984

S. Albeverio, R. Høegh-Krohn and H. Holden, Acta Phys. Austr. Supp. XXVI, p. 211 (1984)

[15] H. Holden, these Proceedings

[16] S. Mandelstam, Ann. Phys. 19, 1 (1962)

I. Bialynicki-Birula, Bull. l'Acad. Pol. Sci. 11, 135 (1963)

[17] J. Lukierski, in Field Theoretical Methods in Particle Physics, W. Rühl, Ed., 1980

[18] O. Babelon and C.M. Viallet, Phys. Lett. 85B, 246 (1979)

I.M. Singer, Physica Scripta 24, 817 (1981)

[19] Z. Haba, BiBoS preprint No.58 , 1985

[20] H. Nicolai, Phys. Lett. 117B, 408 (1982)

[21] M. Asorey and P.K. Mitter, Comm. Math. Phys. 80, 43 (1981)

[22] B. Gaveau and P. Trauber, Journ. Funct. Anal. 38, 324 (1980)

[23] S. Albeverio and R. Høegh-Krohn, Z. Wahr. verw. Gebiete 40, 1 (1977)

[24] C.N. Yang, Phys. Rev. Lett. 38, 1377 (1977)

[25] C.H. Taubes, Comm. Math. Phys. 75, 207 (1980)

PERCOLATION OF THE TWO-DIMENSIONAL ISING MODEL

Yasunari Higuchi
Department of Mathematics
Kobe University
Rokko, Kobe 657 Japan

§1. Introduction; the problem

We consider 2-dimensional Ising model with ferromagnetic interaction; i.e. $\mathbb{Z}^2$ is the square lattice, and $\Omega \equiv \{-1,+1\}^{\mathbb{Z}^2}$ is our configuration space, and the formal Hamiltonian has the form

(1) $$H(\sigma) \sim -\sum_{<x,y>} \sigma(x)\sigma(y) - h\sum_{x} \sigma(x) ,$$

where $<x,y>$ denotes that x and y are nearest neighbour pairs in $\mathbb{Z}^2$, and h is a real parameter, called the external field.
Also we consider a Gibbs state μ for (1), which is a probability measure on Ω, formally having the following form:

(2) $$\mu(\sigma) \sim (\text{normalization}) \exp\{-\beta H(\sigma)\}$$

where $\beta>0$ is a parameter called the inverse temperature. In general, μ is not uniquely determined by (2) for given $(\beta.h)$, but there are at most two distinct Gibbs states μ_+ and μ_-, such that any Gibbs state is a convex combination of these two Gibbs states. Of course μ_+ and μ_- depends on $(\beta.h)$, so in order to stress $(\beta.h)$-dependence, we denote them as $\mu_{+,\beta.h}$ and $\mu_{-,\beta.h}$.
As is well known, there exists $\beta_c>0$ such that

(3) $$\begin{aligned} &\mu_{+,\beta.0} \neq \mu_{-,\beta.0} && \text{if} \quad \beta > \beta_c \\ &\mu_{+,\beta.h} = \mu_{-,\beta.h} && \text{if} \quad \beta \le \beta_c \text{ or } \beta>\beta_c \text{ and } h \neq 0. \end{aligned}$$

Percolation

For every $\omega\in\Omega$, let $\omega^{-1}(+1)$ be the set of all points $x\in\mathbb{Z}^2$, such that $\omega(x)=+1$, and let E_∞^+ be the event that there exists an infinite cluster in $\omega^{-1}(+1)$. The problem is the following:

Problem: For what value of $(\beta.h)$ do we have $\mu_{+,\beta.h}(E_\infty^+)=1$? And how about $\mu_{-,\beta.h}(E_\infty^+)$?

Remark: From tail triviality of μ_+ and μ_-, these probabilities take only values 0 or 1. Also, if $\mu_+=\mu_-$, then we simply write this measure by $\mu_{\beta.h}$, and we only have to look at the value $\mu_{\beta.h}(E_\infty^+)$.

§2. What is known up to now

Dual Graph Argument

It is very convenient to introduce the dual graph $\mathbb{L}$ of $\mathbb{Z}^2$, $\mathbb{L}$ consists of the same set of vertices as $\mathbb{Z}^2$, but the connection in $\mathbb{L}$ is more than in $\mathbb{Z}^2$; i.e. two points $x,y\in\mathbb{L}$ are nearest neighbours in $\mathbb{L}$ if

$$\max(|x^1-y^1|,|x^2-y^2|) = 1.$$

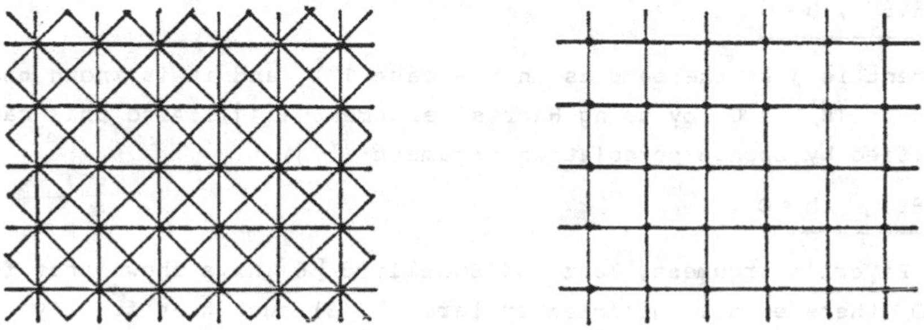

$\mathbb{L}$ $\mathbb{Z}^2$

(Fig. 1)

First, note the following simple observation:

Observation: The following two statements are equivalent.

(i) There is an infinite cluster of $\omega^{-1}(+1)$ in $\mathbb{Z}^2$.

(ii) There exists a finite $\Lambda\subset\mathbb{Z}^2$ such that there is no $\mathbb{L}$-circuit of $\omega^{-1}(-1)$ which surrounds Λ.

Now let us begin with the easier case;

1°) $\beta>\beta_c$, $h=0$.

By the above observation, if $h=0$ we have $\mu_+(E_\infty^+)=1$, for otherwise μ_+ - a.s. we find a $\mathbb{L}$-circuit in $\omega^{-1}(-1)$ surrounding a given finite $\Lambda\subset\mathbb{Z}^2$, which, by Markov property of μ_+, implies that $\mu_+=\mu_-$ which is a contra-

diction to our assumption.
As for $\mu_-(E^+_\infty)$, it was proved in [1] that $\mu_-(E^+_\infty) = 0$ by using Harris' argument ([2]) which can be avoided when we use so-called "Sponge Percolation". ([5])

2°) <u>$\beta > \beta_c$, $h \neq 0$</u>

Once we have a result for $h = 0$, it can be carried into $h \neq 0$ case by using FKG inequality.
Namely it is a direct consequence of 1°) and FKG inequality that

(i) $\mu_{\beta.h}(E^+_\infty) = 1$ for $h > 0$,

(ii) $\mu_{\beta.h}(E^+_\infty) = 0$ for $h < 0$.

So, in the case when $\beta > \beta_c$, we have a complete description of the percolation region.

3°) <u>$\beta \leq \beta_c$, $h = 0$</u>

The argument is just the same as in the case 1°), and it is known that one has $\mu_{\beta.0}(E^+_\infty) = 0$ by using Harris' argument ([1]). Also this can be simplified by sponge percolation argument ([5]).

4°) <u>$\beta \leq \beta_c$, $h \neq 0$</u>

By using Peyerl's argument, Kunz and Souillard [6] have shown that for every $\beta > 0$, there exists sufficiently large $h_o(\beta)$ such that for any $h > h_o(\beta)$ $\mu_{\beta.h}(E^+_\infty) = 1$, which, in conjunction with 3°), implies the existence of critical $h_c(\beta) \geq 0$ such that

$$\mu_{\beta.h}(E^+_\infty) = 1 \quad \text{if } h > h_c(\beta),$$

$$\mu_{\beta.h}(E^+_\infty) = 0 \quad \text{if } h < h_c(\beta) .$$

Therefore we have a kind of phase transition in the parameter region where the free energy of (1) is analytic. This is just like as in the Bernouilli case.
In the Bernoulli percolation, one has a system of no interaction; i.e. the formal Hamiltonian of the system is given by

$$\text{(4)} \qquad H'(\beta) \sim -h\sum_x \sigma(x) ,$$

where h is as before a real parameter. There exists unique Gibbs state $\mu'_{\beta.h}$ for given parameter $(\beta.h)$, having the following form;

(5) $\mu'_{\beta.h}(\sigma(x) = +1 \quad x \in \Lambda) = \left(\frac{1}{1+e^{-2\beta h}}\right)^{|\Lambda|}$.

For simplicity, let us write the right hand side of (5) by $p^{|\Lambda|}$, where $p=p(\beta.h)$ is just defined by

$$p = (1+e^{-2\beta h})^{-1} .$$

Then it is equivalent to say that we have i.i.d. $\{\sigma(x)\}x\in\mathbb{Z}^d$, whose distribution is given by

$$P(\sigma(x)=+1) = p .$$

For Bernoulli percolation, we know that there exists $p_c > \frac{1}{2}$ such that

$$\mu'_{\beta.h}(E^+_\infty) = 1 \qquad \text{if} \quad p(\beta.h) > p_c$$

$$\mu'_{\beta.h}(E^+_\infty) = 0 \qquad \text{if} \quad p(\beta.h) \leq p_c$$

([4],[7]). The best rigorous lowerbound for p_c is recently obtained by Tóth [9];

$$p_c > 0.503478 .$$

Now, there is a very simple and useful criterion to compare our two systems: Ising and Bernoulli.

Lemma. (Russo [8])

If for all $\sigma\in\Omega$,

$$\mu_{\beta.h}(\sigma(x)=+1\,|\,\sigma(y),\ y\in\partial\{x\}) \underset{(\leq)}{\geq} p ,$$

then we have

$$\langle f\rangle_{\beta.h} \underset{(\leq)}{\geq} \langle f\rangle'_p$$

for all increasing functions f of $\sigma\in\Omega$, where $\langle\cdot\rangle_{\beta.h}$ and $\langle\cdot\rangle'_p$ stand for expectations w.r.t. $\mu_{\beta.h}$ and $\mu'_{\beta.h}$ respectively.

Remark. In [8], the statement corresponding to the above lemma is much more general, but for simplicity we presented it in the above form. From the lemma, it is obvious that

$$\mu_{\beta.h}(E^+_\infty) = 0 \qquad \text{if } h < \frac{1}{\beta}\log\frac{p_c}{1-p_c} - 4 ,$$

$$\mu_{\beta.h}(E^+_\infty) = 1 \qquad \text{if } h > \frac{1}{\beta}\log\frac{p_c}{1-p_c} + 4 .$$

([4]).

§3. Sketch of the proof

In this section, we give a sketch of the proof of the statements we gave in §2, 1°)~4°).

First we consider the case when $h = 0$, corresponding to 1°) and 3°) in §2. The only thing remaining to be proved is that

$$\mu_{-,\beta,0}(E^+_\infty) = 0 \qquad \text{for all } \beta > 0, \tag{6}$$

where of course $\mu_{-,\beta,0} = \mu_{\beta,0}$ (the unique Gibbs state) for $\beta \geq \beta_c$.

As we mentioned, this can be proved by Harris' argument given in [2].

Let

$$V_n = \{ x \in \mathbb{Z}^2;\ |x^1| \leq n,\ |x^2| \leq n \},$$

and let E^+_n (E^{+*}_n) be the event that $\{x^1 = -n\}$ is connected (*connected) to $\{x^1 = n\}$ in $V_n \cap \omega^{-1}(+1)$, where *connection means the connection in $\mathbb{L}$. E^-_n and E^{-*}_n are the events defined by the same way for $\omega^{-1}(-1)$.

By duality of $\mathbb{Z}^2$ and $\mathbb{L}$, we have

$$(E^+_n)^c = R(E^{-*}_n), \quad (E^{+*}_n)^c = R(E^-_n) \qquad \text{foe } n \geq 1, \tag{7}$$

where $R: \Omega \to \Omega$ is the rotation around the origin by 90 degrees, i.e.

$$(R\omega)(x^1,x^2) = \omega(-x^2,x^1).$$

Since *connection is easier than the usual connection, we know that

$$E^{+*}_n \supset E^+_n, \quad E^{-*}_n \supset E^-_n.$$

By R-invariance and FKG inequality, we have

$$\mu_{-,\beta,0}(E^{-*}_n) \geq \mu_{-,\beta,0}(E^-_n) = \mu_{+,\beta,0}(E^+_n). \tag{8}$$

Here, we used the fact that $\mu_{-,\beta,h} \circ T = \mu_{+,\beta,-h}$ for any $\beta > 0$ and real h, where $T: \Omega \to \Omega$ reverses all the spins on $\mathbb{Z}^2$, i.e.

$$(T\omega)(x) = -\omega(x).$$

(7) and (8) imply obviously that $\mu_{-,\beta,0}(E^{-*}_n) \geq 1/2$.

Let F^{-*}_n be the event that $\{x^1 = -3n\}$ is *connected to $\{x^1 = 3n\}$ in

$$\{ x \in \mathbb{Z}^2;\ |x^1| \leq 3n,\ |x^2| \leq n \} \cap \omega^{-1}(-1).$$

The key point is to show that there is a n-independent constant $\alpha > 0$, such that

$$\mu_{-,\beta,0}(F^{-*}_n) \geq \alpha \qquad \text{for } n \geq 1,\ \beta > 0. \tag{9}$$

If (9) is proved, then putting $\theta_x: \Omega \to \Omega$ by $(\theta_x\omega)(y) = \omega(x+y)$ and

$$F^{-*}_{n,1} = \theta_{(0,2n)} F^{-*}_n, \quad F^{-*}_{n,2} = \theta_{(-2n,0)} R F^{-*}_n,$$

$$F^{-*}_{n,3} = \theta_{(0,-2n)}F^{-*}_{n}, \quad F^{-*}_{n,4} = \theta_{(2n,0)}RF^{-*}_{n},$$

we obtain

(10) $\mu_{-,\beta,0}($ there exists a *circuit surrounding V_n in $\omega^{-1}(-1)$ $)$

$$\geq \mu_{-,\beta,0}\Big(\bigcap_{1\leq i\leq 4} F^{-*}_{n,i} \Big)$$

$$\geq \alpha^4.$$

The last inequality follows from FKG inequality and the invariance of $\mu_{-,\beta,0}$ under $\{\theta_x;\ x \in \mathbb{Z}^2\}$ and R. As we have observed in §2, (10) implies that

$$\mu_{-,\beta,0}(E^+) \leq 1 - \alpha^4 < 1,$$

which, by the tail triviality of μ_-, proves (6).

Now it remains to show (9). Consider the event that $\{x^2=2n\}$ is *connected to $\{x^2=n\}$, and $\{x^1=2n\}$ is *connected to both $\{x^1=0,\ x^2\geq 0\}$ and $\{x^1=0,\ x^2\leq 0\}$. Denote this event simply by G^{-*}_n.
By FKG inequality and the invariance of μ_-, we obtain

$$\mu_{-,\beta,0}(G^{-*}_n) \geq 4^{-1}\{\mu_{-,\beta,0}(E^{-*}_n)\}^3.$$

But in G^{-*}_n, we know that the maximal *half circuit $\gamma(\omega)$ in $V_n+(n,0) = \{0 \leq x^1 \leq 2n,\ |x^2| \leq n\}$ surrounding the origin is *connected to $\{x^1=2n\}$ in $(V_n+(n,0)) \cap \omega^{-1}(-1)$.
For each *half circuit γ in $V_n+(n,0)$ surrounding the origin, let H^{+*}_γ (H^{-*}_γ) be the event that the origin is surrounded by a *circuit which is *connected to $\hat{\gamma}$ (γ) in $\Theta_\gamma \cap \omega^{-1}(+1)$ ($\Theta_\gamma \cap \omega^{-1}(-1)$), where Θ_γ is the region encircled by γ and its reflection $\hat{\gamma}$ w.r.t. $\{x^1=0\}$.
By the symmetry of the conditional distribution of μ_-, we have

$$\mu_{-,\beta,0}(\ H^{-*}_\gamma \mid \omega(x)=+1 \quad x \in \gamma,\ \omega(y)=-1 \quad y \in \hat{\gamma}\backslash\gamma) \geq 1/2,$$

because

$$\mu_{-,\beta,0}(\ H^{-*}_\gamma \cup H^{+*}_\gamma \mid \omega(x)=+1 \quad x \in \gamma,\ \omega(y) =-1 \quad y \in \hat{\gamma}\backslash\gamma) = 1,$$

and both $\mu_{+,\beta,0}$ and $\mu_{-,\beta,0}$ have the same conditional distribution given $\{\omega(x),\ x \in \gamma \cup \hat{\gamma}\}$.
Hence if $\omega \in G^{-*}_n$, by FKG inequality, the conditional probability that $\mu_{-,\beta,0}(\ H^{-*}_\gamma \mid \gamma(\omega) = \gamma\ )$ is not less than 1/2 for each γ.

Multiplying $\mu_{-,\beta,0}(\ \gamma(\omega) = \gamma\ )$ and summing up for all γ's, we get

$$\mu_{-,\beta,0}\left(\begin{array}{l}\{x^1=2n\} \text{ is *connected to a *circuit surrounding} \\ \text{the origin in } [-2n,2n]\times[-n,n]\cap\omega^{-1}(-1)\end{array}\right)$$

$$\geq 2^{-1}\mu_{-,\beta,0}(\ G_n^{-*}\ ) \geq 2^{-6}.$$

Reflecting this event w.r.t. $\{x^1=0\}$, and using FKG inequality, we get

$$\text{(11)} \qquad \mu_{-,\beta,0}\left[\begin{array}{l}\{x^1=2n\} \text{ is *connected to } \{x^1=-2n\} \\ \text{in } [-2n,2n]\times[-n,n]\cap\omega^{-1}(-1)\end{array}\right]$$

$$\geq 2^{-12}.$$

Denote this event by I_n^{-*}. Then from obvious inclusion

$$\left[\theta_{(-n,0)}I_n^{-*}\right]\cap RE_n^{-*}\cap\left[\theta_{(n,0)}I_n^{-*}\right]\subset F_n^{-*},$$

we obtain

$$\text{(12)} \qquad \mu_{-,\beta,0}(\ F_n^{-*}\ ) \geq \mu_{-,\beta,0}(\ E_n^{-*})\cdot\left[\mu_{-,\beta,0}(\ I_n^{-*})\right]^2 \geq 2^{-25},$$

which proves (9).
For the case $h \neq 0$, the argument is standard for the use of FKG inequality in percolation theory.

§4. Concluding remark.

The known results up to now for the percolation of the two-dimensional Ising model is summarized in Fig.2, where we know that

$$\beta_0 = 4^{-1}\log(p_c/1-p_c) \leq \beta_c.$$

By using FKG inequality, we know that for each $\beta \leq \beta_c$, there exists $h_c(\beta) \geq 0$, such that

$$\mu_{\beta,h}(\ E_\infty^+\ ) = 0 \qquad \text{for} \quad h < h_c(\beta).$$

We expect that $\mu_{\beta,h}(\ E_\infty^+) = 1$ for $h > h_c(\beta)$, too, and it is quite interesting to know the critical value $\beta_c{}'$ defined by

$$\beta_c{}' = \sup\{\ \beta \leq \beta_c;\ h_c(\beta) > 0\ \},$$

which we expect to be equal to β_c. If it is not the case, we have a new critical value for β, whose physical meaning is not quite clear.

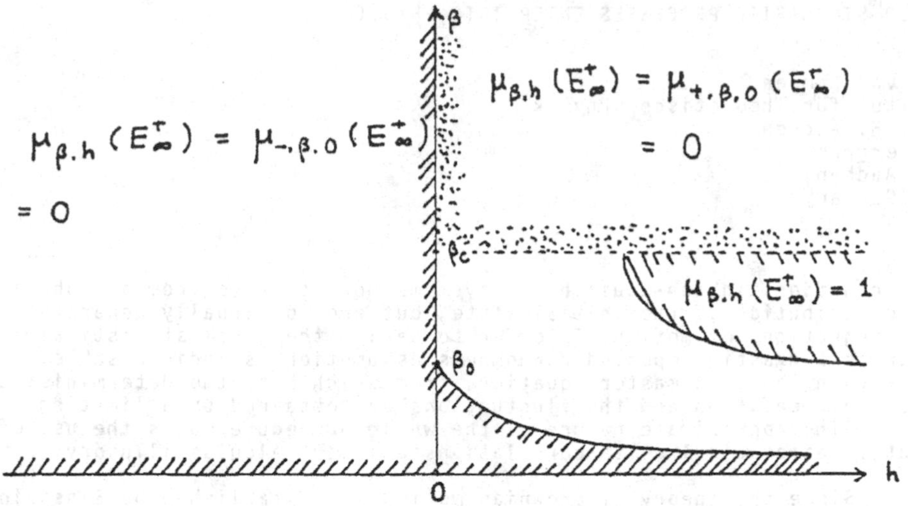

$$\beta_0 = \frac{1}{4} \log \frac{p_c}{1-p_c} \quad .$$

(Fig. 2)

References

[1] Coniglio, A., Nappi, C.R., Peruggi, F., Russo, L.: Comm.Math.Phys. 51, 315-323 (1976).

[2] Harris, T.E.: Proc.Cambridge Philos.Soc. 56, 13-20 (1960).

[3] Higuchi, Y.: Z.Wahrscheinlichkeitstheorie verw. Gebiete 61, 75-81 (1982).

[4] Higuchi, Y.: In Probability Theory and Mathematical Statistics. Proceedings of 4th USSR-Japan Symposium, 1982. Lecture Notes in Math. 1021, 230-237 (1983).

[5] Higuchi, Y.: A weak version of RSW theorem for the two-dimensional Ising model. Preprint.

[6] Kunz, H,. Souillard, B.: J.Stat.Phys. 19, 77-106 (1978).

[7] Russo, L.: Z.Wahrscheinlichkeitstheorie verw. Gebiete 56, 229-237 (1981).

[8] Russo, L.: i.b.i.d. 61, 129-139 (1982).

[9] Tóth, B.: A lower bound for the critical probability of the square lattice site percolation. Preprint.

HOW DO STOCHASTIC PROCESSES ENTER INTO PHYSICS?

N.G. van Kampen
Institut für Theoretische Physik
R.W.T.H. Aachen
Templergraben 55
5100 Aachen
F.R. Germany

Abstract

Fluctuations in non-equilibrium systems do not arise from a probability distribution of the initial state, but are continually generated by the equations of motion. In order to derive them from statistical mechanics a drastic repeated randomness assumption is indispensable. One is then led to a master equation, from which both the deterministic macroscopic equation and the fluctuations are obtained by a limiting process. The approximate nature of the whole procedure makes the use of strictly mathematical delta-correlations and Itô calculus illusory.

1. Since the theory of Brownian motion was established by Einstein and Smoluchowski [1] the role of stochastic processes and stochastic differential equations in physics has grown into that of an indispensable tool. In many cases it is heuristically clear why and how this tool should be utilized, but in other cases it is not. For instance, in the theory of the laser [2,3] stochastic equations are used without the basic understanding that is needed to judge whether the result is reliable. Hence it is not just for intellectual satisfaction that the way in which stochastic processes enter into physics should be investigated.

Of course, like everything in physics, the stochatic description can only be an approximation, but it is necessary to understand precisely which approximations are involved. A test for real understanding is that one can indicate how higher approximations should be obtained. Unfortunately insufficient effort has been devoted to the analysis of these questions.

2. Consider a closed, isolated, classical physical system described by canonical variables $(q_1,\ldots,q_N,p_1,\ldots,p_N)$, and a Hamilton function $H(q,p)$. The equations of motion define a family of trajectories defining a flow in the 2N-dimensional phase space Γ. In an alternative notation: every point $x\in\Gamma$ is carried by a flow into a uniquely defined point $x^t\in\Gamma$ after a time t. The flow preserves the phase space volume. If x is the initial state of the system at $t = 0$, its state at t is $x^t = X(t,x)$ and the Jacobian equals unity: $|dX(t,x)/dx| = 1$ (t fixed).

Statistical mechanics tells us that a physical system in which the number N of degrees of freedom is large should be described by an ensemble of identical replicas. Accordingly, rather than a single initial state x one introduces a probability density $\rho(x)$ of initial states, to be determined by physical considerations. This turns x into a stochastic variable and $x^t = X(t,x)$ into a stochastic process. Its single-time probatility density is

$$P_1(x_1,t_1) = \int \delta(x^{t_1} - x_1)\rho(x)dx = \rho(x_1^{-t_1})$$

and the entire hierarchy of joint probability densities is

$$P_n(x_1,t_1;\ldots;x_n,t_n) = \int \prod_{i=1}^{n} \delta(x^{t_i} - x_i)\rho(x)dx.$$

A physical quantity is a function A(x) in phase space. Its value $A(x^t)$ at time t has also become a stochastic process, fully determined by the choice of the ensemble ρ. A special choice for ρ is a stationary ensemble, that is, a distribution having the property $\rho(x^t) = \rho(x)$; for instance,

$$\rho(x) = e^{-\beta H(q,p)}/Z, \qquad Z = \int e^{-\beta H(q,p)} dq\, dp,$$

with arbitrary positive parameter β. With this choice our random functions X(t,x) and $A(x^t)$ are stationary processes.

3. All this is exact, but of little use when dealing with actual many-body systems since the mapping $x \to x^t = X(t,x)$ is much too complicated to be determined explicitly [4]. What one does in practice is the following. One selects somehow a set of "relevant" variables $A_r(x)$ and makes suitable assumptions concerning the stochastic properties of the associated processes $A_r(x^t)$. These usually amount to assuming that they obey a multivariate Langevin (or Itō)equation.

For instance, in the popular Projection Operator Technique [5] one first formally derives an equation for the A_r alone by eliminating all other variables; this is done purely mathematically and inevitably the equation involves an integral over the preceding values of A from the initial time up to the time considered. In addition there is a term involving the initial values of the eliminated variables. This integral equation is called the "generalized Langevin equation", but is actually merely a different form of the exact microscopic equations of motion. One then turns the additional term into a random force by assuming some probability distribution *for the initial values of the eliminated variables*. Nobody asks what this special initial time is. Subsequently a "Markovian assumption" is used to get rid of the integral and obtain an actual stochastic differential equation of the Langevin type.

In Linear Response Theory [6] the initial time is shifted to $-\infty$, but it is again true that the randomness enters only through the assumed *initial* probability distribution [7].

Thus in these and similar approaches the essential difficulty of statistical mechanics is resolved by assumption; in the absence of anything else one cannot complain, provided that no claim is made that something has been derived.

4. However, the whole idea is wrong. *This is not the way in which stochastic processes enter into physics.* As an illustration take a Brownian particle; together with the surrounding fluid it constitutes a closed, isolated system. The "relevant" variable A is the position of the particle, and constitutes a stochastic process (approximately a Wiener process). Obviously, this is *not* due to our ignorance concerning the state of the system at some initial time. Rather, it is due to the fact that the single variable A does not really satisfy a closed differential equation, but interacts with all fluid molecules. Their variables are not present in the equation for A but their effect shows up in the random Langevin force. Fluctuations in A are constantly being generated by the collisions, and would be there just the same even if I had been able to start the system off in a precise microscopic state at $t = 0$. Another illustration is shuffling a deck of cards.

Generally, the evolution of a many-body system is described exactly on the *microscopic* scale by the flow $x \to x^t$. Experience has taught us that there is also a *macroscopic* description in terms of a few, suitably chosen, macroscopically observable quantities $A_r(x)$, which obey a closed set of differential equations. These equations are not exact, however. The enormous number of eliminated microscopic variables makes itself felt by causing the actual values of the A_r to fluctuate about the values given by those macroscopic equations. The actual values are extremely complicated functions of time, which cannot be found without solving the microscopic equations, but their short-time averages (and other moments) do have simple predictable properties. In practice one replaces these time averages by ensemble averages for convenience.

Summary. Stochastic processes describing fluctuations in physics do not arise from a probability distribution of the initial microstate. Rather, they serve as a tool to describe the irregular motion of the actual trajectory about the smoother evolution determined by the macroscopic equations. That is how stochastic processes enter into physics. Our next task is to describe in somewhat more detail how this happens.

5. The macroscopic variables A_r determine a "coarse-graining" of the phase space Γ by cutting it up in phase cells defined by

$$a_r < A_r(x) < a_r + \Delta a_r \quad \text{(all } r\text{)},$$

where Δa_r is roughly the lack of precision of the observations. An observation or measurement tells me in which cell the point x lies, but no more. The *basic assumption of statistical mechanics of time-dependent processes* is that I don't have to know more, but that I may replace the precise point x by a probability distribution in the cell - with constant density in that cell and zero density outside it.

The flow in Γ carries this density along and after time t a fraction of that density lies in phase cell a'; this fraction may be denoted

$$\Delta a' \, T_t(a'/a).$$

It represents the probability $P(a',t)\,\Delta a'$ that a system starting at *a* will be found (at time t) in phase cell $\Delta a'$ at *a'*.

In our formulation of the basic assumption no restriction has been imposed on how the system originally had arrived in the cell Δa at *a*. Hence we may apply it to the time interval $t, t + \Delta t$ and find

$$P(a'', t + \Delta t) = \int T_{\Delta t}(a''|a')P(a',t)da'. \tag{1}$$

Thus the motion among the cells is a Markov chain. In the limit of small Δt one obtains

$$\frac{\partial P(a,t)}{\partial t} = \int \{W(a|a')P(a',t) - W(a'|a)P(a,t)\}\, da', \tag{2}$$

where $W(a|a')da$ is the transition probability per unit time from *a'* into d*a*. This is the differential form of the Chapman-Kolmogorov or Smoluchowski equation, now usually called the master equation [8]. It describes the evolution of the system, as seen by a macroscopic observer, in terms of a Markov process.

6. With respect to our drastic assumption the following remarks can be made.

(i) All existing treatments relating macroscopic equations to the microscopic ones use such an assumption in the form of a "Stosszahlansatz", "molecular chaos", "random phase", "Markov assumption", or "repeated randomness assumption". Rather than to hide it one should make it explicit so that its validity can be investigated.

(ii) The picture is that the microscopic trajectory is so complicated that it practically covers the whole phase cell during the short time

t. Thus one implicitly uses a kind of local ergodic theorem: the time average during Δt equals the phase cell average.

(iii) As a consequence, (1) cannot be valid when Δt is too small and (2) is not really a limit $\Delta t \to 0$. Rather, (2) holds approximately when it is possible to pick a Δt that is large enough for (1), but still so small that the values of the A_r do not change appreciably.

(iv) It follows that any process in physics is stochastic only if the variables are measured with a sufficient margin. [9] It is Markovian only if one does not look at too small time intervals. In particular, in the Langevin equation the random force is never strictly delta-correlated; at best its auto-correlation time is short compared to the other relaxation times in the system [10]. *From a physical point of view the Itô calculus is based on a misconception,* since it requires strict delta correlations.

(v) The validity of the basic assumption depends on a proper choice of the A_r. This choice is *not* determined by what the experimenter wants to observe. Rather, it is determined by the requirement that the A_r should incorporate *all* slow variations; they must account for *all* long-time correlations, since otherwise the Markov property used in (1) cannot be valid.

(vi) Whether or not such a separation of time scales is possible depends on the system, i.e., on its Hamilton function. For some many-body systems a reduced description in terms of a few A_r may be impossible, such as self-gravitating systems, e.g., stellar clusters. [11]

(vii) In actual physical cases one is often able to guess (on the basis of experience, intuition, or trial and error) what the correct choice is for the A_r. In a simple fluid they are the local density, velocity, and energy density. But the lack of an actual criterion makes it impossible to judge whether higher approximations can be obtained by merely adding terms to the equations for the same A_r [12]; or require the addition of new A_r into the macroscopic equations, such as the heat flux, as in "extended thermodynamics" [13].

7. We have arrived at the master equation (2), which does describe the evolution of the system on a macroscopic scale, but as a stochastic process. The familiar deterministic macroscopic equations (Navier-Stokes, Ohm, etc.) can be extracted from it in the following way [14]. In general, $W(a|a')$ involves a paramter Ω with the property that for large Ω the fluctuations are relatively small. Ω may be the size of the system, the mass of the Brownian particle, or the capacity of a condenser. Expansion of (2) in Ω^{-1} gives to lowest order the desired deterministic

equation

$$\dot{a}_r(t) = f_r(a) = \int (a_r' - a_r)W(a'|a)da'. \qquad (3)$$

Thus the rate of change $\dot{a}_r$ equals the average of the jump per unit time. We therefore call the process "jump-driven".

The next approximation is of order $\Omega^{-1/2}$ and gives the fluctuations in Gaussian approximation. Let $\varphi_r(t)$ be the solution of (3) with some given initial value, and set $\Delta a_r = a_r - \varphi_r(t)$. Then one finds $< \Delta a_r > = 0$ and the matrix $\Xi_{rs} = < \Delta a_r \Delta a_s >$ obeys

$$\dot{\Xi} = A(\varphi)\Xi + \Xi[A(\varphi)]^{\dagger} + B(\varphi), \qquad (4)$$

where the matrices A and B are given by

$$A_{rs}(a) = \partial f_r / \partial a_s$$

$$B_{rs}(a) = \int (a_r' - a_r)(a_s' - a_s)W(a'|a)da'.$$

For the detailed derivation we refer to the literature [8].

For some systems, however, to be called "diffusion-driven", the lowest order (3) happens to be zero. In that case the next order is the leading one. It turns out that this order consists of a Fokker-Planck equation [15,8]:

$$\frac{\partial P(a,t)}{\partial t} = - \sum_r \frac{\partial}{\partial a_r} A_r(a)P + \frac{1}{2} \sum_{r,s} \frac{\partial^2}{\partial a_r \partial a_s} B_{rs}(a)P. \qquad (5)$$

Here B_{rs} is the same as before and $A_r(a)$ is what remains of (3) in the next order of Ω^{-1}, the lowest being zero. An example is an electron in a semiconductor (possibly inhomogeneous), subject to an electric field (possibly not constant).

Evidently for this class of systems the Ω-expansion does not yield a deterministic macroscopic equation, but again a stochastic description. All it does is to substitute a Fokker-Planck approximation for the master equation (2). In order to obtain a deterministic equation an additional expansion is needed, for instance in powers of the temperature. [8,16] One then finds again for the noise in lowest order a Gaussian distribution.

Summary. Statistical mechanics leads, on the macroscopic level, to a stochastic description in terms of the master equation (2). Subsequently, deterministic equations plus fluctuations can be extracted from it by suitable limiting procedures. The view that one should start from the known macroscopic equations and the fluctuations should be somehow tagged on to them is responsible for much confusion in the literature.

8. I call a *stochastic differential equation* any differential equation whose coefficients are random constants or functions of time *with given stochastic properties*, i.e.,

$$\dot{a}_r = F_r(a;\ \xi(t)) \quad \text{with given random } \xi(t). \tag{6}$$

In the mathematical literature one usually restricts the name to the case that ξ is Gaussian and delta-correlated (derivative of the Wiener process), but that is clearly inappropriate in physics, since such a ξ is only an approximation of an actual random force. How do stochastic differential equations arise in physics?

Consider a closed, isolated many-body system consisting of a small subsystem S with few variables Q,P and a large "bath" B with many q,p:

$$H = H_S(Q,P) + H_B(q,p) + H_I(Q,q). \tag{7}$$

An example is a single nuclear spin S in a crystal B; the lattice vibrations are described by the q,p. [17] For the interaction we take the simple, but not unrealistic, form

$$H_I(Q,q) = g(Q) \sum_i h(q_i).$$

The idea is that one first imagines the motion resulting from H_B *in isolation* to be determined, giving rise to complicated and rapidly varying functions $q_i(t)$. Their explicit form is of course unknown, but one puts

$$\sum_i h(q_i(t)) = \xi(t).$$

Then $\xi(t)$ is regarded as a stochastic process, of which the properties are guessed. Usually one assumes $\xi(t)$ to be Gaussian, as it is the sum of many terms (although not independent); and delta-correlated, as the $q_i(t)$ vary rapidly and irregularly (although determinstically). With these assumptions one obtains a stochastic Hamilton function for S

$$H(t) = H_S(Q,P) + g(Q)\,\xi(t), \tag{8}$$

leading to stochastic equation of motion for the subsystem S alone.

One recognizes the same philosophy as before: the Q,P are our previous A_r, the q,p are the eliminated variables, and the fluctuations in the A_r are continually generated by the q,p. In addition, however, one has introduced a *second assumption,* namely that the motion of the q,p may be determined from H_B alone, that is, that *the reaction of the subsystem S on the bath is negligible.* This assumption reduces the interaction with the bath to an external random force with given properties and thereby reduces the master equation (2) to a stochastic differential equation (6).

9. The plausibility of this assumption depends on the system considered.

The earth's troposphere is turbulent so that its density, and hence its refractive index, is a random function in space and time. An electromagnetic wave propagating through it obeys Maxwell's equation involving a random dielectric constant. [18] Clearly the effect that passing radio waves have on the turbulence is negligible, so that in this case a stochastic differential equation is justified.

For a spin embedded in a crystal lattice, however, the assumption cannot be true. (Of course one has to replace Q,P with spin operators, but the argument remains the same.) The reason is that (8) leads inevitably to an ever growing value of the energy H_S, as if the bath had an infinite temperature. What is missing is the spontaneous emission, the fact that transitions from a high level to a lower one are more likely than vice versa. Hence one is forced to add a damping term in the equations of motion, corresponding to the friction term in the equation for the Brownian particle. The magnitude of this term is related to the random term in (8) by the requirement that the outcome of the competition between both must be the known equilibrium distribution ("fluctuation-dissipation theorem"). Yet this damping term is entirely *ad hoc* and demonstates the heuristic nature of stochastic differential equations in physics.

Incidentally, for some simple systems, such as the Brownian particle, a more sophisticated approach is possible [19]. One determines q(t) not from H_B, but from $H_B(q,p) + H_I(Q,q)$, regarding Q as slowly varying and in lowest approximation as constant. As a result the stochastic properties of $\xi(t)$ now depend on Q. It turns out that this automatically leads to the desired damping. The result can again be expressed as a stochastic differential equation; however, it is obtained without the "second assumption" in section 8, but by working out the program of section 5.

Summary. The master equation (2) describes a closed, isolated system. If this system can be subdivided in a subsystem and a bath as in (7), it may be possible to write stochastic equations of motion for the subsystem alone. The fluctuations are then caused by an external random force $\xi(t)$ caused by the bath. This random force is not affected by the subsystem, any effect of the subsystem on the bath must be accounted for by a deterministic damping term in the equations for the subsystem.

10. In systems that cannot be so subdivided the noise is *intrinsic*, it is part and parcel of the evolution itself. Examples are radioactive decay and chemical reactions. They can be described only by the original master equation (2). Yet such systems are often treated in the literature in terms of a Langevin stochastic differential equation. How can this be understood?

First we have seen that there exist *diffusion-driven systems*, in which the dominant term of the Ω-expansion has the form of a Fokker-Planck equation (5). As any Fokker-Planck equation is mathematically equivalent to a suitably chosen Langevin or Itô equation, one is free to use these as a formal device instead of (5). Higher approximations in Ω^{-1}, however, give additional terms to (5), which cannot be incorporated in the Langevin equation.

Secondly, *jump-driven systems* are described in the first two orders of the Ω-expansion by the deterministic macroscopic equation (3) plus Gaussian noise determined by (4). It is possible to construct a Fokker-Planck equation whose solution in the same order of Ω produces the same macroscopic behavior and noise. Thus to this order in Ω such systems can be described by a Fokker-Planck equation and hence by a Langevin equation. [20] It is not clear, however, why one should do this, in particular since these equations are much harder to solve than the first and second order approximations (3), (4) themselves.

REFERENCES

[1] A. Einstein, Ann. Physik (4) 17, 549 (1905); 19, 371 (1906); M. v. Smoluchowski, Ann. Physik (4) 21, 756 (1906).

[2] H. Haken, in *Encyclopaedia of Physics* 25/2c (Springer, Berlin 1970); M. Sargent, M.O. Scully, and W.E. Lamb, *Laser Physics* (Addison-Wesley, Reading, Mass. 1974).

[3] H. Haken, *Synergetics* (Springer, Berlin 1976, 1978); C.W. Gardiner, *Handbook of Stochastic Methods for Physics, Chemistry and the Natural Sciences* (Springer, Berlin 1983).

[4] With the exception of a few soluble cases, viz., the linear harmonic chain, see e.g. G.W. Ford, M. Kac, and P. Mazur, J. Math. Phys. 6, 504 (1965); P. Ullersma, Physica 32, 27, 56, 74, 90 (1966).

[5] S. Nakajima, Prog. Theor. Phys. 20, 948 (1958); R. Zwanzig, J. Chem. Phys. 33, 1338 (1960); M. Mori, Prog. Theor. Phys. 33, 423 (1965).

[6] R. Kubo, J. Phys. Soc. Japan 12, 570 (1957).

[7] N.G. van Kampen, Physica Norvegica 5, 279 (1971).

[8] N.G. van Kampen, *Stochastic Processes in Physics and Chemistry* (North-Holland, Amsterdam 1981).

[9] The vital role of such margins was forcefully argued by P. and T. Ehrenfest, in: Enzyklopädie der mathematischen Wissenschaften 4, Nr. 32 (Teubner, Leipzig 1912); translated by M.J. Moravcsik with the title *Conceptual Foundations of the Statistical Approach in Mechanics* (Cornell Univ. Press, Ithaca 1959).

[10] The founding fathers were of course fully aware of this: G.E. Uhlenbeck and L.S. Ornstein, Phys. Rev. 36, 823 (1930).

[11] T.S. van Albada, Bull. Astr. Inst. Neth. 19, 479 (1968); W. Thirring, Z. Physik 235, 339 (1970); R. Miller, in: Advances in Chemical Physics 26 (Wiley, New York 1970).

[12] N.N. Bogolubov, *Problems of Dynamical Theory in Statistical Physics*, in: Studies in Statistical Mechanics I (G.E. Uhlenbeck and J. de Boer eds., North-Holland, Amsterdam 1962); G.E. Uhlenbeck, in: *Probability and Related Topics in Physical Sciences I* (Proceedings of the Summer Seminar in Boulder, Colorado in 1957; Interscience, London and New York 1959) p. 195 ff.; E.G.D. Cohen, in: *Fundamental Problems in Statistical Mechanics II* (E.G.D. Cohen ed., North-Holland, Amsterdam 1968).

[13] R.E. Nettleton, J. Chem Phys. 40, 112 (1964); I. Müller, Z. Physik 198, 329 (1967); L.S. García-Colin, M. López de Haro, R.F. Rodriguez, and D. Jou, J. Stat. Phys. 37, 465 (1984).

[14] N.G. van Kampen, Can. J. Phys. 39, 551 (1961) and in: Advances in Chemical Physics 34 (Wiley, New York 1976); R. Kubo, K. Matsuo, and K. Kitahara, J. Statis. Phys. 9, 51 (1973).

[15] N.G. van Kampen, Phys. Letters 62A, 383 (1977).

[16] H. Grabert and M.S. Green, Phys. Rev. A19, 1747 (1979); H. Grabert, R. Graham, and M.S. Green, Phys. Rev. A21, 2136 (1980).

[17] C.P. Slichter, *Principles of Magnetic Resonance* (Harper and Row, New York 1963; Springer, Berlin 1978).

[18] V.I. Tatarski, *Wave Propagation in a Turbulent Medium* (McGraw-Hill, New York 1961); U. Frisch, in: *Probabilistic Methods in Applied Mathematics 1* (A.T. Bharucha-Reid ed., Acad. Press, New York 1968); V.I. Klyatskin and V.I. Tatarski, Sov. Phys. Usp. 16, 494 (1974).

[19] P. Mazur and I. Oppenheim, Physica 50, 241 (1970), and literature quoted there.

[20] T.G. Kurtz, J. Appl. Prob. 7, 49 (1970); 8, 344 (1977); J. Chem. Phys. 57, 2976 (1972); Z.A. Akcasu, J. Statis. Phys. 16, 33 (1977); N.G. van Kampen, J. Statis. Phys. 25, 431 (1981).

Estimates on the difference between succeeding eigenvalues and Lifshitz tails for random Schrödinger operators

Werner Kirsch
Institut für Mathematik
Ruhr-Universität
D-4630 Bochum, W.-Germany

1. Introduction: In this note we give estimates on the difference of eigenvalues of second order differential operators. We both treat the one-dimensional and the multi-dimensional case. Moreover, we apply our estimates to certain concrete problems of theoretical physics. In particular we prove Lifshitz behavior of the density of states for a broad class of random Schrödinger operator.

Let us denote by H the second order linear differential operator

$$H = - \sum_{i,j} \frac{\partial}{\partial x_i} a_{ij}(x) \frac{\partial}{\partial x_j} + V(x) \tag{1}$$

where the matrix $[a_{ij}(x)]_{i,j}$ is positive definite for every x, $a_{ij}(x)$ and $V(x)$ are realvalued functions.

To avoid technicalities we assume that a_{ij} and V are bounded, continuous functions and a_{ij} has continuous partial derivatives. The results below, however, are true under much less restrictive assumptions.

Since the operator H is bounded below we may arrange the eigenvalues of H below the essential spectrum in increasing order. We denote by E_o, E_1,... the eigenvalues of H and may suppose that

$$E_o \leq E_1 \leq \ldots \tag{2}$$

The chain (2) may, of course, be empty, finite, or infinite.

Each eigenvalue occurs in (2) according to its multiplicity. For the operator (1) it is known that E_o is always non degenerate, i.e. that

$$E_o < E_1 \leq E_2 \ldots$$

while for d = 1 no eigenvalue is degenerate, that is:

$$E_o < E_1 < E_2 < \ldots$$

In this paper we give estimates (especially from below) on the differences $E_{n+1}-E_n$ in the one-dimensional case and on E_1-E_o in the multidimensional case. The proofs we present are based on the methods developed in [2] and [3]. In those papers, however, only the Schrödinger case (i.e. $a_{ij} \equiv 1$) is considered. We would like to thank M. Hazewinkel for pointing our attention to the more general case we treat here. The result for such operators may be of some interest in engineering. In this note we will concentrate on applications in Quantum Mechanics, i.e. for Schrödinger operators. In particular, we discuss the use of our estimates in connection with the semiclassical limit due to S. Nakamura [27]. In the final section we discuss in details the Lifshitz behavior of the density of states.

2. Exponential bounds for the one-dimensional case

Here we consider an operator H of Sturm-Liouville-type:

$$H = -\frac{d}{dx}\, p(x)\, \frac{d}{dx} + q(x) \tag{3}$$

We ask for estimates on $E_{n+1}-E_n$ from below. Those estimates should be given in "geometric terms" with respect to p and q, they should not depend on the particular "local" behavior, but rather on some rough, "global" characteristics of these functions.

Before clarifying what we mean by this, let us look what we can reasonably expect. Let us consider one dimensional Schrödinger operators with potentials V_L given by

$$V_L(x) = -\chi_{(-1,1)}(x) + \chi_{(-1,1)}(x-L) \tag{4}$$

This is a typical tunneling situation. For L large, one expects by physical reasoning two eigenvalues very near to each other at the bottom of the spectrum. One should come from a particle living near zero, the other from a particle near L. Tunneling prevents those eigenvalues form being degenerate. Harrell [1] proves that, indeed $E_1(L)-E_o(L) \sim e^{-\lambda L}$ for large L ($E_n(L)$ denoting the eigenvalues of $-\frac{d^2}{dx^2} + V_L$). Thus the best we can hope for is to obtain exponentially small lower bounds in the general case.

We define $\mu_n(x) = \sup\limits_{E \in [E_{n-1},E_n]} |V(x)-E|$ and $\lambda_n(x) = \mu_n(x)^{1/2} p(x)^{1/2}$.

We have

Theorem 1: Suppose that $V(x) \geq E_n + \alpha^2 + \alpha p'(x)$ for $x \notin [a,b]$, then if $\mu(x) \geq \mu_n(x)$ $E_n - E_{n-1} \geq C\, e^{-\int_a^b \frac{\mu(x)^{1/2}}{p(x)^{1/2}}dx}$ with an explicitely computable constant C (depending on α, μ and p).

A special case of Theorem 1 was proven in [2] for the case $p \equiv 1$. In sketching the proof we will mainly emphasize the changements to be made in the proof in [2]. Moreover, we restrict ourselves to the case of operators on a finite interval [a,b] rather than on the whole line.

In [2] we introduced a slightly modified Prüfer transform.This transform should be changed to

$$u(x) = r \cos \theta(x)$$

$$pu'(x) = -\lambda(x)\, r \sin \theta(x)$$

The main difference is the x-dependence of the parameter $\lambda(x)$. This makes the first order differential equation for θ more complicated. It reads:

$$\theta' = \lambda^{-1}(E-V) \cos^2 \theta + p^{-1} \lambda \sin^2 \theta - \lambda^{-1}\lambda' \sin \theta \cos \theta \tag{5}$$

Thus the equation for $\phi = \frac{\partial \theta}{\partial E}$ becomes:

$$\phi' = \{(p^{-1}\lambda + \lambda^{-1}(V-E)) \sin 2\theta - \lambda^{-1}\lambda' \cos 2\theta\}\, \phi + \lambda^{-1}(\cos^2 \theta) \tag{6}$$

As in [2] one proves that

$$E_n - E_{n-1} \geq \Big[\sup_{E \in [E_{n-1}, E_n]} \phi(b,E)\Big]^{-1}$$

An upper estimate on ϕ is obtained from (6) by use of Gronwall's lemma. Independently Nakamura [27] used a similar modification of the method of [2] to investigate the semiclassical limit of Schrödinger operators.

3. The multi-dimensional case

The estimates in the case $d > 1$ are quite easy if one uses the "Dirichlet form" representation of H. This technique is used in [8] to define and investigate singular operators and in [9] for estimates related to those presented here. We refer to Fukushima's book [10] for the general theory of Dirichlet forms in a probabilistic context.

Let ψ_o denote the positive ground state of H. Then $U\varphi(x) = \psi_o(x)\varphi(x)$ defines a unitary operator from $L^2(\mathbb{R}^d, \psi_o^2 dx)$ to $L^2(\mathbb{R}^d, dx)$.

Consequently

$$\hat{H} = U^{-1}(H-E_o)U \tag{7}$$

is an operator on $L^2(\psi_o^2 dx)$ unitary equivalent to $H - E_o$. Thus $\hat{E}_o := E_o(\hat{H}) = 0$ and $\hat{E}_1 := E_1(\hat{H}) = E_1(H) - E_o(H)$.

The min-max-principle tells us that

$$\hat{E}_1 = \inf_{f \in Q(H), \langle f,1\rangle_{\psi_o}} \frac{\langle f,\hat{H}f\rangle_{\psi_o}}{\langle f,f\rangle_{\psi_o}} \tag{8}$$

where $Q(\hat{H})$ denotes the form domain of $\hat{H}$ and $\langle f,g\rangle_{\psi_o} = \int f(x)g(x)\,\psi_o^2(x)dx$ denote the scalar product on $L^2(\psi_o^2 dx)$. Note that the constant 1 is the (normalized) ground state of $\hat{H}$.

Moreover, we compute:

$$\langle f,\hat{H}f\rangle_{\psi_o} = \int\Big(\sum_{j,j} a_{ij}(x) \frac{\partial f}{\partial x_i} \frac{\partial \bar{f}}{\partial x_j} \Big)\, \psi_o(x)dx.$$

Following Agmon [26] we set

$$|\nabla_A f|^2 = \Sigma a_{ij}(x) \frac{\partial f}{\partial x_i} \frac{\partial \bar{f}}{\partial x_j} .$$

From (8) we get:

Theorem 2 (Variational principle):

$$E_1(H) - E_o(H) = \inf_{f:\ \int f\psi_o^2 dx=0} \frac{\int |\nabla_A f(x)|^2 \psi_o^2(x)dx}{\int |f(x)|^2 \psi_o^2(x)dx} \tag{9}$$

This variational principle enables us to compare the distance $E_1 - E_o$ for different operators H and $\tilde{H}$.

Denote by E_n (resp. $\tilde{E}_n$) the eigenvalues of $H = \sum_{i,j} \frac{\partial}{\partial x_i} a_{ij}(x) \frac{\partial}{\partial x_i} + V(x)$ (resp. $\tilde{H} = \sum_{i,j} \frac{\partial}{\partial x_i} \tilde{a}_{ij}(x) \frac{\partial}{\partial x_j} + \tilde{V}(x)$) and by ψ_o and $\tilde{\psi}_o$ the corresponding positive normalized ground states.

We set $a_+ = \sup \psi_o(x)\, \tilde{\psi}_o(x)^{-1}$, $a_- = \sup \psi_o(x)\, \tilde{\psi}_o(x)^{-1}$.

Furthermore, we denote the matrix $a_{ij}(x)$ by $A(x)$.

Theorem 3 (Comparison theorem): If $\gamma\tilde{A}(x) \le A(x) \le \Gamma\tilde{A}(x)$ for all x ($\le$ in the sense of a matrix inequality) then

$$\left(\frac{a_-}{a_+}\right)^2 \gamma(\tilde{E}_1 - \tilde{E}_o) \le E_1 - E_o \le \left(\frac{a_+}{a_-}\right)^2 \Gamma(\tilde{E}_1 - \tilde{E}_o) \tag{10}$$

To prove (10) we set $Tf(x) = f(x) - \langle f, \tilde{\psi}_o^2\rangle$. ($\langle\cdot,\cdot\rangle$ denotes the inner product of $L^2(dx)$). So, we have:

$$E_1 - E_o = \inf_{\langle f,\psi_o^2\rangle=0} \frac{\int|\nabla_A f|^2\psi_o^2 dx}{\int|f|^2\psi_o^2 dx} = \inf_{\langle f,\psi_o^2\rangle=0} \frac{\int|\nabla_A(Tf)|^2\psi_o^2 dx}{\int|Tf+\langle f,\tilde{\psi}_o^2\rangle|^2\psi_o^2 dx}$$

$$\le \frac{a_+^2}{a_-^2}\Gamma \inf_{\langle f,\psi_o^2\rangle=0} \frac{\int|\nabla_{\tilde{A}}(Tf)|^2\tilde{\psi}_o^2 dx}{\int|Tf+\langle f,\tilde{\psi}_o^2\rangle|^2\tilde{\psi}_o^2 dx} = \frac{a_+^2}{a_-^2}\Gamma \inf_{\langle f,\psi_o^2\rangle=0} \frac{\int|\nabla_{\tilde{A}}(Tf)|^2\tilde{\psi}_o^2 dx}{\int|Tf|^2\tilde{\psi}_o^2 dx+\langle f,\tilde{\psi}_o^2\rangle^2}$$

(since $\langle Tf, \tilde{\psi}_o^2\rangle = 0$)

$$\le \frac{a_+^2}{a_-^2}\Gamma \inf_{\langle g,\tilde{\psi}_o^2\rangle=0} \frac{\int|\nabla_{\tilde{A}}g|^2\tilde{\psi}_o^2 dx}{\int|g|^2\tilde{\psi}_o^2 dx} = \frac{a_+^2}{a_-^2}\Gamma(\tilde{E}_1 - \tilde{E}_o)$$ which proves the theorem.

In the forthcoming paper [3] we use the above theorems to prove exponentially small lower bounds on E_1-E_o in the multi-dimensional case. Here we turn to applications in the context of periodic and random operators.

4. Estimates for periodic Schrödinger operators

Now, we consider Schrödinger operators with periodic potentials satisfying:

$$V(x + i) = V(x) \text{ for all } i \in \mathbb{Z}^d \tag{11}$$

We set $C_L := \{x \in \mathbb{R}^d \mid -1/2\, L \le x_i \le 1/2\, L \text{ for } i = 1,\dots,d\}$. So C_1 is "the" unit cell of the lattice $\mathbb{Z}^d$. It is common to introduce the reduced Hamiltonian

$$H(k) = -\Delta + V,\ k \in [0,2\pi]^d \tag{12}$$

on C_1 with boundary conditions

$$\varphi(-1/2, x_2,\dots,x_d) = e^{ik_1}\varphi(1/2, x_2,\dots,x_d)$$

and similar conditions for the other coordinates. The operators H(k) have purely discrete spectrum, we denote their eigenvalues by $\varepsilon_n(k)$ (with the usual convention of the ordering of eigenvalues). The operator $H = -\Delta + V$ on $L^2(\mathbb{R}^d)$ is a direct integral of the H(k) (see [11]). In particular, the spectrum $\sigma(H)$ is given by

$$\sigma(H) = \bigcup_{n=0}^{\infty} \{\varepsilon_n(k) \mid k \in [0,2\pi]^d\} \tag{13}$$

The function $\varepsilon_n(\cdot)$ is called the n-th band function, the set $B_n = \{\varepsilon_n(k) \mid k \in [0,2\pi]^d\}$ is the n-th band.

So, the spectrum $\sigma(H)$ of H consists of bands B_n with possibly (but not necessarily) gaps between the bands.

It is known that the ground band B_o - and thus the spectrum $\sigma(H)$ - starts with $\varepsilon_o(0)$. Thus $\varepsilon_o(0)$ is the minimum of $\varepsilon_o(\cdot)$. The following theorem from [3] tells us more precisely how $\varepsilon_o(k)$ behaves near the bottom of the spectrum.

Theorem 4: The band function ε_o is parabolic near $k = 0$, i.e.

$$ck^2 \leq \varepsilon_o(k) - \varepsilon_o(0) \leq k^2 \qquad (14)$$

This theorem can be proven from the variational principle (theorem 2) - see [3] for details.

As a corollary to the above result we obtain:

Corollary: The density of states $\rho(\lambda)$ for periodic Schrödinger operators behaves near the bottom λ_o of the spectrum like $(\lambda - \lambda_o)^{d/2}$, more precisely:

$$0 < \lim_{\lambda \downarrow \lambda_o} \frac{\rho(\lambda)}{(\lambda-\lambda_o)^{d/2}} < \infty \qquad (15)$$

For the definition of ρ see e.g. [11].

We will see in the next section that the behavior (15) is in sharp contrast to the case of random potentials.

Now, we look at the periodic operator H on the hypercube C_L with periodic boundary conditions. We denote the resulting operator by H_L^{per} and its eigenvalues by $E_n^{per}(L)$.

Consider the ground state ψ_o of H_1^{per}. If is known that ψ_o can be taken strictly positive (see [11] or [12]). We take ψ_o positive and normalized so that $\|\psi_o\|_2 = 1$. Since ψ_o is periodic we may extend it to the whole space $\mathbb{R}^d$. So, we see that (the extended) ψ_o is also an eigenfunction for H_L^{per} with arbitrary L. Moreover, since ψ_o is positive it has to be a positive ground state eigenfunction. Hence $E_o^{per}(L) = E_o^{per}(1) = \inf \sigma(H)$ for all L.

It is intuitively clear (and not difficult to prove) that $E_1^{per}(L)$ tends to $E_o^{per}(L) = \inf \sigma(H)$ as L tends to infinity. The following theorem tells us how fast this convergence is:

<u>Theorem 5:</u> $\alpha L^{-2} \leq E_1^{per}(L) - E_o^{per}(L) \leq \beta L^{-2}$ (16)

To prove this result we use the comparison theorem (th. 3) where the free operator $H_o{}^{per}_L = (-\Delta)_L^{per}$ serves as $\tilde{H}$. The essential point is that $\sup_{x \in C_L} \psi_o(x) \tilde{\psi}_o^{-1}(x)$ (as well as the inf) do <u>not</u> depend on L since both ground state function are Z^d-periodic.

However, in the final section 5, we have to deal with Neumann boundary conditions instead of periodic ones. So define H_L^N to be the operator H on C_L with Neumann boundary conditions and denote by $E_n^N(L)$ the corresponding eigenvalues. In this case the ground states will not be periodic in general, so the above argument will not work ($\sup_{x \in C_L} \psi_o(x) \tilde{\psi}_o^{-1}(x)$ may depend on L and we have no control on this dependence).

Therefore, we will restrict the class of potentials we deal with. We suppose that V in addition to being periodic is also <u>reflection invariant</u>, that is we suppose that $V(x_1, x_2, ---, x_d) = V(-x_1, x_2, ---, x_d)$ and similar for the other components. This condition is satisfied for example if $V(x) = \sum_{i \in Z^d} f(x-i)$ and f is spherically symmetric.

Let ψ_o be the positive, normalized ground state of H_L^{per}. By reflection symmetry the "reflected" ψ_o is also a positive, normalized ground state of H_1^{per} and by uniqueness must agree with ψ_o. Thus ψ_o satisfies Neumann boundary conditions and by positivity must be the Neumann ground state. Therefore in the case of reflection symmetry we may apply the procedure to prove Theorem 5 also to $E_1^N(L) - E_o^N(L)$ to obtain:

<u>Theorem 6:</u> If V is reflection invariant, then

$$\alpha L^{-2} \geq E_1^N(L) - E_o^N(L) \geq \beta L^{-2} \tag{17}$$

<u>5. Lifshitz tails</u>

In this final section we consider random potentials V_ω. For simplicity, we suppose that V_ω is of the form

$$V_\omega(x) = \Sigma q_i(\omega) f(x-i) \tag{18}$$

with q_i independent, identically distributed with distribution P_o. Furthermore we suppose that the support supp P_o of P_o is bounded. Moreover, we assume that f is a reflection invariant function satisfying $|f(x)| = O(|x|^{-(d+\varepsilon)})$ near infinity.

We emphasize that many other random potentials can be handled by the method below (see [5], [7] and [4]).

We denote by $H_L^D(\omega)$ and $H_L^N(\omega)$ the operator $H(\omega) = -\Delta + V_\omega$ on C_L with Dirichlet and Neumann boundary conditions respectively.

By $\rho_L^D(\omega)$ (resp. $\rho_L^N(\lambda)$) we denote the number

$\#\{n | E_n(H_L^D(\omega)) \leq \lambda\}$ and

$\#\{n | E_n(H_L^N(\omega)) \leq \lambda\}$ respectively.

The thermodynamic limit

$$\rho(\lambda) = \lim_{L\to\infty} \frac{1}{L^d} \rho_L^D(\lambda) = \lim_{L\to\infty} \frac{1}{L^d} \rho_L^N(\lambda) \tag{19}$$

is called the integrated density of states for $H(\omega)$ or the density of states, for short.

The existence of the above limits and the independence of the boundary conditions is proven in [13] and [14]. We refer to the literature cited in [7] for other papers on the density of states mentioning only the pioneering work of Pastur (e.g. [15]). For the physical as well as the mathematical background on ρ we refer the reader to [7].

In a famous paper [16] Lifshitz argued on the basis of physical considerations that the density ρ should behave near $\lambda_o = \inf \sigma(H_\omega)$ like

$$\rho(\lambda) \sim e^{-(\lambda-\lambda_o)^{-d/2}} \quad \text{for } \lambda \downarrow \lambda_o. \tag{20}$$

This behavior is known as the Lifshitz behavior. Lifshitz behavior was proven for certain discretized Schrödinger operators on $\ell^2(Z^d)$ by Fukushima [17], Nagai [18] and Romerio-Wrezinski [19] and for Schrödinger operators on $L^2(\mathbb{R}^d)$ by Benderskii-Pastur [20], Friedburg-Luttinger [21], Luttinger [22], Nakao [23] and Pastur [24]. The latter authors prove that

$$\lim_{\lambda\downarrow\lambda_o} (\lambda-\lambda_o)^{d/2} \ln \rho(\lambda) = C \tag{21}$$

where C is a certain nonzero constant.

The potentials they considered where of the type

$$V_\omega(x) = \Sigma f(x - \xi_i(\omega)) \tag{22}$$

where $\xi_i(\omega)$ are Poisson distributed points in $\mathbb{R}^d$, $f \geq 0$ and $f(x) = O(|x|^{-(d+2)})$ near infinity (for details see [23]). Nakao and Pastur use the Donsker-Varadhan-maschinery [25] to obtain (21). Their methods seem tobe bounded to their special potentials.

Our theorem below shows a weaker assertion than (21). However, the potentials that can be treated cover also those as in (22) as well as many others.

We assume that V_ω is as in (18) with the assumption made there. In addition we assume that with $r := \inf \operatorname{supp} P_o$ we have $P_o([r,r+\varepsilon]) \geq C\,\varepsilon^m$ for some $C,m > 0$.

<u>Theorem 6:</u> (i) If $|f(x)| \leq O(|x|^{-(d+2)})$ near infinity, than

$$\lim_{\lambda\downarrow\lambda_o} \frac{\ln(-\ln \rho(\lambda))}{\ln(\lambda-\lambda_o)} = -\frac{d}{2} \tag{23}$$

(ii) If $C|x|^{-\alpha} \leq |f(x)| \leq C'|x|^{-\alpha}$

for some α with $d < \alpha < d+2$ then

$$\lim_{\lambda\downarrow\lambda_o} \frac{\ln(-\ln \rho(\lambda))}{\ln(\lambda-\lambda_o)} = -\frac{d}{\alpha-d} \tag{24}$$

As it stands the theorem is due to Kirsch-Simon [4]. However this paper relies on the methods developed in Kirsch-Martinelli [5] and Simon [6].

To conclude this paper we sketch the proof of one half of (23).

The <u>first step</u> is a simple observation. If $\tilde{V}_\omega(x) \leq V_\omega(x)$ then $\tilde{\rho}(\lambda) \geq \rho(\lambda)$ if $\tilde{\rho}$ denotes the density of states corresponding to $\tilde{H} = -\Delta + \tilde{V}_\omega$. So for an upper estimate we may suppose that $\operatorname{supp} f \subset C_1$, so that different parts of the sum in (18) do not overlap.

The <u>second step</u> uses Neumann-Dirichlet-bracketing to obtain

$$\rho(\lambda) \leq \frac{1}{L^d} E(\rho_L^N(\lambda)) \tag{25}$$

where E denotes the expectation (integral) with respect to the probability measure $P = \prod_{i\in Z^d} P_o$.

The inequality (25) is valid for any choice of the side length L. In particular we may choose $L = L(\lambda)$ depending on λ. We will make this choice in the final part of the proof keeping L arbitrary up to this step.

Set $r = \inf \operatorname{supp} P_o$ ("the smallest possible q_i") and $V_o(x) = \sum r\, f(x-i)$ (26). V_o is a refelction invariant periodic potential. We denote by $\rho_{oL}^N(\lambda)$ the density of states for $H_o = -\Delta + V_o$. It is known that $\lambda_o = \inf \sigma(H(\omega)) = \inf \sigma(H_o)$. By adding a constant we can assume that $\lambda_o = 0$.

Now we estimate the right hand side of (25) further:

$$E(\rho_L^N(\omega) \leq \rho_{oL}^N(\lambda) \quad P(E_o(H_L^N(\omega)) < \lambda) \tag{27}$$

Since $\frac{1}{L^d} \rho_{oL}^N(\lambda) \leq C_1 \lambda^k$ for some k we are left with the problem to estimate $P(E_o(H_L^N(\omega) < \lambda)$.

The idea behind the <u>third step</u> is that $E_o(H_L^N(\omega))$, the ground state of $H_L^N(\omega)$, can be small only if the potential V_ω inside C_L is small and hence the q_i are small for $i \in C_L$. To make this idea precise we employ Temple's inequality.

<u>Temple's inequality:</u> If H is a selfadjoint operator bounded below such that $E_o(H) < E_1(H)$ and if ψ is a normalized vector in the domain of H such that $<\psi, H\psi> < \mu \leq E_1(H)$, then:

$$E_o(H) \geq <\psi, H\psi> - \frac{<\psi, H^2\psi> - <\psi, H\psi>^2}{\mu - <\psi, H\psi>}$$

A proof of Temple's inequality can be found in [11].

We are going to apply Temple's inequality to $H_L^N(\omega)$. For μ we take $E_1(H_{o\,L}^{\,N}) \leq E_1(H_L^N(\omega))$. (see (26))

Theorem 5 supplies us with an estimate of $E_1(H_{oL}^N)$ from below ($\lambda_o = 0!$):

$$E_1(H_{oL}^N) \geq \beta L^{-2} \tag{28}$$

For ψ we take the ground state $\psi_o^{(L)}$ of $H_L^N(\omega)$. If ψ_o denotes the ground state $\psi_o^{(1)}$ extended by periodicity to all of $\mathbb{R}^d$ we have that

$$\psi_o^{(L)} = \frac{1}{L^{d/2}} \psi_o \tag{29}$$

as was explained in section 4. Since $H_o\,\psi_o^{(L)} = 0$ we compute easily:

$$<\psi_o^{(L)}, H_L^N(\omega)\,\psi_o^{(L)}> = \frac{1}{L^d} \sum_{i\in C_L} q_i(\omega) \int f(x)\,\psi_o^2(x)dx \tag{30}$$

and

$$<\psi_o^{(L)}, (H_L^N(\omega))^2\psi_o^{(L)}> = \frac{1}{L^d} \sum_{i\in C_L} q_i^2(\omega) \int f^2(x)\psi_o^2(x)dx \tag{31}$$

However, in order to apply Temple's inequality we need $<\psi_o^{(L)}, H_L^N(\omega)\,\psi_o^{(L)}> < E_1(H_{oL}^N)$ which is not the case in general. To remedy this situation we make the potential smaller (as we already did in step 1) by setting:

$$\tilde{q}_i(\omega) = \min(q_i(\omega), \gamma\frac{\beta}{f_1} L^{-2}) \tag{32}$$

with $f_1 = \int f(x)\,\psi_o^2(x)dx$ for a $\gamma < 1$.

Defining $\tilde{H}_L^N(\omega)$ with $\tilde{q}_i$ instead of q_i we obtain from (30):

$$<\psi_o^{(L)}, \tilde{H}_L^N(\omega)\,\psi_o^{(L)}> \leq \gamma\beta L^{-2} < \beta L^{-2} \leq E_1(H_{oL}^N) \tag{33}$$

So that we can apply Temple's inequality to $\tilde{H}_L^N(\omega)$.

Observing that with $f_2 = \int f^2(x)\,\psi_o^2(x)dx$ we have

$$<\psi_o^{(L)}, (\tilde{H}_L^N(\omega))^2\,\psi_o^{(L)}> \leq (\gamma\frac{\beta f_2}{f_1} L^{-2})\cdot \frac{1}{L^d} \sum_{i\in C_L} \tilde{q}_i(\omega) \tag{34}$$

we obtain:

$$E_o(H_L^N(\omega)) \geq E_o(\tilde{H}_L^N(\omega)) \geq (f_1 - \frac{\gamma}{1-\gamma}\frac{f_2}{f_1})\,\frac{1}{L^d} \sum_{i\in C_L} \tilde{q}_i(\omega) > (\tfrac{1}{2}f_1)\,\frac{1}{L^d} \sum_{i\in C_L} \tilde{q}_i(\omega) \tag{35}$$

if $\gamma > 0$ is taken small enough.

From (35) we get as the net result of the third step:

<u>Proposition 1:</u> $P(E_o(H_L^N(\omega)) < \lambda) \leq P(\#\{i\in C_L \mid \tilde{q}_i < \frac{5}{f_1}\lambda\} \geq \frac{1}{2} L^d)$ (36)

To prove the Proposition suppose that less than half of the points $i\in C_L$ have

$\tilde{q}_i < \frac{5}{f_1}\lambda$ then $\tilde{q}_i \geq \frac{5}{f_1}\lambda$ for at least $\frac{1}{2} L^d$ points i, so $(\frac{1}{2}f_1) \frac{1}{L^d} \sum_{i\in C_L} \tilde{q}_i > \lambda$. Thus

$E_o(H_L^N(\omega) > \lambda)$ by (35).

The <u>fourth</u> step consists in estimating the probability on the right hand side of (36). To get the Lifshitz behavior we want this probability to be exponentially small. This is true only if $E(\tilde{q}_i) > \frac{5}{f_1}\lambda$ since otherwise the probability in question will not go to zero. But $E(\tilde{q}_i) \leq \frac{\beta}{f_1} L^{-2}$ by definition (32). This suggests to take $L = L(\lambda)$ proportional to $\lambda^{-1/2}$. Setting $L = [\alpha\lambda^{-1/2}]$ where [-] denotes the integer part (L has to be an integer) we see that $\frac{\beta}{f_1} L^{-2} > \frac{5}{f_1}\lambda$ if α is taken sufficiently small. So, $\tilde{q}_i < \frac{5}{f_1}\lambda$ if and only if $q_i < \frac{5}{f_1}\lambda$.

Hence, for L sufficiently large, i.e. λ small we have by a standard probabilistic estimate:

$$P(\#\{i \in C_L \mid \tilde{q}_i < \frac{5}{f_1}\lambda\} \geq 1/2L^d)$$

$$= P(\#\{i \in C_L \mid q_i < \frac{5}{f_1}\lambda\} \geq 1/2L^d)$$

$$\leq e^{-CL^d} \leq e^{-C'\lambda^{-d/2}}$$

(see [4] for details).

Collecting our estimates we obtain

$$\rho(\lambda) \leq C_1\lambda^k \quad e^{-e'\lambda^{-d/2}}$$

as desired.

This ends the proof of one part of (i). We refer the reader to [4] for a complete proof.

Acknowledgement

The auther is grateful to the organizing comitee for the kind invitation and financial support. Financial support by the DFG is also acknowledged.

References:

[1] E. Harrell: Double wells; Comm. Math. Phys. 75, 239-261 (1980)

[2] W. Kirsch, B. Simon: Universal lower bounds on eigenvalue splittings for one dimensional Schrödinger operators; Comm. Math. Phys. 97, 453-460 (1985)

[3] W. Kirsch, B. Simon: Comparison theorems for the gap of Schrödinger operators; to appear

[4] W. Kirsch, B. Simon: Lifshitz tails for periodic plus random potentials; Preprint

[5] W. Kirsch, F. Martinelli: Large deviations and Lifshitz singularity of the integrated density of states of random Hamiltonians; Comm. Math. Phys. 89, 27-40 (1983)

[6] B. Simon: Lifshitz tails for the Anderson model, Journ. Stat. Phys.

[7] W. Kirsch: Random Schrödinger operators and the density of states; in: Lect. Notes Math. 1109, 68-102 (1985)

[8] S. Albeverio, R. Høegh-Krohn, L. Streit: Energy forms, Hamiltonians, and distorted Brownian paths, J. Math. Phys. 18, 907-917 (1977)

[9] E.B. Davies, B. Simon: Ultracontractivity and the heat kernel for Schrödinger operators and Dirichlet Laplacians, Preprint

[10] M. Fukushima: Dirichlet forms and Markov processes North-Holland, 1980

[11] M. Reed, B. Simon: Method of modern mathematical physics, Vol. IV, Academic Press, 1978

[12] M.S.P. Eastham: The spectral theory of periodic differential operators, Scottish Academic Press, 1973

[13] W. Kirsch, F. Martinelli: On the density of states of Schrödinger operators with a random potential, J. Phys. A15, 2139-56 (1982)

[14] J. Droese, W. Kirsch: The effect of boundary conditions on the density of states for random Schrödinger operators; Preprint

[15] L.A. Pastur: Spectra of random selfadjoint operators, Russian Math. Surv. 28, 1-67 (1973)

[16] I.M. Lifshitz: Energy spectrum structure and quantum states of disordered condensed systems, Sov. Phys. Usp. 7, 549 (1965)

[17] M. Fukushima: On the spectral distribution of a disordered system and the range of a random walk, Osaka J. Math. 11, 73-85 (1974)

[18] H. Nagai: On an exponential character of the spectral distribution function of a random difference operator; Osaka J. Math. 14, 111-116 (1977)

[19] M. Romerio, W. Wreszinski: On the Lifshitz singularity and the tailing in the density of states for random lattice systems; J. Stat. Phys. 21, 169-179 (1979)

[20] M. Benderskii, L.A. Pastur: On the spectrum of the one-dimensional Schrödinger equation with random potential, Mat. Sb. 82, 245-256 (1970)

[21] R. Friedberg, J. Luttinger: Density of electronic energy levels in disordered systems; Phys. Rev. B12, 4460 (1975)

[22] J. Luttinger: New variatonal method with applications to disordered systems; Phys. Rev. Lett. 37, 609 (1976)

[23] S. Nakao: On the spectral distribution of the Schrödinger operator with random potential; Japan J. Math. 3, 111-139 (1977)

[24] L.A. Pastur: Behavior of some Wiener integrals as $t \to \infty$ and the density of states of Schrödinger equation with a random potential, Teo. Mat. Fiz. 32, 88-95 (1977)

[25] M. Donsker, S.R.S. Varadhan: Asymptotics for the Wiener sausage, Comm. Pure Appl. Math. 28, 525-565 (1975)

On Identification

for Distributed Parameter Systems*

Timo Koski
Åbo Akademi
Department of Mathematics
SF-20500 Åbo 50
Finland

Wilfried Loges
Ruhr-Universität Bochum
Mathematisches Institut
4630 Bochum
West Germany

Introduction.

In this note we consider the problem of parameter identification for a system given as solution of a linear stochastic differential equation (SDE) in a Hilbert space. By means of the minimum contrast method we derive an (explicit) estimator (MCE) for a real parameter occuring in the infinitesimal generator of the underlying SDE. The MCE is shown to satisfy certain desirable asymptotic properties, i.e. strict consistency and asymptotic normality. Finally we apply this method to estimate the thermal conductivity parameter in a stochastic heat flow problem.
The minimum contrast method (MCM) was originally designed by Pfanzagl [1968] for observations of independent, identically distributed random variables. Gänßler [1972] extended this theory to discrete time Markov processes.

The basic ideas of Pfanzagl have been also adapted for continuous-time real valued stochastic processes by Lánska [1979] and Mühleis/Wittwer [1980].

The MCM includes maximum likelihood estimation (MLE) as a particular case and can for this and other reasons be regarded as a contribution to a unified (asymptotic) theory of statistical estimation. A statistical estimation method beyond the maximum likelihood method is relevant and necessary for the statistical models which are considered here. In fact, in many cases where the system is generated by an unbounded operator the measures induced by the system processes w.r.t. different parameter values are orthogonal. In consequence, the maximum likelihood estimator has been successfully found only for some, as such interesting,

*Talk presented by W. Loges

special cases of Hilbert space-valued stochastic differential equations (cf. Loges [1984], Koski/Loges [1984]). Contrary to this situation it will turn out that our MCE works for an extensive class of the portrayed models.

2. Construction of a minimum constrast estimator and its consistency

Before defining the family of contrast functions and our minimum contrast estimator we give a precise description of the underlying dynamical system. We study infinite-dimensional stochastic differential equations (SDE) of the type

$$(2.1)\qquad dx(t) = \theta Ax(t)dt + dw(t),$$

where θ belongs to a parameter set Θ contained in the positive reals. A is the infinitesimal generator of a strongly continuous semigroup $(T_t)_{t\geq 0}$ acting on a real separable Hilbert space H with scalar product $\langle\cdot,\cdot\rangle$ and norm $\|\cdot\|$; $(w(t))_{t\geq 0}$ is an H-valued Wiener process; i.e. $(w(t))_{t\geq 0}$ is a stochastic process defined on a probability space $(\Omega,\mathcal{A},P)$ with independent and stationary increments such that the associated (incremental) covariance operator W is nuclear and w(t) has zero mean for all $t \geq 0$. We use the theory of infinite dimensional SDE's as (e.g.) developed in Curtain/Pritchard [1978]. It should also be observed that, in fact, only the mild solution of (2.1) is needed in the sequel (in the sense discussed in Curtain/Pritchard [1978] ch. 5.3).

To deal with (at least asymptotically) ergodic processes we only study exponentially stable semigroups $(T_t)_{t\geq 0}$; i.e. we assume that there exist constants $M \geq 1$ and $\mu > 0$ s.t. ($\mathcal{L}(H)$ is the operator norm)

$$(2.2)\qquad \|T_t\|_{\mathcal{L}(H)} \leq Me^{-\mu t} \text{ for all } t \geq 0.$$

For a discussion of stability of a semigroup determined by the generator, see Slemrod [1976]. In this situation we can define (for each $\theta > 0$) the (nuclear) operator

$$(2.3)\qquad S_\infty(\theta) = \int_0^\infty T_{\theta t}WT^*_{\theta t}dt = \frac{1}{\theta}\int_0^\infty T_tWT^*_t dt = \frac{1}{\theta}\cdot S_\infty.$$

It is easy to show (cf. Lemmas 4 and 5 Loges [1984]) that the Gaussian measure with mean zero and covariance operator $S_\infty(\theta)$ is the unique invariant distribution of the system defined by $dx(t) = \theta Ax(t)dt + dw(t)$. Moreover, the process $x(t) \equiv x(t;\theta)$ starting with a random element having this invariant distribution (and being independent of $(w(t))_{t\geq 0}$) is ergodic. We have of course $E(\|x(t)\|^2) = \mathrm{tr} S_\infty(\theta) = \frac{1}{\theta}\mathrm{tr} S_\infty$, where tr designates the trace of a (nuclear) operator.

Thus we define our contrast function as

$$(2.4) \qquad L_T(\theta) := \int_0^T (\|x(t)\|^2 - \frac{1}{\theta}\mathrm{tr}\, S_\infty)^2 dt.$$

That this function indeed possesses the properties of a meaningfully defined family (as function of θ) of contrast functions (cf. Lånska [1979], Mühleis/Witwer [1980]) can easily be seen by the fact that for an arbitrary square integrable random variable Y $E((Y - c)^2)$ is minimal for $c = E(Y)$; i.e. $(Y - E(Y))^2$ is a contrast function, as has been pointed out already by Pfanzagl [1968].

To the end of making statistical inference for a time-continuous process we integrate this functional over $[0,T]$, the period of observation, ending up with (2.4). One should remark that in the case of a one-dimensional space H (2.4) coincides with a contrast function given in the example of Lånska [1979].

Minimizing $\frac{1}{T} \cdot L_T(\theta)$ w.r.t. θ we obtain by elementary calculations the estimator

$$(2.5) \qquad \hat{\theta}_T = (\frac{1}{T}\int_0^T \|x(t)\|^2 dt)^{-1}\mathrm{tr} S_\infty,$$

where S_∞ is supposed to be known by the very construction of the statistical model.

Theorem 1.

The minimum contrast estimator $\hat{\theta}_T = (\frac{1}{T}\int_0^T \|x(t)\|^2 dt)^{-1}\mathrm{tr} S_\infty$ is strictly consistent; i.e. $\hat{\theta}_T \to \theta$ P-a.s. as $T \to \infty$.

Proof: As already mentioned, $(x(t))_{t\geq 0}$ is an ergodic process in H and hence also $(\|x(t)\|^2)_{t\geq 0}$ in $\mathbb{R}$. The ergodic theorem ensures $\lim_{T\to\infty} \frac{1}{T}\int_0^T \|x(t)\|^2 dt = E(\|x(0)\|^2) = \frac{1}{\theta}\mathrm{tr} S_\infty$.

3. Asymptotic normality of the estimator

In this section we assume that the generator A is selfadjoint which does not present a severe restriction in view of most common applications. Then we can state the main result.

Theorem 2.

The estimator $\hat{\theta}_T$ is asymtotically normal, i.e. $\sqrt{T}(\hat{\theta}_T - \theta)$ converges weakly to a centered Gaussian distribution with variance $\theta \mathrm{tr}(WA^{-1}S_\infty A^{-1})(\mathrm{tr}S_\infty)^{-2}$.

Proof: Evidently,

$$\sqrt{T}(\hat{\theta}_T - \theta) = \left(\frac{1}{T}\int_0^T \|x(t)\|^2 dt\right)^{-1} \sqrt{T}\left(\mathrm{tr}\, S_\infty - \theta \cdot \frac{1}{T}\int_0^T \|x(t)\|^2 dt\right). \tag{3.1}$$

In order to derive the desired asymptotic result we have to find a suitable representation of the numerator. The simple observation effecting this is that $\sqrt{T}(\mathrm{tr}\, S_\infty - \theta \cdot \frac{1}{T}\int_0^T \|x(t)\|^2 dt)$ can be transformed (up to terms of order $T^{-\frac{1}{2}}$) into a stochastic integral for which an appropriate central limit theorem is true.

Indeed, if $g(x) := \langle -A^{-1}x, x\rangle$ for $x \in H$ (it follows by the stability requirement that A^{-1} is a bounded operator in H), then Ito's lemma, in the form proved in Curtain [1981] (see lemma 3.3 on p. 146 and also lemma 3.2) entails

$$\begin{aligned} dg(x(t)) &= -2\theta\langle A(A^{-1}x(t)), x(t)\rangle dt + \sum_{i=1}^{\infty} \lambda_i \langle -A^{-1}e_i, e_i\rangle dt \\ &\quad -2\langle A^{-1}x(t), dw(t)\rangle \\ &= -2\theta\langle x(t), x(t)\rangle dt + \sum_{i=1}^{\infty} \lambda_i \langle -A^{-1}e_i, e_i\rangle dt - 2\langle A^{-1}x(t), dw(t)\rangle \end{aligned} \tag{3.2}$$

where $\{e_i\}$ is an orthonormal basis consisting of the eigenvectors of the covariance operator W and $\{\lambda_i\}$ is the corresponding sequence of (positive) eigenvalues.

Hence we may compute the integral of the squared norm as

$$\begin{aligned} \int_0^T \|x(t)\|^2 dt &= \frac{1}{2\theta}\left(\langle A^{-1}x(T), x(T)\rangle - \langle A^{-1}x(0), x(0)\rangle\right) \\ &\quad + \frac{T}{\theta}\sum_{i=1}^{\infty} \frac{\lambda_i}{2}\langle -A^{-1}e_i, e_i\rangle \\ &\quad - \frac{1}{\theta}\int_0^T \langle A^{-1}x(t), dw(t)\rangle. \end{aligned} \tag{3.3}$$

Next we verify that

$$\mathrm{tr}\, S_\infty = \sum_{i=1}^{\infty} \frac{\lambda_i}{2} \langle -A^{-1}e_i, e_i\rangle . \tag{3.4}$$

In this regard we note first that (3.4) is true, if A has a pure point spectrum i.e. $A\phi_i = -k_i\phi_i$ for an orthonormal basis $\{\phi_i\}$, since then $T_t\phi_i = e^{-k_i t}\phi_i$ and (3.4) follows by straightforward computations using the basic representational formula (2.3) above.

In order to establish (3.4) for an arbitrary selfadjoint generator A of a stable semigroup we select an approximating sequence $\{A_\varepsilon\}$, $\varepsilon > 0$, defined by $A_\varepsilon := A + B_\varepsilon$, where B_ε is a bounded selfadjoint operator with

$$(3.5) \qquad \|B_\varepsilon\|_{\mathcal{L}(H)} \leq \varepsilon$$

such that A_ε has pure point spectrum. That this can be done for any ε is guaranteed by the wellknown Weyl-von Neumann theorem (cf. Kato [1976], Thm. 2.1, p. 525, in fact B_ε can be taken as a Hilbert-Schmidt-operator). If $(T_t^\varepsilon)_{t\geq 0}$ denotes the semigroup generated by A_ε, it becomes a stable semigroup, too, by choosing ε small enough or for any $\varepsilon < \mu\cdot M^{-1}$, since then according to e.g. Curtain/Pritchard [1978] Thm. 2.3.1.

$$(3.6) \qquad \|T_t\|_{\mathcal{L}(H)} \leq Me^{-(\mu - M\varepsilon)t}$$

(and hence $S_\infty^\varepsilon = \int_0^\infty T_t^\varepsilon W T_t^\varepsilon dt$). Moreover, the perturbed semigroup is known to satisfy the equation

$$(3.7) \qquad T_t^\varepsilon x = T_t x + \int_0^t T_{t-s} B_\varepsilon T_s^\varepsilon x ds$$

for all $x \in H$.

Writing now $S_t^\varepsilon := T_t^\varepsilon - T_t$ it holds obviously that

$$(3.8) \qquad S_\infty^\varepsilon - S_\infty = \int_0^\infty S_t^\varepsilon W T_t dt + \int_0^\infty T_t W S_t^\varepsilon dt + \int_0^\infty S_t^\varepsilon W S_t^\varepsilon dt$$

Hence (here $\|\cdot\|_1$ denotes "the trace-class"-norm)

$$(3.9) \qquad |\mathrm{tr}\, S_\infty^\varepsilon - \mathrm{tr}\, S_\infty| = |\mathrm{tr}(S_\infty^\varepsilon - S_\infty)|$$

$$\leq \|S_\infty^\varepsilon - S_\infty\|_1 \leq \int_0^\infty \|S_t^\varepsilon W T_t\|_1 dt + \int_0^\infty \|T_t W S_t^\varepsilon\|_1 dt$$

$$+ \int_0^\infty \|S_t^\varepsilon W S_t^\varepsilon\|_1 dt$$

$$\leq 2\|W\|_1 \int_0^\infty \|S_t^\varepsilon\|_{\mathcal{L}(H)} \|T_t\|_{\mathcal{L}(H)} dt + \|W\|_1 \int_0^\infty \|S_t^\varepsilon\|^2_{\mathcal{L}(H)} dt$$

by some standard properties of $\|\cdot\|_1$. Now it follows with the aid of (3.7), (3.6), (3.5), (2.2) and by some elementary integrations that

$$\|S_t^\varepsilon\|_{\mathcal{L}(H)} \leq \int_0^t \|T_{t-s}\|_{\mathcal{L}(H)} \|B_\varepsilon\|_{\mathcal{L}(H)} \|T_s\|_{\mathcal{L}(H)} ds$$

$$\leq Me^{-\mu \cdot t}(e^{M\varepsilon t}-1) \leq M .$$

Due to (3.9) and the definition of S_t^ε we get consequently

$$\text{(3.10)} \qquad |\mathrm{tr}\ S_\infty^\varepsilon - \mathrm{tr}\ S_\infty| \leq 3M\|W\|_1 \int_0^\infty \|S_t^\varepsilon\|_{\mathcal{L}(H)} dt$$

$$\leq \left(\frac{1}{\mu - M\varepsilon} - \frac{1}{\mu}\right) 3M^2 \|W\|_1 \leq \frac{6M^3}{\mu^2} \|W\|_1 \cdot \varepsilon$$

in view of the stability bound and since we can also select $\varepsilon < \frac{1}{2}M^{-1}\mu$ without restriction. In other words, since A_ε has a pure point spectrum, and thus fulfills (3.4),

$$\text{(3.11)} \qquad \mathrm{tr}\ S_\infty = \lim_{\varepsilon \to 0} \mathrm{tr}\ S_\infty^\varepsilon = \lim_{\varepsilon \to 0} \frac{1}{2} \sum_{i=1}^\infty \lambda_i \langle -A_\varepsilon^{-1} e_i, e_i \rangle$$

$$= \frac{1}{2} \sum_{i=1}^\infty \lambda_i \langle -A^{-1} e_i, e_i \rangle$$

as was to be shown. The last equality in (3.11) is explained by the fact that

$$\|A_\varepsilon^{-1} - A^{-1}\|_{\mathcal{L}(H)} \leq \int_0^\infty \|T_t^\varepsilon - T_t\|_{\mathcal{L}(H)} dt$$

an inequality which is derived by an application of the resolvent identity $(\lambda I - A)^{-1} x = \int_0^\infty e^{-\lambda t} T_t x dt$ where $x \in H$, λ lies in the resolvent set of A and I denotes the identity operator in H. Evidently we obtain now

$$\|A_\varepsilon^{-1} - A^{-1}\|_{\mathcal{L}(H)} \leq 2M^2 \mu^{-2} \cdot \varepsilon$$

by an estimation procedure already used to derive the bound (3.10) above.

Hence (3.3) and (3.4) yield that

$$(3.12)\quad \sqrt{T}(\operatorname{tr} S_\infty - \theta T^{-1}\int_0^T \|x(t)\|^2 dt) = (2T)^{-1/2}(\langle -A^{-1}x(T),x(T)\rangle - \langle -A^{-1}x(0),x(0)\rangle) + T^{-1/2}\int_0^T \langle A^{-1}x(t),dw(t)\rangle \quad .$$

Using a standard rule of representation of H-valued Wiener processes we write

$$z(T) := T^{-1/2}\int_0^T \langle A^{-1}x(t),dw(t)\rangle = T^{-1/2}\sum_1^\infty \lambda_i^{-1/2}\int_0^T \langle W^{1/2}A^{-1}x(t),e_i\rangle d\langle w(t),e_i\rangle \quad .$$

But then it holds evidently by lemma 3 of Loges [1984] that z(T) converges weakly to a Gaussian distribution with mean zero and variance equal to

$$\lim_{T\to\infty}\frac{1}{T}\int_0^T \|W^{1/2}A^{-1}x(t)\|^2 dt = E\|W^{1/2}A^{-1}x(0)\|^2 = \frac{1}{\theta}\operatorname{tr}(WA^{-1}S_\infty A^{-1})$$

Also, since the process x(T) is stationary, the first term in the r.h.s. of (3.12) is of order $T^{-1/2}$ in probability and as the denominator in (3.1) converges to $\frac{1}{\theta}\operatorname{tr} S_\infty$ by thm. 1 above, the desired result follows by a standard property of weak convergence.

Remark. The asymptotic properties of the estimator do not depend on the assumption that the initial value of the process has stationary distribution. Due to the stability of the semigroup $(T_t)_{t\geq 0}$ the influence of the starting variable decays exponentially in t. Thus the limit behaviour of $T^{-\beta}\int_0^T \|x(t)\|^2 dt, \beta = 1$ or $1/2$, will not be changed.

Example. Estimating the thermal conductivity in a stochastic heat flow. Although there exists a wide class of problems to which the minimum contrast estimation procedure above can be applied (since our only condition of restrictive nature on the generator A is the stability of its semigroup), we confine ourselves to the well known heat equation with stochastic disturbance.

For that reason we consider $H = L_2[0,1]$ endowed with the usual scalar product and A the (one-dimensional) Laplacian $\partial^2/\partial\xi^2$ with domain of definition

$$D(A) = \{h \in H \mid h_\xi, h_{\xi\xi} \in H;\ h_\xi(0) = h(1) = 0\} \quad .$$

The set of eigenvalues $(-k_i)_{i\in\mathbb{N}}$ is given by $k_i = (\pi i)^2$, i = 1,2,..., which obviously implies the stability of the corresponding semigroup. Moreover, the Neumann-boundary conditions make A selfadjoint. Thus we are able to consistently estimate the parameter θ, i.e. the thermal conductivity, in the stochastic heat equation

$$\partial_t x(t,\xi) = \theta \partial^2/\partial\xi^2 x(t,\xi)dt + dw(t,\xi)$$

with the aid of $\hat{\theta}_T$ in (2.5).

REFERENCES

Curtain R.F.: Markov processes generated by linear stochastic evolution euqations. Stochastics 5 (1 & 2), 1981, 135-165

Curtain R.F. and Pritchard A.: Infinite dimensional linear systems theory. Springer Verlag, Berlin, 1978

Gänssler P.: Note on Minimum Contrast Estimates for Markov processes. Metrika 19, 1972, pp. 115-130

Kato T.: Perturbation theory for linear operators. Springer Verlag, Berlin, 1976

Koski T. and Loges W.: Asymptotic statistical inference for a stochastic heat flow problem. Preprint Ruhr-Universität Bochum, 1984

Lânska V.: Minimum contrast estimation in diffusion processes. Journal of Applied Probability 16, 1979, 65-75

Loges W.: Girsanov's theorem in Hilbert space and an application to the statistics of Hilbert space-valued stochastic differential equations. Stochastic Processes and their Applications 17, 1984, 243-263

Mühleis W. and Witwer G.: Minimum-Kontrast-Schätzungen für stochastische Prozesse. Mathematische Operationsforschung und Statistik. Ser. A. Statistics 11, 1980, 219-227

Pfanzagl J.: On the measurability and consistency of minimum contrast estimates. Metrika 14, 1968, 249-272

Slemrod M.: Asymptotic behaviour of C_0-semigroups as determined by the spectrum of the generator. Indiana University Journal 25, 1976, 738-792

FOCK SPACE AND PROBABILITY THEORY

by P.A. Meyer

The first part of this paper is an exposition, meant to be understandable to probabilists with no knowledge of quantum mechanics, of some of the simplest and most attractive parts of the Hudson-Parthasarathy work on non commutative stochastic calculus. The second part is a discrete (Bernoulli) approximation to boson calculus, which is possibly (but not probably) new. It is, at least, elementary and amusing.

INTRODUCTION

1. In the classical (von Neumann) form of non-commutative probability theory, the space Ω is replaced by a complex Hilbert space H, the probability law P is replaced by a normalized vector ϕ, and random variables are replaced by <u>self-adjoint operators</u>, possibly unbounded. We need no technique concerning these operators : just as in classical probability we might <u>define</u> a r.v. X as the family of subsets $\{X\leq\lambda\}$ of Ω (increasing, right continuous, with intersection $\emptyset$ and union Ω), and the probability distribution μ_X is defined on $\mathbb{R}$ by $\mu_X(]-\infty,\lambda])=P\{X\leq\lambda\}$, here we may <u>define</u> X as a family $\{X\leq\lambda\}$ of closed subspaces of H , increasing, right continuous, with intersection $\{0\}$ and closed union H, and define the probability distribution of X under ϕ as $\mu_X(]-\infty,\lambda]) = <\phi, I_\lambda\phi>$ where I_λ is the projector on $\{X\leq\lambda\}$. More generally, given a r.v. X we may define a closed subspace $\{X\in A\}$ for every Borel subset A of $\mathbb{R}$, satisfying some obvious properties, and define a r.v. f(X) for every Borel function f on $\mathbb{R}$ ($\{f(X)\leq\lambda\}=\{X\in\{f\leq\lambda\}\}$).

More generally, a r.v. X with values in a measurable space $(E,\underline{\underline{E}})$ may be defined as a family of subspaces $\{X\in A\}$, $A\in\underline{\underline{E}}$. The main difference with ordinary probability theory is the fact that, given two r.v. with values in $\mathbb{R}$ X and Y, the pair (X,Y) may have no meaning as a r.v. with values in $\mathbb{R}^2$: it is a r.v. <u>if and only if</u> the projections $I_{\{X\leq\lambda\}}$ and $I_{\{Y\leq\mu\}}$ all <u>commute</u>. More generally - this is the key to the understanding of the sequel - a family of r.v. (X_t) whose associated projectors all commute may be considered as a stochastic process in the ordinary sense.

If X is a r.v., we may also define f(X) for f <u>complex</u> Borel, and this will be a $\mathbb{C}$-valued r.v.. This leads us to the most practical way of defining random variables : via the bounded operators $e^{itX}=U_t$, which turn out to be the most general <u>unitary group</u> (strongly continuous). Then the distribution μ_X is known through its Fourier transform, which is just $<\phi,U_t\phi>$. This is sufficient to understand the language below.

2. Fock space. We start with some complex Hilbert space H, which we may assume to be $L^2_{\mathbb{C}}(E,\lambda)$ for some positive measure λ (for us, it will be $L^2_{\mathbb{C}}(\mathbb{R}_+,dt)$). Then the Hilbert space $L^2_{\mathbb{C}}(E^n,\lambda^n)$ for the n-fold product measure can be described in purely algebraic terms as the Hilbert tensor product $H^{\otimes n}$. Also, we have obvious mappings $L^2_{\mathbb{C}}(E^n,\lambda^n)\times L^2_{\mathbb{C}}(E^m,\lambda^m) \to L^2_{\mathbb{C}}(E^{n+m},\lambda^{n+m})$ which can be described algebraically. There is no need to insist.

In practice, one works only with the subspaces of $L^2_{\mathbb{C}}(E^n,\lambda^n)$ consisting of symmetric or antisymmetric functions . These subspaces will be denoted by $H^{\circ n}$, $H^{\wedge n}$ respectively. The above mappings must be symmetrized or antisymmetrized to become respectively the symmetric product $H^{\circ n}\times H^{\circ m}\to H^{\circ n+m}$ and the exterior product $H^{\wedge n}\times H^{\wedge m}\to H^{\wedge n+m}$. These are associative, and there is a huge algebraic literature on symmetric and exterior algebra ! We need very little of it, except that we'll make a very convenient norm convention : we take the scalar product on $H^{\circ n}$ or $H^{\wedge n}$ to be n! times the scalar product induced by $L^2_{\mathbb{C}}(E^n,\lambda^n)$. This will spare us many clumsy constants.

We make the convention that for n=0, $H^{\circ 0}=H^{\wedge 0}=\mathbb{C}$, and extend the $\circ,\wedge$ product to n=0 in the obvious way .

The symmetric or boson Fock space, the antisymmetric or fermion Fock space, are then the complex Hilbert spaces $\oplus_n H^{\circ n}$, $\oplus_n H^{\wedge n}$ respectively. We simplify our notation by calling H_n the n-th summand in any one of them, and Φ the sum, whether it is boson or fermion. The vector $1\in H_0$ is the vacuum vector. In the whole paper, we shall use it to define our quantum probability law on Fock space.

There is a unique integer valued r.v. on Fock space such that $\{N=n\}=H_n$. This is called number of particles. In particular, N=0 a.s. in the vacuum state.

Given $h\in H$, we define an unbounded operator on Fock space, the creation operator a^+_h , as the closure of the linear operator given by the following recipe

$$a^+_h(x_1\circ\ldots\circ x_n) = h\circ x_1\circ\ldots\circ x_n \quad\text{(boson)} \qquad ; \quad a^+_h(x_1\wedge\ldots\wedge x_n) = h\wedge x_1\ldots\wedge x_n \quad\text{(fermion)}$$

its adjoint a^-_h is the annihilation operator, defined similarly by

$$\begin{aligned} a^-_h(y_1\circ\ldots\circ y_{n+1}) &= \langle h,y_1\rangle y_2\circ\ldots\circ y_{n+1} \\ &+ \langle h,y_2\rangle y_1\circ y_3\ldots\circ y_{n+1} \\ +\ldots &+ \langle h,y_{n+1}\rangle y_1\circ\ldots\circ y_n \end{aligned} \quad ; \quad \begin{aligned} a^-_h(y_1\wedge\ldots\wedge y_{n+1}) &= \langle h,y_1\rangle y_2\wedge\ldots\wedge y_{n+1} \\ &- \langle h,y_2\rangle y_1\wedge y_3\ldots\wedge y_{n+1} \\ +\ldots &+(-1)^n\langle h,y_{n+1}\rangle y_1\wedge\ldots\wedge y_n . \end{aligned}$$

In the symmetric case, a very convenient set of « test vectors » in Fock space is constituted of the so called coherent, or exponential, vectors

$$\text{for } h\in H , \qquad \mathcal{E}(h) = \Sigma_n \frac{1}{n!} h^{\circ n} \quad (\|h^{\circ n}\|^2 = n!\|h\|^{2n}_H , \text{ hence } \|\mathcal{E}(h)\|^2=e^{\|h\|^2})$$

More generally, $<\mathcal{E}(h),\mathcal{E}(k)> = e^{\langle h,k\rangle}$.

PROBABILISTIC INTERPRETATION OF FOCK SPACE (1).

The material in this section is quite classical, and we describe it rapidly. I suppose the main ideas are all due to Segal.

Let Ω be the set of all continuous mappings from $\mathbb{R}_+$ to $\mathbb{R}$, equal to 0 for $t=0$, let (B_t) be the coordinate process, let μ be Wiener measure, i.e. (B_t) under μ is standard brownian motion. Let C_n be the subset $\{s_1<\dots<s_n\}$ of $\mathbb{R}_+^n$. Every probabilist knows the Ito-Wiener multiple integrals

(1) $I_n(f) = \int_{C_n} f(s_1,\dots,s_n)dB_{s_1}\dots dB_{s_n}$ if $\|f\|^2=\int_{C_n} |f(s_1,\dots,s_n)|^2 ds_1\dots ds_n < \infty$.

and the following facts : 1) $\|I_n(f)\|^2=\|f\|^2$ 2) For $m\neq n$, $I_m(f)$ and $I_n(g)$ are orthogonal 3) Denoting by K_n the space of all n-fold stochastic integrals, and by K_0 the space of constants, $L^2(\Omega)=\oplus_n K_n$.

Now make the convention that for a <u>symmetric</u> L^2 function f on $\mathbb{R}_+^n$

$$J_n(f) = \int f(s_1,\dots,s_n)dB_{s_1}\dots dB_{s_n} = n!\int_{C_n} f(s_1,\dots,s_n)dB_{s_1}\dots dB_{s_n} = n!I_n(f)$$

Then $\|J_n(f)\|^2=n!\int|f(s_1,\dots,s_n)|^2ds_1\dots ds_n$. If we take the basic Hilbert space H to be $L^2_{\mathbb{C}}(\mathbb{R}_+,dt)$, this is exactly the norm our convention gave to f as an element of $H^{\circ n}$. Hence we have an isomorphism J_n from $H^{\circ n}$ to K_n , and <u>we may identify boson Fock space with</u> $L^2(\Omega)$.

Let us now describe some features of the structure of Wiener space.

a) The vacuum vector $1\in\Phi$ has been mapped into the constant $1\in L^2(\Omega)$.

b) Elements of the Wiener chaos K_n and K_m belong to every $L^p(\Omega)$, hence can be multiplied (as random variables in $L^2(\Omega)$), the product being in L^2. This multiplication isn't the same as the product $\circ$ in Fock space. To emphasize this, we call it <u>Wiener multiplication</u> in Fock space.

Given a (real) r.v. X on Ω, we may let X act on $L^2(\Omega)$ as a Wiener multiplication operator : then X is a r.v. in the quantum sense, and its distribution function under the quantum law 1 is exactly its distribution function under the classical law μ . In particular, for $h\in L^2_{\mathbb{R}}(\mathbb{R}_+,dt)$, let $\tilde{h}$ denote $\int h(s)dB_s\in K_1$, and Q_h denote the corresponding multiplication operator : for $h=I_{[0,t]}$ Q_h is simply multiplication by B_t . We have

$$< 1,e^{iQ_h}1 > = \exp(-\|h\|^2/2) .$$

c) Assume again h is real, and denote by $\underset{\sim}{h}$ the Cameron-Martin function $\int_0^{\cdot} h_s ds$. From Girsanov's theorem, we know that the Girsanov transform

$$G_t f(\omega) = f(\omega-t\underset{\sim}{h})\exp[t\tilde{h} - \tfrac{1}{2}t^2\|h\|^2]$$

preserves Wiener measure, and therefore

$$T_t f(\omega) = f(\omega-2t\underset{\sim}{h})\exp[t\tilde{h}-t^2\|h\|^2]$$

is an unitary group, which can be written as $T_t=e^{itP_h}$ for a s.a. operator P_h . This can be made explicit : denoting by ∇_h the derivation operator in Wiener space along the Cameron-Martin function $\underset{\sim}{h}$, we have

$$P_h = i(Q_h - 2\nabla_h) \ .$$

Then we have $< 1, e^{itP_h} 1 > = < 1, \exp(t\tilde{h} - t^2\|h\|^2) > = \exp(-t^2\|h\|^2/2)$, a gaussian characteristic function. Since these r.v. all commute, they have a joint gaussian distribution, which is exactly the same as that of the Q_h. In particular, for $h = I_{[0,t]}$, P_h may be denoted by $(\hat{B}_t)$, and these quantum r.v. form a second brownian motion « hidden » in classical Wiener space.

d) The random variable corresponding to the element h^{on} of Fock space, where h is a normed real element of H, is $\mathcal{H}_n \circ \tilde{h}$, where $\mathcal{H}_n$ is the n-th Hermite polynomial defined by the generating function

$$e^{tx - t^2/2} = \Sigma_n \frac{t^n}{n!} \mathcal{H}_n(x)$$

It follows that, for real $h \in H$, the exponential vector $\mathcal{E}(h)$ is read as the stochastic exponential $\exp[\tilde{h} - \frac{1}{2}\|h\|^2]$.

e) The Ornstein-Uhlenbeck semi-group on Wiener space is defined (for everywhere defined moderately large functions f) by

$$\Pi_t f(w) = \int f(we^{-t} + \omega(1-e^{-2t})^{1/2}) \mu(d\omega)$$

This is a Markov semi-group on Ω, symmetric with respect to Wiener measure, hence inducing positive contractions on $L^2(\Omega) = \Phi$. It can be shown that it acts on the n-th Wiener chaos simply by

$$\Pi_t f = e^{-nt} f \quad \text{if} \quad f \in K_n$$

Therefore $\Pi_t = e^{-tN}$. This can be extended to complex values of the parameter t such that $\Re t \geq 0$. Of special importance is $\Pi_{-i\pi/2} = \mathcal{F}$, which corresponds to the multiplier i^n on the n-th Wiener chaos K_n : this turns out to be an unitary operator on $L^2(\Omega)$, called the Fourier-Wiener transform. The following formulas can be proved (see Sem. Prob. XVI, p.115)

$$\nabla_h \Pi_t = e^{-t} \Pi_t \nabla_h \ , \quad \Pi_t Q_h = e^{-t} Q_h \Pi_t + (1-e^{-2t}) \Pi_t \nabla_h \quad (t \geq 0)$$

They can be extended by analytic continuation, and give

$$\mathcal{F} Q_h \mathcal{F}^{-1} = P_h \ , \quad \mathcal{F} P_h \mathcal{F}^{-1} = -Q_h \ .$$

f) Let us give now (without proof) the explicit form of the creation and annihilation operators on Wiener space, for h real

$$a_h^- = \nabla_h \ , \quad a_h^+ = Q_h - \nabla_h$$

from which we can deduce the expressions on Fock space of our two brownian motions :

$$Q_h = a_h^+ + a_h^- \ , \quad P_h = i(a_h^+ - a_h^-) \ .$$

We can now state some (formal) commutation relations, for h,k real

$$[a_h^-, a_k^+] = < h,k > I \ , \quad [P_h, Q_k] = -2i < h,k > I$$

(the first one may be extended to h,k complex).

One may see from all this the very strong analogy between the pair (Q,P) and the classical (position, momentum) pair.

COMMENT. There is one point which seems to me worthy of interest. As in any probability space, one should distinguish between elements which belong to the measurable structure and elements which depend on the choice of a probability law - here, the choice of the vacuum vector. All the operators $Q_h, P_h, N\ldots$ belong to the measurable structure, as does the identification as Hilbert spaces of Φ with $L^2(P)$. If we choose another quantum probability law, the « processes » (B_t) or $(\hat{B}_t)$ will no longer be brownian motions, but will still be stochastic processes in the classical sense.

PROBABILISTIC INTERPRETATION OF FOCK SPACE (2).

The existence of the multiple Wiener integrals, the computation of their norm, the orthogonality of the Wiener chaos, didn't depend at all on the fact that (B_t) was brownian motion : everything can be done in the same way replacing (B_t) by any square integrable martingale (X_t) such that $\langle X,X\rangle_t$ (the usual quadratic variation of a real valued martingale) is equal to t. Therefore, if $(\Omega,\mathcal{F},P)$ is some probability space carrying (X_t), we may realize an imbedding of the boson Fock space Φ into $L^2_{\mathbb{C}}(\Omega)$ which maps $1\in H_0$ into the constant 1, and $h\in H$ into the stochastic integral $\int h(s)dX_s$. Therefore, it should somehow be possible to « see » (X_t) on Fock space. The only case I am going to describe, however, is that which Hudson-Parthasarathy have studied : (X_t) is a compensated Poisson process with intensity $\lambda>0$, and jumps of size $\sqrt{\lambda}$. In this case, the imbedding we have just indicated is known to map Φ onto $L^2(\Omega)$, as in the brownian motion case.

First of all, we are going to define some generalized number operators. Given a real valued function $\alpha(t)$, we define T_α by its action on the n-th chaos

$$T_\alpha(\int f(s_1,\ldots,s_n)dB_{s_1}\ldots dB_{s_n}) = \int e^{i\alpha(s_1)+\ldots+i\alpha(s_n)}f(s_1,\ldots,s_n)dB_{s_1}\ldots dB_{s_n}$$

and in particular, using complex exponential vectors $\mathcal{E}(h) = \Sigma_n \frac{h^{\circ n}}{n!} = \exp[\tilde{h} - \frac{1}{2}\int_0^\infty h_s^2 ds]$ we have

$$T_\alpha \mathcal{E}(h) = \mathcal{E}(e^{i\alpha}h)$$

These operators are unitary, so we may define a self-adjoint operator N_α by $e^{itN_\alpha} = T_{t\alpha}$: N_α acts on the n-th chaos by

$$N_\alpha(\int f(s_1,\ldots,s_n)dB_{s_1}\ldots dB_{s_n}) = \int(\alpha(s_1)+\ldots+\alpha(s_n))f(s_1,\ldots,s_n)dB_{s_1}\ldots dB_{s_n}$$

(no problem of domain if α is bounded). If $\alpha=1$, we get the number operator N. If $\alpha=I_{[0,t]}$, we get an operator « number of particles in $[0,t]$» which we denote by N_t. Now, following Hudson-Parthasarathy, we set

$$C = N_\alpha + Q_h + (\int \alpha_s h_s^2 ds) I \qquad \alpha \text{ real bounded for instance} \quad h \in H \text{ real}$$

One must be careful when adding two self-adjoint operators which are unbounded, and do not commute. The way H-P use to bypass this difficulty consists in giving an explicit computation for the unitary group $K_t = e^{itC}$ as follows

$$K_t \mathcal{E}(f) = \exp[\int (e^{it\alpha_s} - 1) h_s f_s ds + (\cos(t\alpha_s) - 1) h_s^2 ds] \mathcal{E}(\Theta_t f)$$
$$\Theta_t f = e^{it\alpha} f + h(e^{it\alpha} - 1)$$

Checking that (K_t) is an unitary group with generator extending C would be cumbersome (and H-P have some tools to do it which haven't been presented here), so we leave it aside. As a by-product, we get the characteristic function of C :

$$< 1, K_t 1 > = \exp[\int (e^{it\alpha_s} - 1) h_s^2 ds]$$

which shows that the distribution of C is Poisson. One can show also that all the operators C corresponding to the same function h commute. Finally, we get that the random variables

(2) $$Z_t^h = N_t + \int_0^t h_s dQ_s$$

constitute a stochastic process in the classical sense (i.e. they all commute) which is a compensated Poisson process of intensity $h_s^2 ds$. This is quite paradoxical : we have added a gaussian process and a process (N_t) which is a.s. equal to 0, and got a compensated Poisson !

If h never vanishes, the process

(3) $$X_t^h = \int_0^t \frac{1}{h_s} dZ_s^h$$

turns out to be a martingale with $<X,X>_t = t$, and it can be shown that conversely, if we start from a martingale which has this distribution, map Fock space into its L^2 space by the procedure described at the beginning of this section, then the mapping carries Φ onto L^2 and the operators (3) are exactly the multiplication operators in L^2 corresponding to position at time t of the martingale. So what we get is really a probabilistic interpretation of Fock space.

In particular, taking $h = \sqrt{\lambda}$, we see that the processes

(4) $$X_t^\lambda = \frac{1}{\sqrt{\lambda}} N_t + Q_t$$

are compensated Poisson processes of intensity λ and jumps of size $1/\sqrt{\lambda}$. As expected, they converge to a brownian motion when $\lambda \to \infty$.

Let us now apply Fourier-Wiener transform : it can be easily seen that N_t remains fixed, while Q_t is changed into P_t. Therefore the conjugate process of (4) is the similar process constructed from P_t instead of Q_t .

We may interpret these results in a different way : we have one fixed

Hilbert space (Fock space Φ). Each time we map Fock space onto some space $L^2_{\mathbb{C}}(\Omega,P)$ in such a way that the vacuum vector is mapped into 1, we are giving to Φ some additional structure : essentially a conjugation, and an associative, commutative, not everywhere defined multiplication (hence we may also talk of bounded elements of Φ, positive elements of Φ, L^p norm in Φ, etc.). To emphasize the relative character of these notions, we may talk about << Wiener multiplication, λ-Poisson multiplication >> in Fock space, or << Wiener positivity, λ-Poisson positivity >>, etc. For instance, the <u>same</u> coherent vector in Fock space $\mathcal{E}(h)$, for real h, can be read as

$$\exp[h - \tfrac{1}{2}\|h\|^2] \quad \text{(Wiener exponential)}$$

or as

$$\exp(-\textstyle\int f_s h_s ds)\prod_{s\in S} (1+ \tfrac{f}{h}(s)) \quad \text{(h-Poisson products (3), S is the set of jumps of } X_\bullet \text{)}$$

As at the end of section 1, all these << processes >> exist on Fock space as elements of the <u>measurable</u> structure, and their laws may be studied whenever the state is the vacuum state or not.

The vague idea one may get from the preceding sections is that somehow, to go from Wiener to Poisson, we may (if we wish to) remain on Wiener space, with the random variables (X_t) we all know, but having them act on L^2 by a <u>different multiplication</u> . Given the expansion of a function along the chaos

$$f = f_0 + \int f_1(s)dX_s + \int_{s_1<s_2} f_2(s_1,s_2)dX_{s_1}dX_{s_2} + \dots$$

the computation of the product of two such expansions follows from the two elementary rules

(5) $$dX_s dX_t = dX_t dX_s \text{ for } s\neq t \quad , \quad dX_s^2 = ds + \frac{1}{\sqrt{\lambda}}\, dX_s$$

($\lambda=+\infty$ corresponds to Wiener multiplication, λ finite to Poisson).
For instance,

$$(\textstyle\int f(s)dX_s)(\int g(t)dX_t) = \int_{s<t} f(s)g(t)dX_s dX_t + \int_{s>t} \dots + \int_{s=t} \dots$$

$$= \int_{s<t} [(f(s)g(t)+ g(s)f(t)]dX_s dX_t \quad \text{(first rule)}$$

$$+ \quad \frac{1}{\sqrt{\lambda}} \int f(s)g(s)dX_s \;+ \int f(s)g(s)ds \quad \text{(second rule).}$$

So one may wonder whether there are <u>other rules for multiplication of stochastic integrals</u>, associative, but not necessarily commutative. It came as a surprise to me that the anticommutative Wiener rule

(6) $$dX_s dX_t + dX_t dX_s = 0 \quad \text{for } s\neq t \quad , \quad dX_s^2 = dt$$

defines on ordinary Wiener space a kind of continuous Clifford algebra, and that ordinary brownian motion (X_t), acting by this Clifford multiplication, is just what physicists call fermion brownian motion. Thus antisymmetric Fock space too has some connections with probability theory.

If we perform in this Clifford algebra the same computation as above, we get for the product

$$(\int f(s)dX_s).(\int g(t)dX_t) = \int_{s<t} [f(s)g(t)-g(s)f(t)]dX_s dX_t + \int f(s)g(s)ds$$

One may wonder whether Poisson-like non commutative processes can be constructed from fermion brownian motion. The rule $dX_t^2 = dt + dX_t/\sqrt{\lambda}$, however, isn't compatible with associativity and anticommutativity. Possibly slightly more complicated anticommutation rules are compatible (anyhow, we are led to this question by mere curiosity, and it is probably unimportant !).

TOY FOCK SPACE

The results in this section are the result of discussions on martingale representation with Jean-Lin Journé, mostly at the Café, rue des Ursulines. See also the references at the end of the paper.

Let M be some integer, and Ω be $\{0,1\}^M$, with its coordinate mappings ν_k , and the probability law P under which the ν_k are independent and $P\{\nu_k=1\}=p$. As usual with Bernoulli r.v.'s, we set $q=1-p$, and we normalize

$$X_k = \frac{\nu_k - p}{\sqrt{pq}}$$

Given a subset A of $\{1,\ldots,M\}$, with number of elements $|A|$, we set $X_A = \prod_{i\in A} X_i$. It is well known that these r.v.'s constitute an o.n. basis of the finite dimensional Hilbert space $H=L^2(\Omega)$. The subspace of H generated by all X_A with $|A|=n$ is denoted by H_n and called the n-th Bernoulli chaos. If $p=1/2$, P is Haar measure on the group Ω (the group operation being addition mod.2), the r.v. $(-1)^{|A|}X_A$ being all the characters of Ω .

Let us now define on H symmetric and antisymmetric creation and annihilation operators : in the antisymmetric case, we get the correct rules for finite systems of fermions, but in the symmetric case, we do not get boson Fock space, which is <u>never</u> finite dimensional : we get some kind of toy Fock space, without the unpleasant features of Weyl algebra, but really quite close to the true one. We are quite sure this isn't new[1]: it should at least be known to non-standardists !

toy bosons		fermions
$a_k^- X_A = 0$ if $k\notin A$, $X_{A\setminus\{k\}}$ if $k\in A$	;	$b_k^- X_A = 0$ if $k\notin A$, $(-1)^{n(k,A)} X_{A\setminus\{k\}}$ if $k\in A$
$a_k^+ X_A = X_{A\cup\{k\}}$ if $k\notin A$, 0 if $k\in A$	;	$b_k^+ X_A = (-1)^{n(k,A)}$ if $k\notin A$, 0 if $k\in A$

Here $n(k,A)$ is the number of elements of A strictly smaller than k .

1. See indeed the additional references.

We also introduce the <u>occupation number</u> at k

$$N_k = a_k^+ a_k^- = b_k^+ b_k^- \quad : \quad N_k X_A = X_A \text{ if } k\in A, \ 0 \text{ if } k\notin A$$

Note that $N_k^2 = N_k$, and that $\Sigma_k N_k$ is the usual number operator, $NX_A = |A| X_A$. With this notation, we can get the commutation rule for toy bosons : looking explicitly at the action on the basis elements X_A, we find that

$$(8) \qquad [a_k^-, a_k^+] = I - 2N_k$$

which mimics the Heisenberg c.r. in the sense that, if M is large and we remain in chaos of bounded order, then it is very likely that one given state k will be inoccupied and the commutator is I . If we define a (unitary) Fourier transform $\mathcal{F}$ by $\mathcal{F}x = i^n x$ for $x\in H_n$, we have as in true Fock space

$$\mathcal{F} a_k^+ \mathcal{F}^{-1} = i a_k^+ \ , \ \mathcal{F} a_k^- \mathcal{F}^{-1} = -i a_k^- \quad (\text{ same relations for } b_k^+, b_k^-)$$

We also define the analogs of boson and fermion brownian motions

$$q_k = a_k^+ + a_k^- \ , \ p_k = i(a_k^+ - a_k^-) \quad ; \quad r_k = b_k^+ + b_k^- \ , \ s_k = i(b_k^+ - b_k^-)$$

which are conjugate under Fourier transform : $\mathcal{F} q_k = p_k \mathcal{F}$, $\mathcal{F} p_k = -q_k \mathcal{F}$,etc.

Up to now, we have just expressed the combinatorial properties of the basis X_A . We are going to express now the Bernoulli multiplication : we first remark that

$$(9) \qquad X_k^2 = 1 + cX_k \quad , \text{ with } \ c = (1-2p)/\sqrt{pq}$$

and then, computing explicitly the product $X_k . X_A$ for $k\notin A$, $k\in A$, we get

$$(10) \qquad X_k . X_A = (q_k + cN_k) X_A$$

which is strikingly similar to (4). Hence, we may realize the different Bernoulli processes on toy Fock space just as we realized the Poisson processes in the first sections. An immediate consequence is the fact that, if we consider toy Fock space as an algebraic object in itself, the operators $q_k + cN_k$ for given c all commute, and may be considered as random variables (for an <u>arbitrary</u> state now). By Fourier transform, the same is true for the conjugate operators $p_k + cN_k$.

For the use of non-standardists, let us write a small dictionary of correspondences between toy Fock space and boson Fock space :

$a_k^{\pm}$ is read as $da_t^{\pm}/\sqrt{dt}$, q_k, p_k are read as $dQ_t/\sqrt{dt}$, $dP_t/\sqrt{dt}$ (the two conjugate brownian motions) ; c is read as $1/\sqrt{\lambda dt}$, $p\neq 1/2$ is read as λdt ($p=1/2$ corresponds to $\lambda=\infty$) ; N_k is read as dN_t (number operator process) ; X_k (the Bernoulli r.v.) is read as $dX_t/\sqrt{dt}$, (X_t) being the compensated Poisson $Q_t \mp N_t/\sqrt{\lambda} = X_t = \frac{1}{\sqrt{\lambda}}(-\nu_t + \lambda t)$, and finally ν_k is read as $d\nu_t$. The formula $[da_t^-, da_t^+] = dt - 2dN_t dt$ turns out to be an useful second order approximation.

Finally, let us mention the problem of martingales on toy Fock space. In the boson stochastic calculus developed by Hudson, Parthasarathy (and now many other authors), operator martingales are represented as stochastic integrands w.r. to three basic integrators : da_t^-, da_t^+, dN_t. Or rather, three are needed, and it is conjectured that they suffice, but one is plagued with unbounded operators, and it isn't even quite clear in which sense the representation takes place. Journé's remarks on toy Fock space will explain at once why three and not more should be necessary and sufficient.

We order our basis of toy Fock space $H([1,M])$ on $\{1,..,M\}$ as follows :

$$\underline{X_\emptyset|}\ \underline{X_1|}\ \underline{X_2,X_{12}|}\ \underline{X_3,X_{13},X_{23},X_{123}|}\ \ldots$$

Going from the left to the successive vertical lines determines segments of the basis, of sizes $1,2^1,2^2,2^3..$, which may be identified with the corresponding basis in $H(\emptyset)$, $H(\{1\})$, $H([1,2])$... . On the other hand, any element of the large basis can be written as X_AX_B, where A is a subset of $\{1,..,n\}$ and B a subset of $\{n+1,...,M\}$. This means that $H([0,M])$ has a natural tensor product structure, and we say that a matrix U on $H([0,M])$ is $\mathcal{F}_n$-measurable if it is the ampliation to $H([0,M])$ of a matrix V on $H([0,n])$, i.e., in the above representation, $U(X_AX_B)=V(X_A)X_B$. To help the reader we take n=2 : a matrix (4,4) is $\mathcal{F}_0$-measurable if it is written as a diagonal

$$\begin{vmatrix} a & & & 0 \\ & a & & \\ & & a & \\ 0 & & & a \end{vmatrix}$$

and $\mathcal{F}_1$ measurable if it is a block diagonal

$$\begin{vmatrix} a & b & & 0 \\ c & d & & \\ & & a & b \\ 0 & & c & d \end{vmatrix}$$

We can now define the conditional expectation of any matrix U w.r. to $\mathcal{F}_n$ (in which sense this may be interpreted as a conditional expectation will not be explained here : this depends on the choice of the vacuum state as the basic probability law). We just take the $(2^n,2^n)$ square matrix in the left upper corner, repeat it diagonally until filling the diagonal and write zeros outside of the region we have filled. For instance, the two matrices above are the successive conditional expectations of any (4,4) matrix $U = \begin{vmatrix} A & B \\ C & D\end{vmatrix}$ such that the upper left block is $A=\begin{vmatrix} a & b \\ c & d\end{vmatrix}$.

The martingale representation problem consists in representing a martingale difference $E[U|\mathcal{F}_n]-E[U|\mathcal{F}_{n-1}]$ as a linear combination , with coefficients which are $\mathcal{F}_{n-1}$-measurable, of basic martingale differences. The concrete case n=2 will make the whole situation clear.
Our martingale difference $U-E[U|\mathcal{F}_2]$ here is a block matrix $\begin{vmatrix} 0 & B \\ C & D-A\end{vmatrix}$. On the other hand, the three matrices $\begin{vmatrix} 0 & I \\ 0 & 0\end{vmatrix}$, $\begin{vmatrix} 0 & 0 \\ I & 0\end{vmatrix}$, $\begin{vmatrix} 0 & 0 \\ 0 & I\end{vmatrix}$ are orthogonal to $\mathcal{F}$, and we have the (unique) previsible representation

$$\begin{vmatrix} 0 & B \\ C & D-A\end{vmatrix} = \begin{vmatrix} B & 0 \\ 0 & B\end{vmatrix}\begin{vmatrix} 0 & I \\ 0 & 0\end{vmatrix} + \begin{vmatrix} C & 0 \\ 0 & C\end{vmatrix}\begin{vmatrix} 0 & 0 \\ I & 0\end{vmatrix} + \begin{vmatrix} D-A & 0 \\ 0 & D-A\end{vmatrix}\begin{vmatrix} 0 & 0 \\ 0 & I\end{vmatrix}$$

Now it turns out that our three martingale differences are just the matrices of a_2^-, a_2^+ and N_2 . The general situation is the same : one just allows A, B,C,D,I to denote $(2^n,2^n)$ matrices instead of (2,2) ones. The extension to the continuous case should now be left to the expert in non-standard analysis.

REFERENCE

R.L. Hudson, K.R. Parthasarathy. Quantum Ito's formula and stochastic evolutions. Comm. Math. Phys. 93, 1984, p. 301-323.

ADDITIONAL REFERENCES

As expected, toy Fock space isn't a new object among theoretical physicists. Discussions at the BiBoS meeting have produced the following references, which are quite interesting since they emphasize the similarities between toy Fock space and true Fock space from a group theoretical (and not probabilistic) point of view.

Ph. Combe, R. Rodriguez, M. Sirugue Collin, M. Sirugue.

1) On the quantization of spin systems and Fermi systems. J. Math. Phys. 20, 1979, p. 611-616.

2) Weyl quantization of classical spin systems, quantum spins and Fermi systems. In Feynman path integrals, Proc. Marseille 1978. Lecture Notes in Physics (Springer) 106 .

I.R.M.A., Université de Strasbourg
67084 Strasbourg-Cedex, France.

ON A TRANSFORMATION OF SYMMETRIC MARKOV PROCESS AND RECURRENCE PROPERTY

Yoichi-OSHIMA
Department of Mathematics
Faculty of Engineering
Kumamoto University
Kumamoto 860, Japan

Masayoshi-TAKEDA
Department of Mathematics
Faculty of Science
Osaka University
Toyonaka, Osaka 560, Japan

§1. Introduction

Let ψ be a function belonging to H^1_{loc}. We consider the Dirichlet form $\mathcal{E}_\psi$ on $L^2(\psi^2 dx)$

(1.1) $$\mathcal{E}_\psi(u,v) = \frac{1}{2}\int(\mathrm{grad}u,\mathrm{grad}v)_{R^d}\,\psi^2 dx, \text{ for } u,v \in \mathcal{F}_\psi,$$

where $\mathcal{F}_\psi$ is the closure of $C_0^\infty(R^d)$ with respect to $\mathcal{E}_{\psi,1} = \mathcal{E}_\psi + (\ ,\ )_{\psi^2 dx}$. By Fukushima's theory, we can construct the diffusion process associated with the Dirichet space $(\mathcal{E}_\psi,\mathcal{F}_\psi)$ ([3:Chap. 6]).

From the probabilistic point of view, Fukushima and Takeda [5] proved the following. Let (X_t,P_x) be the d-dimensional Brownian motion (in fact, they discussed general symmetric Markov processes) and ψ be the function $R_\alpha g(=E.[\int_0^\infty e^{-\alpha t}g(X_t)dt])$ for $\alpha > 0$ and a bounded positive function g in $L^2(dx)$. Denote by (X_t^ψ,P_x^ψ) the transformation of Brownian motion (X_t,P_x) by the multiplicative functional L_t given by

(1.2) $$L_t = \frac{\psi(X_t)}{\psi(X_0)}\exp\left(-\int_0^t \frac{1}{2}\frac{\Delta\psi}{\psi}(X_s)ds\right).$$

Then, (X_t^ψ,P_x^ψ) becomes a $\psi^2 dx$-symmetric conservative diffusion process whose Dirichet space is an extension of (1.1).

Let

(1.3) $$\psi(X_t) - \psi(X_0) = M_t^{[\psi]} + N_t^{[\psi]}$$

be the Fukushima's decomposition of $\psi(X_t) - \psi(X_0)$ into the martingale additive functional $M_t^{[\psi]}$ and the additive functional $N_t^{[\psi]}$

of zero energy. Then L_t of (1.2) is represented as

$$(1.4) \qquad L_t = \exp\left(\int_0^t \frac{1}{\psi(X_s)}\, dM_s^{[\psi]} - \frac{1}{2}\int_0^t \frac{1}{\psi^2(X_s)}\, d\langle M^{[\psi]}\rangle_s\right).$$

The object of this article is to extend the results in [3] for sufficiently wide class of ψ by using expression (1.4).

Now, we shall introduce the basic notations. We consider a locally compact smooth manifold X and a positive Radon measure m on X such that $\mathrm{supp}[m] = X$. Let $M = (X_t, P_x, \zeta)$ be a m-symmetric diffusion process associated with regular Dirichlet space $(\mathcal{E}, \mathcal{F})$ which is of the form

$$(1.5) \qquad \mathcal{E}(u,v) = \frac{1}{2}\sum_{i,j=1}^{d} \int \frac{\partial u}{\partial x_i}(x)\frac{\partial v}{\partial x_j}(x)\,\nu_{ij}(dx) \left(= \frac{1}{2}\int d\mu^c_{\langle u,v\rangle}\right).$$

Let us fix a non-trivial m-a.e. non-negative function $\psi \in \mathcal{F}$. We may suppose that ψ is quasi-continuous and denote by $M^{(0)}$ the part process of M on the fine open set $X^{(0)} = \{x;\ \psi(x) > 0\}$. When we set the sequence of fine open sets, $K_n = \{x; \frac{1}{2^n} < \psi(x) < 2^n\}$ our basic hypothesis is

<u>Hypothesis H:</u> For q.e. $x \in K_n$

$$(1.6) \qquad E_x[L^2_{t\wedge\tau_n}] \le S_{n,t}, \quad E_x[L^{-2}_{t\wedge\tau_n}] \le S_{n,t},$$

where τ_n is the first exit time from K_n.

Let $M^{(n)}$ be the part process of $M^{(0)}$ on K_n. Denote by $\tilde{M}$, $\tilde{M}^{(0)}$ and $\tilde{M}^{(n)}$ the transformed process of M, $M^{(0)}$ and $M^{(n)}$ by the multiplicative functional L_t in (1.4), for example, $\tilde{M}^{(n)}$ is the process with transition function

$$(1.7) \qquad \tilde{P}_t^{(n)} f(x) = E_x[L_t f(X_t); t < \tau_n].$$

Then, we obtain

<u>Theorem 1.1.</u> Under the hypothesis H, the following results hold.

i) $\tilde{M}^{(0)}$ is $\psi^2 m$-symmetric

ii) Denoting by $(\mathcal{E}^{(n)}, \mathcal{F}^{(n)})$ and $(\tilde{\mathcal{E}}^{(0)}, \tilde{\mathcal{F}}^{(0)})$ the Dirichlet spaces associated with $M^{(n)}$ and $\tilde{M}^{(0)}$ respectively, it holds that

$\bigcup_n \mathcal{F}^{(n)} \subset \tilde{\mathcal{F}}^{(0)}$ and for u, $v \in \bigcup_n \mathcal{F}^{(n)}$

(1.8) $$\tilde{E}^{(0)}(u,v) = \frac{1}{2}\int \psi^2 \, d\mu_{<u,v>}.$$

By using (1.8), we see that the function 1 belongs to the domain of the extended Dirichlet space of $(\tilde{E}^{(0)}, \tilde{F}^{(0)})$ (Lemma 4.1) and $\tilde{M}^{(0)}$ is conservative. In particular the process $\tilde{M}$ does not hit the set $\{x; \psi(x) = 0\}$. In this connection, we mention the paper by Meyer and Zheng [6]. Further we show a stronger assertion that $\tilde{M}^{(0)}$ is recurrent provided that $(\mathcal{E}^{(0)}, \mathcal{F}^{(0)})$ is irreducible.

§2. Some AFs locally of zero energy

Let $(E^{(0)}, F^{(0)})$ be the Dirichlet form on $L^2(dm)$ associated with $M^{(0)}$. Then $\log\psi \in F^{(0)}_{loc}$. Let

$$\log\psi(X_0) - \log\psi(X_0) = M_t^{[\log\psi]} + N_t^{[\log\psi]} \quad t < \tau^{(0)},$$

$\tau^{(0)} = \inf\{t;\ X \notin X^{(0)}\}$, be the Fukushima's decomposition, then

(2.1) $$M_t^{[\log\psi]} = \int_0^t \frac{1}{\psi(X_s)} dM^{[\psi]} \quad \text{for} \quad t < \tau^{(0)}$$

Let N_t be the AF locally of zero energy defined by

(2.2) $$N_t = N_t^{[\log\psi]} + \frac{1}{2}\int_0^t \frac{1}{\psi^2(X_s)} d\langle M^{[\psi]}\rangle_s.$$

Then L_t is written as

(2.3) $$L_t = \frac{\psi(X_t)}{\psi(X_0)} e^{-N_t} \quad t < \tau^{(0)}$$

Note that, if the function ψ belongs the domain of the generator, then $N_t = \int_0^t \frac{1}{\psi(X_s)} dN_s^{[\psi]}$. So that (2.3) becomes (1.2) in the Brownian motion case. For each $n \geq 1$, there exists a bounded function $\psi_n \in F$ such that $\log\psi = \psi_n + \varepsilon_n$ on K_n, where ε_n is some constant. Let

(2.4) $$\psi_n(X_t) - \psi_n(X_0) = M_t^{[\psi_n]} + N_t^{[\psi_n]}$$

be the decomposition. Then

$$(2.5)\qquad \psi_n(X_{t\wedge\tau_n}) = \log\psi(X_{t\wedge\tau_n}),\ M_{t\wedge\tau_n}^{[\psi_n]} = M_{t\wedge\tau_n}^{[\log\psi]},\quad N_{t\wedge\tau_n}^{[\psi_n]} = N_{t\wedge\tau_n}^{[\log\psi]}$$

Lemma 2.1. For non-negative functions f and g on X which vanish on $X-(K_n)^c$ and a non-negative function F on R^1, we have

$$(2.6)\qquad E_m^{(n)}[f(X_0)F(N_t)g(X_t)] = E_m^{(n)}[f(X_t)F(N_t)g(x_0)].$$

Proof. It is enough to show the case that f and g are bounded and F is bounded continuous. Let $\Delta^k = \{0 = t_0^k < t_1^k < \cdots < t_{2^k}^k = t\ (t_i^k = \frac{i}{2^k}t)$ be the partition of $[0,t]$ into subintervals of length $\Delta t^k = \frac{t}{2^k}$. Set $m_n = I_{K_n}m$. Then there exists a subsequence of Δ^k (which is denoted by Δ^k again) such that

$$<M^{[\psi_n]}>_t = \lim_{k\to\infty}\sum_{i=0}^{2^k-1}(\psi_n(X(t_{i+1}^k))-\psi_n(X(t_i^k)))^2$$

$$\text{a.s } P_{m_n} \quad \text{on} \quad t < \tau_n.$$

Thus,

$$\begin{aligned}E_m^{(n)}[f(X_0)F(N_t)g(X_t)] &= E_m[f(X_0)F(N_t)g(X_t)\ ;\ t<\tau_n]\\ &= E_m[f(X_0)F(N_t^{[\psi_n]}+\tfrac{1}{2}<M^{[\psi_n]}>_t)g(X_t)\ ;\ t<\tau_n]\\ &= \lim_{k\to\infty}E_m[f(X_0)F(N_t^{[\psi_n]}+\tfrac{1}{2}\sum_{i=0}^{2^k-1}(\psi_n(X(t_{i+1}^k))-\psi_n(X(t_i^k))^2\ g(X_t)\\ &\qquad ;\ t<\tau_n].\end{aligned}$$

By [3, Theorem 5.2.2], $N_t^{[\psi_n]}$ can be written as the P_{m_n}-a.e. limit of AFs which are of the form $\int_0^t f(X_s)ds$. Thus, using the symmetry of X_t, the right hand side is equal to

$$\begin{aligned}&\lim_{k\to\infty}E_m[f(X_t)F(N_t^{[\psi_n]}+\tfrac{1}{2}\sum_{i=0}^{2^k-1}(\psi_n(X_{t_{i+1}^k})-\psi_n(X_{t_i^k}))^2)g(X_0)\ ;\ t<\tau_n]\\ &= E_m^{(n)}[f(X_t)F(N_t)g(X_0)].\end{aligned}$$

q.e.d

Concerning the Hypothesis (H), we shall remark the following

<u>Lemma 2.2.</u> (1.6) is satified if

$$(2.7)\qquad E_x[\exp(10\int_0^{t\wedge\tau_n}\frac{1}{\psi^2(X_s)}d\langle M^{[\psi]}\rangle_s)] \le S_{n,t}^2, \quad \text{for q.e. } x \in K_n.$$

Proof. Since $M_t^{[\psi]}$ is a local martingale relative to P_x, for q.e. x,

$$E_x[\exp(\int_0^{t\wedge\tau_n}\frac{4}{\psi(X_s)}\,dM_s^{[\psi]} - \frac{1}{2}\int_0^{t\wedge\tau_n}(\frac{4}{\psi})^2(X_s)d\langle M^{[\psi]}\rangle_s)] \le 1$$

for all $t \ge 0$. Thus,

$$E_x[L_{t\wedge\tau_n}^{-2}] = E_x[\exp(-2\int_0^{t\wedge\tau_n}\frac{1}{\psi(X_s)}dM_s^{[\psi]} + \int_0^{t\wedge\tau_n}\frac{1}{\psi^2(X_s)}d\langle M^{[\psi]}\rangle_s]$$

$$= Ex[\exp(-2\int_0^{t\wedge\tau_n}\frac{1}{\psi(X_s)}dM_s^{[\psi]} - 4\int_0^{t\wedge\tau_n}\frac{1}{\psi^2(X_s)}d\langle M^{[\psi]}\rangle_s$$
$$+5\int_0^{t\wedge\tau_n}\frac{1}{\psi^2(X_s)}d\langle M^{[\psi]}\rangle)]$$

$$\le E_x[\exp(-\int_0^{t\wedge\tau_n}(\frac{4}{\psi})(X_s)dM_s^{[\psi]} - \frac{1}{2}\int_0^{t\wedge\tau_n}(\frac{4}{\psi})^2(X_s)d\langle M^{[\psi]}\rangle_s)]^{1/2}$$
$$\cdot E_x[\exp(10\int_0^{t\wedge\tau_n}\frac{1}{\psi^2(X_s)}d\langle M^{[\psi]}\rangle_s)]^{1/2} \le S_{n,t}.$$

The other inequality follows similarly.

§3. Dirichlet form of M

The purpose of this section is to prove Theorem 1.1.

<u>proof of i)</u>

Denote by $\tilde{P}_t^{(n)}$ the transition function of $\tilde{M}^{(n)}$. Let f and g be bounded non-negative functions in $F^{(n)}$, then, by Lemma 2.1, we obtain

$$(\tilde{P}_t^{(n)}g,f)_{\psi^2 m} = (E_\cdot^{(n)}[L_t g(X_t)],f)_{\psi^2 m}$$
$$= E_m[\psi(X_0)f(X_0)\psi(X_t)g(X_t)e^{-N_t};\ t < \tau_n]$$
$$= E_m[\psi(X_t)f(X_t)\psi(X_0)g(X_0)e^{-N_t};\ t < \tau_n]$$

$$= (g, \tilde{P}_t^{(n)} f)_{\psi^2 m}.$$

Letting $n \to \infty$, we get

$$(\tilde{P}_t^{(0)} f, g)_{\psi^2 m} = (f, \tilde{P}_t^{(0)} g)_{\psi^2 m}.$$

<u>proof of ii)</u> We shall suppose that $u = R_1^{(n)} g \quad (g \in C_0^+)$.

$$(3.1) \qquad \frac{1}{t}(u-\tilde{P}_t^{(n)} u, u)_{\psi^2 m} = \frac{1}{t}(u-E^{(n)}[L_t u(X_t)],\ u)_{\psi^2 m}$$

$$= \frac{1}{t}(u-P_t^{(n)} u, u)_{\psi^2 m} - \frac{1}{t}(E^{(n)}[(L_t-1)u(X_t)], u)_{\psi^2 m}$$

Since $u \in F^{(n)}$ and ψ^2 is bounded on K_n, $\psi^2 u$ belongs to $F^{(n)}$. Hence, we have

$$\lim_{t\to 0} \frac{1}{t}(u-P_t^{(n)} u, u)_{\psi^2 m} = \lim_{t\to 0} \frac{1}{t}(u-P_t^{(n)} u, \psi^2 u)_m$$

$$= \mathcal{E}^{(n)}(u, \psi^2 u)$$

$$= \frac{1}{2} \int d\mu^c_{\langle u, \psi^2 u\rangle}$$

$$= \frac{1}{2} \int \psi^2 d\mu^c_{\langle u,u\rangle} + \int \psi u d\mu^c_{\langle u,\psi\rangle}.$$

To treat the second term of the right hand side of (3.1), we need the following two lemmas.

<u>Lemma 3.1.</u> It holds that

$$(3.2) \qquad \lim_{t\to 0} \frac{1}{t} E_{u\psi^2 m}[(L_{t\wedge\tau_n}-1)u(X_t); t \geq \tau_n] = 0.$$

<u>Lemma 3.2.</u> It holds that

$$(3.3) \qquad \lim_{t\to 0} \frac{1}{t} E_{um}\Big[\int_0^{t\wedge\tau_n} \Big(\frac{\psi^2(X_0)}{\psi^2(X_s)} L_s - 1\Big) \psi(X_s)\, d\langle M^{[\psi]}, M^{[\psi]}\rangle_s\Big] = 0.$$

If the lemmas were proved, then the second term of (3.1) approaches to

$$\lim_{t\to 0} \frac{1}{t}\ (E_\cdot^{(n)}[(L_t-1)u(X_t)], u)_{\psi^2 m}$$

$$= \lim_{t\to 0} \frac{1}{t} E_{u\psi^2 m}[(L_t-1)u(X_t); t<\tau_n]$$

$$= \lim_{t\to 0} \frac{1}{t} E_{u\psi^2 m}[(L_{t\wedge\tau_n}-1)u(X_t)] \quad (\because \text{ Lemma 3.1})$$

$$= \lim_{t\to 0} \frac{1}{t} E_{u\psi^2 m}[(L_{t\wedge\tau_n}-1)(u(X_t)-u(X_0))]$$

$$= \lim_{t\to 0} \frac{1}{t} E_{u\psi^2 m}[\int_0^{t\wedge\tau_n} \frac{L_s}{\psi(X_s)} d\langle M^{[\psi]},M^{[u]}\rangle_s]$$

$$= \lim_{t\to 0} \{\frac{1}{t} E_{um}[\int_0^{t\wedge\tau_n}(\frac{\psi^2(X_0)}{\psi^2(X_s)} L_s-1)\psi(X_s)d\langle M^{[\psi]},M^{[u]}\rangle_s]$$

$$+ \frac{1}{t} E_{um}[\int_0^{t\wedge\tau_n} \psi(X_s)d\langle M^{[\psi]},M^{[u]}\rangle_s]\}$$

$$= \int_{K_n} u\psi d\mu^c_{\langle\psi,u\rangle} \quad (\because \text{ Lemma 3.2}).$$

Hence, we have

$$\lim_{t\to 0} \frac{1}{t}(u-\tilde{P}_t^{(n)}u,u)_{\psi^2 m} = \frac{1}{2} \int_X \psi^2 d\mu^c_{\langle u,u\rangle},$$

which implies that $u \in \tilde{F}^{(0)}$ and $\tilde{E}^{(0)}(u,u)$ is given by (1.8). Extension of this result for all $u \in \bigcup_n F^{(n)}$ can be done similary to [5; Theorem 3.1].

<u>Proof of Lemma 3.1.</u> By Schwarz's inequality

$$E_{u\psi^2 m}[(L_{t\wedge\tau_n}-1)u(X_t);t \geq \tau_n]^2$$

$$\leq E_{u\psi^2 m}[(L_{t\wedge\tau_n}-1)^2]\cdot E_{u\psi^2 m}[u^2(X_t);\ t \geq \tau_n].$$

Since

$$\lim_{t\to 0} \frac{1}{t} E_{u\psi^2 m}[u^2(X_t);\ t \geq \tau_n]$$

$$= \lim_{t\to 0} [\frac{1}{t} (u\psi^2,P_t u^2-u^2)_m - \frac{1}{t}(u\psi^2,P_t^{(n)}u^2-u^2)_m]$$

$$= E(u\psi^2,u^2)-E(u\psi^2,u^2)$$

$$= 0,$$

it is enough to show that

$$\overline{\lim_{t\to 0}} \frac{1}{t} E_{u\psi^2 m}[(L_{t\wedge\tau_n}-1)^2] < \infty. \tag{3.4}$$

Let $t_i^k = \frac{i}{2^k}t$ and $\Delta t^k = \frac{t}{2^k}$, then

$$E_{u\psi^2 m}[(L_{t\wedge\tau_n}-1)^2] = E_{u\psi^2 m}[\int_0^{t\wedge\tau_n} \frac{L_s^2}{\psi^2(X_s)} d\langle M^{[\psi]}\rangle_s]$$
$$= E_m[u(X_0)\int_0^{t\wedge\tau_n} e^{-2N_s} d\langle M^{[\psi]}\rangle_s]$$
$$= \lim_{j\to\infty} E_m[u(X_0)\int_0^{t\wedge\tau_n}(e^{-2N_s}\wedge j)d\langle M^{[\psi]}\rangle_s]$$
$$= \lim_{j\to\infty}\lim_{k\to\infty}\sum_{i=0}^{2^k-1} E_m[u(X_0)\left(\exp(-2N_{t_i^k\wedge\tau_n})\wedge j\right)\left(\langle M^{[\psi]}\rangle_{t_{i+1}^k\wedge\tau_n} - \langle M^{[\psi]}\rangle_{t_i^k\wedge\tau_n}\right)]$$
$$= \lim_{j\to\infty}\lim_{k\to\infty}\sum_{i=0}^{2^k-1} E_m[u(X_0)\left(\exp(-2N_{t_i^k})\wedge j\right)E_{X_{t_i^k}}[\langle M\rangle_{\Delta t^k\wedge\tau_n}] ; t_i^k < \tau_n]$$
$$= \lim_{j\to\infty}\lim_{k\to\infty}\sum_{i=0}^{2^k-1} E_m[u(X_{t_i^k}) \exp(-2N_{t_i^k})\wedge j\ E_{X_0}[\langle M\rangle_{\Delta t^k\wedge\tau_n}] ; t_i^k < \tau_n].$$

Hence, we obtain

$$\frac{1}{t}E_{u\psi^2 m}[L_{t\wedge\tau_n}-1)^2]$$
$$= \frac{1}{t}\lim_{j\to\infty}\lim_{k\to\infty}\sum_{i=0}^{2^k-1}\int E_x[u(X_{t_i^k})\left(\exp(-2N_{t_i^k})\wedge j\right);t_i^k<\tau_n]\cdot E_x[\langle M^{[\psi]}\rangle_{\Delta t^k\wedge\tau_n}]dm$$
$$= \frac{1}{t}\lim_{j\to\infty}\lim_{k\to\infty}\sum_{i=0}^{2^k-1}\int E_x[u(X_{t_i^k})\left(\frac{\psi^2(X_0)}{\psi^2(X_{t_i^k})}L^2_{t_i^k\wedge\tau_n}\right)\wedge j;t_i^k<\tau_n]\cdot E_x[\langle M^{[\psi]}\rangle_{\Delta t^k\wedge\tau_n}]dm$$
$$\le 2^{2n}\|u\|_\infty \lim_{k\to\infty}\frac{1}{t}\sum_{i=0}^{2^k-1}\int E_x[L^2_{t_i^k};t_i^k<\tau_n]\ E_x[\langle M^{[\psi]}\rangle_{\Delta t^k\wedge\tau_n}]dm$$
$$\le 2^{2n}\|u\|_\infty S_{n,t}\lim_{k\to\infty}\frac{1}{\Delta t^k}\int E_x[\langle M^{[\psi]}\rangle_{\Delta t^k\wedge\tau_n}]dm$$
$$\le 2^{2n}\|u\|_\infty \mu_{\langle\psi\rangle}(K_n)S_{n,t} \quad \text{for all} \quad t \le T. \quad \text{(q.e.d.)}$$

Proof of Lemma 3.2.

It is enough to show that

$$\lim_{t\to 0}\frac{1}{t} E_{um}[\int_0^{t\wedge\tau_n}\left|\frac{\psi^2(X_0)}{\psi^2(X_s)}L_s-1\right|d|\langle M^{[\psi]},M^{[u]}\rangle_s|] = 0.$$

By Kunita-Watanabe's inequality

$$E_{um}[\int_0^{t\wedge\tau_n}\left|\frac{\psi^2(X_0)}{\psi^2(X_s)}L_s-1\right|d|\langle M^{[\psi]},M^{[u]}\rangle_s|]^2$$

$$\leq E_{um}[\int_0^{t\wedge\tau_n}|\frac{\psi^2(X_0)}{\psi^2(X_s)}L_s-1|d\langle M^{[\psi]}\rangle_s]\cdot E_{um}[\int_0^{t\wedge\tau_n}|\frac{\psi^2(X_0)}{\psi^2(X_s)}L_s-1|d\langle M^{[u]}\rangle_s],$$

Let t_i^k and Δt^k be those in the proof of Lemma 3.1, then, as in the proof of Lemma 3.1, we have

$$\frac{1}{t}E_{um}[\int_0^{t\wedge\tau_n}\frac{\psi^2(X_0)}{\psi^2(X_s)}L_s-1|d\langle M^{[u]}\rangle_s]$$

$$=\frac{1}{t}\lim_{j\to\infty}\lim_{k\to\infty}\sum_{i=0}^{2^k-1}E_m[u(X_0)(|\frac{\psi(X_0)}{\psi(X_{t_i^k})}\exp(-N_{t_i^k})-1|\wedge j)\cdot E_{X_{t_i^k}}[\langle M^{[u]}\rangle_{\Delta t^k\wedge\tau_n}];t_i^k<\tau_n]$$

$$\leq\frac{1}{t}\lim_{j\to\infty}\lim_{k\to\infty}\sum_{i=0}^{2^k-1}E_m[u(X_{t_i^k})(|\frac{\psi(X_{t_i^k})}{\psi(X_0)}\exp(-N_{t_i^k})-1|\wedge j)$$

$$E_{X_0}[\langle M^{[u]}\rangle_{\Delta t^k\wedge\tau_n}];t_i^k<\tau_n]$$

$$\leq\frac{1}{t}\lim_{k\to\infty}\sum_{i=0}^{2^k-1}\int\|u\|_\infty E_x[L_{t_i^k}-1;t_i^k<\tau_n]\cdot E_x[\langle M^{[u]}\rangle_{\Delta t^k\wedge\tau_n}]dm$$

$$\leq\lim_{k\to\infty}\frac{1}{t}\sum_{i=0}^{2^k-1}\|u\|_\infty\int E_x[L_{t_i^k\wedge\tau_n}+1]\cdot E_x[\langle M^{[u]}\rangle_{\Delta t^k\wedge\tau_n}]dm$$

$$=2\|u\|_\infty\lim_{k\to\infty}\frac{1}{\Delta t_k}\int E_x[\langle M^{[u]}\rangle_{\Delta t^k\wedge\tau_n}]dm$$

$$=2\|u\|_\infty\,\mu_{\langle u\rangle}(K_n)<\infty.$$

Similarly,

$$\frac{1}{t}E_{um}[\int_0^{t\wedge\tau_n}|\frac{\psi^2(X_0)}{\psi^2(X_s)}L_s-1|d\langle M^{[\psi]}\rangle_s]\tag{3.5}$$

$$\leq\lim_{k\to\infty}\frac{1}{t}\sum_{i=0}^{2^k-1}\int\|u\|_\infty E_x^{(n)}[|L_{t_i^k}-1|;t_i^k<\tau_n]\cdot E_x^{(n)}[\langle M^{[\psi]}\rangle_{\Delta t^k\wedge\tau_n}]dm$$

Since $\langle M^{[\psi]}\rangle_{t\wedge\tau_n}$ is a positive continuous AF of $M^{(n)}$ with associated smooth measure $\mu_n(=I_{K_n}\cdot\mu_{\langle\psi\rangle})$ by [3; Lemma 5.15], using Lemma 5.14 (iii) of [3] and Lemma 2.2, we have

(The right hand side of (3.5))

$$\leq\lim_{k\to\infty}\frac{\|u\|_\infty}{t}\sum_{i=0}^{2^k-1}\int_0^{\Delta t^k}\langle\mu_n,P_s^{(n)}E_\cdot[|L_{t_i^k\wedge\tau_n}-1|;t_i^k<\tau_n]\rangle ds$$

$$=\lim_{k\to\infty}\frac{\|u\|_\infty}{t}\sum_{i=0}^{2^k-1}\int_0^{\Delta t^k}\langle\mu_n,E_\cdot[L_{s\wedge\tau_n}^{-1}|L_{(t_i^k+s)\wedge\tau_n}-L_{s\wedge\tau_n}|;t_i^k+s<\tau_n]\rangle ds$$

$$\leq \lim_{k\to\infty} \frac{\|u\|_\infty}{t} \sum_{i=0}^{2^k-1} \int_0^{\Delta t^k} \langle \mu_n, E.[L^{-2}_{s\wedge\tau_n}; s<\tau_n]\rangle^{1/2}$$

$$\cdot \langle \mu_n, E.[(L_{(t_i^k+s)\wedge\tau_n} - L_{s\wedge\tau_n})^2; t_i^k+s<\tau_n]\rangle^{1/2} ds$$

$$\leq \lim_{k\to\infty} \frac{\|u\|_\infty}{t} \sum_{i=0}^{2^k-1} \int_0^{\Delta t^k} S_{n,T}\sqrt{\mu_n(K_n)}\, E_{\mu_n}[(L_{t\wedge\tau_n}-1)^2]^{1/2} ds$$

$$= S_{n,T}\|u\|_\infty \sqrt{\mu_{\langle\psi\rangle}(K_n)}\, E_{\mu_n}[(L_{t\wedge\tau_n}-1)^2]^{1/2}$$

$$= S_{n,T}\|u\|_\infty \sqrt{\mu_{\langle\psi\rangle}(K_n)}\, E_{\mu_n}[\int_0^{t\wedge\tau_n} \frac{L_s^2}{\psi^2(X_s)} d\langle M^{[\psi]}\rangle_s].$$

When $t \downarrow 0$, $E_{\mu_n}[\int_0^{t\wedge\tau_n} \frac{L_s^2}{\psi^2(X_s)} d\langle M^{[\psi]}\rangle_s]$ decreases to zero.

§4. Recurrence of $\tilde{M}^{(0)}$

Since $\psi \in F$ and $\psi = 0$ on $X - X^{(0)}$, ψ belongs to $F^{(0)}$. Let $\{a_n(x)\}_{n=1}^\infty$ be a sequence of $C_0^\infty(R^1)$-functions satisfying

i) $0 \leq a_n(x) \leq 1$, ii) $a_n(x) = \begin{cases} 1 & \text{for } \frac{1}{2^n} \leq x \leq 2^n \\ 0 & \text{for } x \leq \frac{1}{2^{n+1}},\ x \geq 2^{n+1}, \end{cases}$

iii) $a'(x) \leq \begin{cases} c\ 2^{n+1} & \text{on } V_n = \{x : \frac{1}{2^{n+1}} \leq \psi(x) \leq \frac{1}{2^n}\} \\ c\ \frac{1}{2^n} & \text{on } U_n = \{x : 2^n \leq \psi(x) \leq 2^{n+1}\}, \end{cases}$

where c is a constant.

Lemma 4.1. The identity function 1 belongs to the extended Dirichlet space of $(\tilde{E}^{(0)}, \tilde{F}^{(0)})$ and $\tilde{E}^{(0)}(1,1) = 0$.

Proof. By Theorem 3.1, we see that $\phi_n = a_n \circ \psi \in F^{(n+1)} \subset \tilde{F}^{(0)}$. Since $\psi_n \uparrow 1$ and

$$\tilde{E}^{(0)}(\phi_n-\phi_m, \phi_n-\phi_m) = \frac{1}{2}\int \psi^2\, d\mu^c_{\langle\phi_n-\phi_m, \phi_n-\phi_m\rangle}$$

$$= \frac{1}{2}\int \psi^2 (a_n'(\psi) - a_m'(\psi))^2 d\mu^c_{\langle\psi\rangle}$$

$$\leq \frac{c^2}{2}\{\int_{V_n}\psi^2(2^{n+1})^2 d\mu^c_{<\psi>} + \int_{U_n}\psi^2\left(\frac{1}{2^n}\right)^2 d\mu^c_{<\psi>} + \int_{V_m}\psi^2 (2^{m+1})^2 d\mu^c_{<\psi>} + \int_{U_m}\psi^2\left(\frac{1}{2^n}\right)^2 d\mu^c_{<\psi>}\}$$

$$\leq 2c^2\{\int_{M_n} d\mu^c_{<\psi>} + \int_{N_n} d\mu^c_{<\psi>} + \int_{M_m} d\mu^c_{<\psi>} + \int_{N_m} d\mu^c_{<\psi>}\} \to 0,$$

as $m,n \to \infty$. Furthermore,

$$\begin{aligned}\tilde{\mathcal{E}}^{(0)}(1,1) &= \lim_{n\to\infty}\tilde{\mathcal{E}}^{(0)}(\phi_n,\phi_n) = \lim_{n\to\infty}\frac{1}{2}\int\psi^2 d\mu^c_{<\phi_n,\phi_n>} \\ &\leq \lim_{n\to\infty} 2c^2\{\int_{M_n} d\mu^c_{<\psi>} + \int_{N_n} d\mu^c_{<\psi>}\} \\ &= 0.\end{aligned}$$

Theorem 4.2. $\tilde{M}^{(0)}$ is conservative. In particular, if $\tilde{M}^{(0)}$ is irreducible, then $\tilde{M}^{(0)}$ is recurrent.

Proof. Lemma 4.1 implies that $\tilde{M}^{(0)}$ is conservative. The irreducibility of $\tilde{M}^{(0)}$ comes from that of $M^{(0)}$. If $\tilde{M}^{(0)}$ is transient, then there exists a non-negative function g such that $\int g\, dm > 0$ and $Rg(x) = E_x[\int_0^\infty g(X_t)dt]$ belongs to the extended Dirichlet space of $(\tilde{\mathcal{E}}^{(0)},\tilde{\mathcal{F}}^{(0)})$. For such function g,

$$0 < \int g\, dm = \tilde{\mathcal{E}}^{(0)}(1,Rg) \leq (\tilde{\mathcal{E}}^{(0)}(1,1))^{1/2}(\tilde{\mathcal{E}}^{(0)}(Rg,Rg))^{1/2} = 0,$$

which is a contradiction.

Since $\tilde{M}$ coincides with $\tilde{M}^{(0)}$ before the life time of $\tilde{M}^{(0)}$ and $\tilde{M}^{(0)}$ is conservative, $\tilde{M}$ is nothing but $\tilde{M}^{(0)}$ on $X^{(0)}$. In particular, we have the following.

Corollary 4.3. The process $\tilde{M}$ does not hit the set $\{x:\psi(x) = 0\}$ for q.e. starting points $x \in X^0$.

We shall give two examples which satisfy our hypothesis H. Those were treated in [5] and [1].

Example 1. Let $\psi = R_\alpha g$ for a non-negative bounded function g. Then,

$$L_t = \frac{\psi(X_t)}{\psi(X_0)} e^{-N_t^{[\psi]}} = \frac{\psi(X_t)}{\psi(X_0)} \exp\left(-\int_0^t \left(\frac{\alpha\psi - g}{\psi}\right)(X_s)\,ds\right).$$

Since $\alpha\psi - g$ is bounded on K_n, L_t satisfies Hypothesis H.

Example 2. Let X_t be a d-dimensional Brownian motion and ψ be a non-negative function such that $\psi \in H_2^1(R^d)$. Moreover, suppose that $\frac{\partial\psi}{\partial x_i} \in L^p(R^d)$ for some $p > d$, then the Hypothesis H is fulfilled. In fact, by Carmona [2]

$$\sup_{x\in K_n} E_x\left[\exp\left(10\int_0^{t\wedge\tau_n} \frac{1}{\psi^2(X_s)} d\langle M^{[\psi]}\rangle_s\right)\right]$$

$$\le \sup_{x\in K_n} E_x\left[\exp\left(10\int_0^{t\wedge\tau_n} \frac{\|\mathrm{grad}\ \psi\|^2_{R^d}}{\psi^2}(X_s)\,ds\right)\right]$$

$$\le \sup_{x\in K_n} E_x\left[\exp\left(10\cdot 2^{2n}\int_0^t \|\mathrm{grad}\ \psi\|^2_{R^d}(X_s)\,ds\right)\right]$$

$$\le \sum_{k\ge 0} \frac{C(p)\ \|\mathrm{grad}\ \psi\|_p \Gamma(1-\frac{d}{p})(1-\frac{d}{p})^k}{\Gamma(1+k(1-\frac{d}{p}))} < \infty.$$

Remark 4.4.

In [4], Fukushima has raised the question whether for $\psi \in H_2^1$, the diffusion process corresponding to (1.1) and $\tilde{M}$ coincide or not. If ψ satisfies $\psi \ge c(K) > 0$ for any compact set K, then Theorem 3.1 and Example 2 tell us that $C_0^\infty(R^d)$-functions is included in $\tilde{F}$ under the condition that $\frac{\partial\psi}{\partial x_i} \in L^p(R^d)\ (p>d)$. But in this case, we can prove the uniqueness of Markovian self-adfoint extension of operator $Lu = \frac{1}{2}\Delta u + \frac{1}{\psi}(\mathrm{grad}\ \psi, \mathrm{grad}\ u)_{R^d}$ ([8; Remark 1]). In particular, in case that $\psi \in H_2^2(R^d)\ (d\le 3)$, we can identify the processes because $H_2^2(R^d) \subset H_4^1(R^d)$.

Acknowledgements.

We gratefully acknowledge the hospitality at ZiF, Universitat Bielefeld and Forschungszentrum BiBoS.

Reference

[1] S. Albeverio, R. Høegh-Krohn and L. Streit, Energy forms, Hamiltonians and distorted Brownian paths, J. Math. Phys., 18, 907-917, 1977.

[2] R. Carmona, Regularity properties of Schrödinger and Dirichlet semigroups, J. Funct. Anal. 33, 259-296, 1979.

[3] M. Fukushima, Dirichlet forms and Markov processes, North Holland, Kodansha, 1980.

[4] M. Fukushima, Energy forms-recent developments, ZiF-preprint.

[5] M. Fukushima, M. Takeda, A transformation of symmetric Markov Processes and the Donsker-Varadhan theory, Osaka J. Math. 21, 311-326, 1984.

[6] P.A. Meyer, W.A. Zheng, Construction de processus de Nelson reversibles, Springer Lect. Note, 1123, 12-26, 1985.

[7] Y. Oshima, T. Yamada, On some representations of continuous additive functionals locally of zero energy, J. Math. Soc. Japan, 36, 315-339, 1984.

[8] M. Takeda, On the uniqueness of Markovian self-adjoint extension, in this proceedings.

ON ABSOLUTE CONTINUITY OF TWO SYMMETRIC DIFFUSION PROCESSES

Y. Oshima

Department of Mathematics,
Faculty of Engineering,
Kumamoto University,
Kumamoto, 860, Japan

§1. Introduction

Let D be a domain in R^d and $\Omega = C([0,\infty)\to D)$ be the space of all continuous functions on $[0,\infty)$ with value in D. For $\omega \in \Omega$, let $X_t(\omega)$ be the t-th coordinate of ω and $\mathcal{B}_t$ be the σ-field generated by X_s, $s \leqq t$. For two probability measures P and P' on Ω, we write $P \sim P'$ if P and P' are mutually absolutely continuous on $\mathcal{B}_t$ for all t. Let $M = (\Omega, \mathcal{B}_t, X_t, P_x)$ be a conservative diffusion process which is symmetric relative to an everywhere dense positive Radon measure m on D. Let us denote by $(\mathcal{E}, \mathcal{F})$ the Dirichlet form of M, that is,

$$(1.1)\qquad \mathcal{F} = \{u \in L^2(D;m)\ ;\ \lim_{t\to 0} \frac{1}{t} \int u(x)(I-P_t)u(x)m(dx) < \infty\},$$

$$(1.2)\qquad \mathcal{E}(u,v) = \lim_{t\to 0} \frac{1}{t} \int u(x)(I-P_t)v(x)m(dx) \qquad \text{for } u,\ v \in \mathcal{F}.$$

We say that M is C_0^1-regular if the space $C_0^1(D)$ is dense in $\mathcal{F}$ relative to the metric induced by the inner product

$$(1.3)\qquad \mathcal{E}_1(u,v) = \mathcal{E}(u,v) + (u,v)_{L^2(D;m)}, \qquad u,\ v \in \mathcal{F}.$$

In this case, $\mathcal{E}$ can be represented as

$$(1.4)\qquad \mathcal{E}(u,v) = \frac{1}{2} \sum_{i,j=1}^{d} \int_D \frac{\partial u}{\partial x_i} \frac{\partial v}{\partial x_j} \nu_{ij}(dx) \qquad u,\ v \in C_0^1(D),$$

for suitable Radon measure $(\nu_{ij})_{i,j=1,\dots,d}$ such that

$$\sum_{i,j=1}^{d} \xi_i \xi_j \nu_{ij}(B) \geqq 0$$

for all $\xi = (\xi_1,\dots,\xi_d) \in R^d$ and Borel set $B \subset D$.

In this paper, we shall be concerned with the following problem: Let $M = (\Omega, \mathcal{B}_t, X_t, P_x)$ and $M' = (\Omega, \mathcal{B}_t, X_t, P'_x)$ be, respectively, m and m'-symmetric irreducible conservative C_0^1-regular diffusion process on D with the Dirichlet forms $(\mathcal{E},\mathcal{F})$ and $(\mathcal{E}',\mathcal{F}')$. If $P_x \sim P'_x$ then how $(m,\mathcal{E})$ and $(m',\mathcal{E}')$ are related ?

In §2, under the hypothesis imposed there, we shall prove that there exists a function ρ such that $dm' = \rho dm$ and

$$(1.5)\qquad \mathcal{E}'(u,v) = \frac{1}{2}\sum_{i,j=1}^{d} \int_D \frac{\partial u}{\partial x_i}\frac{\partial v}{\partial x_j}\,\rho d\nu_{ij} \quad \text{for } u,\ v \in C_0^1(D),$$

if $\mathcal{E}$ is given by (1.4). The function ρ satisfies some regularity conditions. In §3, under the additional hypothesis (3.1), we shall show that there exists a version of ρ which is differentiable and $\frac{\partial}{\partial x_i}\log\rho \in L^2(\{K_n\};(1+\rho)dx)$ for suitable nest $\{K_n\}$. In this case, a converse result holds.

Such the problem was discussed by Orey [5] in the case that M is 1-dimensional Brownian motion and M' is a conservative diffusion process. His result says that $P_x \sim P'_x$ if and only if there exists an absolutely continuous function $\rho > 0$ such that $\rho' \in L^2_{loc}(R^1)$, $dm = \rho dx$ and

$$\mathcal{E}'(u,v) = \frac{1}{2}\int u'(x)v'(x)\rho(x)dx.$$

Fukushima [3] treated the case that M is a multi-dimensional Brownian motion and M' is a conservative diffusion process. Our case is a generalization of his case.

§2. Relation of the Dirichlet forms

Let m and m' be everywhere dense positive Radon measures and $M = (\Omega, \mathcal{B}_t, X_t, P_x)$ and $M' = (\Omega, \mathcal{B}_t, X_t, P'_x)$ be m and m'-symmetric C_0^1-regular diffusion processes on D associated with the Dirichlet forms $(\mathcal{E},\mathcal{F})$ and $(\mathcal{E}',\mathcal{F}')$, respectively, where

$$(2.1)\qquad \mathcal{E}(u,v) = \frac{1}{2}\sum_{i,j=1}^{d} \int_D \frac{\partial u}{\partial x_i}\frac{\partial v}{\partial x_j}\,d\nu_{ij} \quad \text{and}$$

$$(2.2)\qquad \mathcal{E}'(u,v) = \frac{1}{2}\sum_{i,j=1}^{d} \int_D \frac{\partial u}{\partial x_i}\frac{\partial v}{\partial x_j}\,d\nu'_{ij}$$

for $u, v \in C_0^1(D)$. Our basic assumptions are the followings.

Hypothesis

(i) M and M' are irreducible and conservative.

(ii) $P_x \sim P'_x$ for M-q.e.x.

(iii) m (resp. m') does not charge the M'-polar (resp. M-polar) sets.

(iv) (ν_{ij}) is locally uniformly elliptic, that is, for any compact set K there exist positive constants C_1 and C_2 depending on K such that

$$(2.3)\qquad C_1 \sum_{i=1}^{d} \xi_i^2 \, \nu_{ii}(B) \leqq \sum_{i,j=1}^{d} \xi_i \xi_j \, \nu_{ij}(B) \leqq C_2 \sum_{i=1}^{d} \xi_i^2 \, \nu_{ii}(B)$$

for all $\xi = (\xi_1, \ldots, \xi_d) \in R^d$ and Borel set $B \subset K$.

Lemma 1. The measures m and m' are mutually absolutely continuous.

Proof. Suppose that a non-negative bounded function f satisfies $\int_D f(x)m(dx) = 0$. Then

$$0 = \int_D f(x)P_t g(x)m(dx) = \int_D P_t f(x)g(x)m(dx)$$

for all $g \geqq 0$, where P_t is the transition function of M. This implies that $P_t f(x) = 0$ m-a.e. Since $P_t f$ is M-quasi-continuous, it follows that $P_t f = 0$ M-q.e. and hence m'-a.e. By the hypothesis, since $P_t(x, \ )$ and the transition function $P'_t(x, \ )$ of M' are mutually absolutely continuous for a.a.x relative to m', we have $P'_t f(x) = 0$ m'-a.e. Thus

$$\int_D f(x)m'(dx) = \lim_{t \to 0} \int_D P'_t f(x)m'(dx) = 0,$$

which implies that m' is absolutely continuous relative to m. The converse absolute continuity follows similarly.

Lemma 2. A Borel set is M-polar if and only if M'-polar.

Proof. Let B be a M-polar set. Then $P_x(\sigma_B < \infty) = 0$ for M-q.e.x, where $\sigma_B = \inf\{t \geqq 0 \,;\, X_t \in B\}$. By the hypothesis,

$P'_x(\sigma_B < \infty) = 0$ M-q.e. and hence m'-a.e. Since $P'_\cdot(\sigma_B < \infty)$ is M'-quasi-continuous, $P'_x(\sigma_B < \infty) = 0$ M'-q.e. Which implies that B is M'-polar.

An increasing sequence $\{K_n\}$ of compact sets is called M-nest if $\mathrm{Cap}_M(K-K_n) \to 0$ as $n \to \infty$ for all compact set K, where Cap_M is the capacity relative to M. By using Lemma 2, it is easy to see the following

Lemma 3. $\{K_n\}$ is M-nest if and only if M'-nest.

In the followings, we shall use the terminologies polar set, q.e. and nest not specifying the processes. By Lemma 3, any Radon measure which does not charge polar sets is a smooth measure.

Let L_t be the Radon-Nikodym derivative of P'_x relative to P_x, then it is a continuous martingale multiplicative functional of M such that $L_0 = 1$ and $0 < L_t < \infty$ (see [5]). By Kunita-Watanabe [4], L_t admits the expression $L_t = \exp(M_t - \frac{1}{2}\langle M\rangle_t)$ for suitable local martingale additive functional (M_t). By virtue of [2 , 8 Lemma 3.2], there exist a nest $\{K_n\}$ and functions f_i (i=1,...,d) such that $f_i \in L^2(\{K_n\};\nu_{ii})$ and

$$(2.4) \qquad M_t = \sum_{i=1}^{d} \int_0^t f_i(X_s)\,dM_s^i,$$

where (M_t^i) is the martingale part of $X_t^i - X_0^i$. Thus L_t is represented as

$$(2.5) \qquad L_t = \exp\Big(\sum_{i=1}^{d} \int_0^t f_i(X_s)\,dM_s^i - \frac{1}{2}\sum_{i,j=1}^{d} \int_0^t f_i f_j(X_s)\,d\langle M^i,M^j\rangle_s\Big).$$

Let $d\nu = dm + \sum_{i=1}^{d} d\nu_{ii}$. Then it is a Radon measure which does not charge polar sets. Hence it is a smooth measure of both processes. Let (A_t) be the continuous additive functional (CAF) of M associated with ν. Then it satisfies $A_t < \infty$ a.s. P_x. Since $d\nu \geqq dm$, we have $dA_t \geqq dt$ and which implies that A_t is strictly increasing and $A_\infty = \infty$ a.s. P_x.

Let $\tilde{M} = (\Omega, \tilde{B}_t, Y_t, P_x)$ be the time changed process of M by A_t. Then it is a ν-symmetric diffusion process. Denote by $(\tilde{E}, \tilde{F})$ the Dirichlet form of $\tilde{M}$. Then $C_0^1(D) \subset \tilde{F}$ and $\tilde{E} = E$ on $C_0^1(D)$ (see

[9], Proof of Theorem 5.1 of [7]). Moreover, there exists a set $\mathcal{D} \subset \mathcal{F} \cap \tilde{\mathcal{F}}$ such that $\tilde{\mathcal{E}} = \mathcal{E}$ on $\mathcal{D} \times \mathcal{D}$ and $\mathcal{D}$ is dense in $\mathcal{F}$ and $\tilde{\mathcal{F}}$ relative to $\mathcal{E}_1(\ ,\ ) = \mathcal{E}(\ ,\ ) + (\ ,\ )_{L^2(m)}$ and $\tilde{\mathcal{E}}_1(\ ,\ ) = \tilde{\mathcal{E}}(\ ,\ ) + (\ ,\ )_{L^2(\nu)}$, respectively. Since $\mathcal{E}_1(u,u) \leqq \tilde{\mathcal{E}}_1(u,u)$ for $u \in \mathcal{D}$, it follows that $\tilde{\mathcal{F}} \subset \mathcal{F}$ and which implies that the function of $\tilde{\mathcal{F}}_{loc}$ has a M-quasi-continuous modification.

According to the equivalence of M and M', A_t is a CAF of M'. Let ν' be the Revuz measure of A_t relative to M' and $\tilde{M}' = (\Omega, \tilde{\mathcal{B}}_t, Y_t, P'_x)$ be the time changed process of M' by A_t. Then it is a ν'-symmetric diffusion process and whose Dirichlet form is given by (2.2) on $C_0^1(D)$. Let $\tilde{P}_t$ ($\tilde{P}'_t$) be the transition function of $\tilde{M}$ ($\tilde{M}'$) then $\tilde{P}_t f$ ($\tilde{P}'_t f$) is M (M') -quasi-continuous. Since $dm \leqq d\nu$, $dm' \leqq d\nu'$ and $\tilde{P}_t(x,\ )$ and $\tilde{P}'_t(x,\ )$ are mutually absolutely continuous for q.e. M and M', by a similar argument to the proof of Lemma 1, we have

<u>Lemma 4.</u> ν and ν' are mutually absolutely continuous.

Let ρ be the density of ν' relative to ν, then $0 < \rho < \infty$ a.e. (ν) and $\int_{K_n} \rho\, d\nu < \infty$ for all nest $\{K_n\}$. For $u \in \mathcal{F}_{loc}$, let

$$u(X_t) - u(X_0) = M_t^{[u]} + N_t^{[u]}$$

be the Fukushima's decomposition of $u(X_t) - u(X_0)$ into the local martingale AF and CAF locally of zero energy of M-process ([1]). Let us define $M_t'^{[u]}$ and $N_t'^{[u]}$ similarly relative to $u \in {}'_{loc}$ and M'-process. $\tilde{M}_t^{[u]}$ and $\tilde{M}_t'^{[u]}$ are defined as the time changed functional of $M_t^{[u]}$ and $M_t'^{[u]}$ by A_t. In particular, if $u(x_1,\dots,x_d) = x_i$, then we shall denote by M_t^i, $M_t'^i$ and so on for $M_t^{[u]}$, $M_t'^{[u]}$ and so on.

<u>Lemma 5.</u> If $u \in \mathcal{F}_{loc} \cap \mathcal{F}'_{loc}$, then

$$M_t'^{[u]} = M_t^{[u]} - \langle M^{[u]}, M\rangle_t, \quad N_t'^{[u]} = N_t^{[u]} + \langle M^{[u]}, M\rangle_t. \tag{2.6}$$

Proof. It is enough to suppose that $u \in \mathcal{F} \cap \mathcal{F}'$. Denote by $\hat{M}_t$ and $\hat{N}_t$ the right-hand sides of (2.6), respectively. Then $\hat{M}_t$ is a martingale AF of M' and

$$u(X_t) - u(X_0) = \hat{M}_t + \hat{N}_t = M_t'^{[u]} + N_t'^{[u]}.$$

Hence $\hat{M}_t - M_t'^{[u]} = N_t'^{[u]} - \hat{N}_t$. For any $g \in L^1(D;m)$ such that $g \geqq 0$, there exists a sequence $n_k \to \infty$ such that

$$(2.7) \qquad \lim_{k\to\infty} \sum_{j=0}^{n_k-1} \{(\hat{M}_{(j+1)t/n_k} - M'^{[u]}_{(j+1)t/n_k}) - (\hat{M}_{jt/n_k} - M'^{[u]}_{jt/n_k})\}^2$$

$$= \langle \hat{M} - M'^{[u]} \rangle_t \qquad \text{a.s. } P'_{gm}.$$

On the other hand, since $\langle M'^{[u]}, M\rangle_t$ is an additive functional of M' locally of zero energy, there exist a sequence $\tau_n \uparrow \infty$ of exit times and additive functionals $N_t^{(n)}$ of zero energy such that $\langle M'^{[u]}, M\rangle_t = N_t^{(n)}$ for $t < \tau_n$. Hence, by choosing a subsequence of n_k if necessary, we have

$$P'_{gm}[\lim_{k\to\infty} \sum_{j=0}^{n_k-1} (\langle M'^{[u]}, M\rangle_{(j+1)t/n_k} - \langle M'^{[u]}, M\rangle_{jt/n_k})^2 = 0,\ t<\tau_n] = 1,$$

that is,

$$(2.8) \qquad \lim_{k\to\infty} \sum_{j=0}^{n_k-1} (\langle M'^{[u]}, M\rangle_{(j+1)t/n_k} - \langle M'^{[u]}, M\rangle_{jt/n_k})^2 = 0 \quad \text{a.s.} P'_{gm}.$$

As for $N^{[u]}$, by choosing a subsequence, we have

$$(2.9) \qquad \lim_{k\to\infty} \sum_{j=0}^{n_k-1} (N^{[u]}_{(j+1)t/n_k} - N^{[u]}_{jt/n_k})^2 = 0 \qquad \text{a.s. } P_{gm},$$

and hence a.s. P'_{gm}. The similar equality holds for $N'^{[u]}$. Combining (2.7), (2.8) and (2.9), we have

$$\langle M - M'^{[u]} \rangle_t = 0 \qquad \text{a.s. } P'_{gm}.$$

Which implies that $M_t = M_t'^{[u]}$.

<u>Corollary.</u> For $u, v \in F_{loc} \cap F'_{loc}$,

$$(2.10) \qquad \langle M^{[u]}, M^{[v]}\rangle_t = \langle M'^{[u]}, M'^{[v]}\rangle_t.$$

Suppose that $u \in F$, then there exists a E-Cauchy sequence $\{u_n\}$ of $C_0^1(D)$ functions such that $u_n \to u$ in $L^2(D;m)$. From (2.3), $\{\partial u/\partial x_i\}$ forms an $L^2(D;\nu_{ii})$-Cauchy sequence for $i = 1,\ldots,d$. Define $\partial u/\partial x_i = \lim_{n\to\infty} \partial u_n/\partial x_i$. By the closability, this definition is independent of the choice of the sequence $\{u_n\}$. Under this notation, the expression of $\mathcal{E}$ on (2.1) remains valid for all $u \in \mathcal{F}$ (see [8 ; Example 1]).

<u>Theorem 6.</u> If $u, v \in \mathcal{F}'$, then

$$(2.11) \qquad \mathcal{E}'(u,v) = \frac{1}{2} \sum_{i,j=1}^{d} \int_D \frac{\partial u}{\partial x_i} \frac{\partial v}{\partial x_j} \rho(x) \nu_{ij}(dx).$$

Proof. Let $u, v \in C_0^1(D)$, then, by the Corollary of Lemma 5, $\langle M^{[u]}, M^{[v]} \rangle_t = \langle M'^{[u]}, M'^{[v]} \rangle_t$. Changing the time, we have

$$(2.12) \qquad \langle \tilde{M}^{[u]}, \tilde{M}^{[v]} \rangle_t = \langle \tilde{M}'^{[u]}, \tilde{M}'^{[v]} \rangle_t .$$

Set $a_{ij} = \nu_{ij}(dx)/\nu(dx)$, then the form $\tilde{\mathcal{E}}$ of $\tilde{M}$ is written as

$$\tilde{\mathcal{E}}(u,v) = \frac{1}{2} \sum_{i,j=1}^{d} \int_D a_{ij}(x) \frac{\partial u}{\partial x_i} \frac{\partial v}{\partial x_j} \nu(dx).$$

From this form, we can see that the associated Revuz measure of the CAF $(\langle \tilde{M}^{[u]}, \tilde{M}^{[v]} \rangle_t)$ of $\tilde{M}$ is equal to

$$\sum_{i,j=1}^{d} a_{ij}(x) \frac{\partial u}{\partial x_i} \frac{\partial v}{\partial x_j} \nu(dx).$$

Which implies that

$$(2.13) \qquad \langle \tilde{M}^{[u]}, \tilde{M}^{[v]} \rangle_t = \int_0^t (a_{ij} \frac{\partial u}{\partial x_i} \frac{\partial v}{\partial x_j})(X_s) ds.$$

From (2.12) and (2.13), we can see that the associated Revuz measure of the CAF $(\langle \tilde{M}'^{[u]}, \tilde{M}'^{[v]} \rangle_t)$ of $\tilde{M}'$ is equal to

$$\sum_{i,j=1}^{d} a_{ij} \frac{\partial u}{\partial x_i} \frac{\partial v}{\partial x_j} \nu'(dx) = \sum_{i,j=1}^{d} \frac{\partial u}{\partial x_i} \frac{\partial v}{\partial x_j} \rho(x) \nu_{ij}(dx).$$

This equality shows that the form of $\tilde{M}'$ is given by

$$\tilde{\mathcal{E}}'(u,v) = \frac{1}{2} \sum_{i,j=1}^{d} \int_D \frac{\partial u}{\partial x_i} \frac{\partial v}{\partial x_j} \rho(x) \nu_{ij}(dx)$$

for $u, v \in C_0^1(D)$. Changing the time, we have (2.12) for $u, v \in C_0^1(D)$. Since $C_0^1(D)$ is dense in $\mathcal{F}'$, we have the result.

By the relation $d\nu' = \rho d\nu$, ν' is the Revuz measure associated with the CAF $\int_0^t \rho(X_s) ds$ of $\tilde{M}$. Changing the time by A_t^{-1}, the associated Revuz measure of $\int_0^t \rho(X_s)ds$ of M is ρdm. On the other hand, which is equal to m'. Thus we have

$$dm' = \rho dm. \tag{2.14}$$

Remark. By (2.4), we have

$$P_x[\sum_{i,j=1}^{d} \int_0^t f_i f_j(X_s)d\langle M^i,M^j\rangle_s < \infty] = 1 \qquad \text{q.e. } x. \tag{2.15}$$

Hence, by (2.6) and the absolutely continuity hypothesis,

$$P'_x[\sum_{i,j=1}^{d} \int_0^t f_i f_j(X_s)d\langle M'^i,M'^j\rangle_s < \infty] = 1 \qquad \text{q.e. } x. \tag{2.16}$$

§3. Regularity of ρ

In this section, we shall suppose that

$$m(dx) = m(x)dx, \qquad \nu_{ij}(dx) = a_{ij}(x)dx \tag{3.1}$$

for $m(x) \in C^1(D)$, $m(x) > 0$, $a_{ij}(x) \in C^1(D)$ and $(a_{ij}(x))$ is locally uniformly elliptic in D. As in §2, let M be an m-symmetric irreducible conservative diffusion process associated with the $L^2(D;m)$-Dirichlet form determined by the smallest closed extension of

$$E(u,v) = \frac{1}{2}\sum_{i,j=1}^{d} \int_D a_{ij}(x)\frac{\partial u}{\partial x_i}\frac{\partial v}{\partial x_j} m(x)dx, \qquad u,\ v \in C_0^1(D). \tag{3.2}$$

Under the hypothesis in §2, M' is $\rho(x)m(x)dx$-symmetric and whose Dirichlet form is given by the smallest closed extension of

$$E'(u,v) = \frac{1}{2}\sum_{i,j=1}^{d} \int_D a_{ij}(x)\frac{\partial u}{\partial x_i}\frac{\partial v}{\partial x_j} \rho(x)m(x)dx, \quad u,\ v \in C_0^1(D). \tag{3.3}$$

In this section, under the above hypothesis, we shall derive a regularity property of ρ. When there exists a version $\hat{\rho}$ of ρ such that, for every k $(1 \leq k \leq d)$ and for a.a. fixed $(x_1,\dots,x_{k-1}, x_{k+1},\dots,x_d) \in R^{d-1}$, $\hat{\rho}$ is absolutely continuous in x_k such that $(x_1,\dots,x_d) \in D$, we shall call ρ is differentiable.

Theorem 7. Under the above hypothesis, ρ is differentiable and $(\partial/\partial x_i)\log\rho \in L^2(\{K_n\};dx)$. Moreover, the density L_t is given by

$$(3.4)\qquad L_t = \exp\Big(\frac{1}{2} \sum_{i=1}^{d} \int_D \frac{\partial}{\partial x_i}\log\rho(X_s)\,dM_s^i - \frac{1}{8} \sum_{i,j=1}^{d} \int_0^t a_{ij} \frac{\partial}{\partial x_i}\log\rho \times \frac{\partial}{\partial x_j}\log\rho(X_s)\,ds\Big).$$

Proof. Let $u \in C^1(D)$ and $\eta \in C_0^1(D)$, then $u\eta \in F \cap F'$. According to (3.1),

$$M_t^{[u\eta]} = \int_0^t \sum_{i=1}^{d} \frac{\partial(u\eta)}{\partial x_i}(X_s)\,dM_s^i ,$$

$$\langle M^i, M^j\rangle_t = \int_0^t a_{ij}(X_s)\,ds \qquad \text{and}$$

$$(3.5)\qquad N_t^{[u\eta]} = \frac{1}{2}\int_0^t \sum_{i,j=1}^{d} \frac{1}{m}\frac{\partial}{\partial x_i}\Big(ma_{ij}\frac{\partial(u\eta)}{\partial x_j}\Big)(X_s)\,ds.$$

These equality combined with (2.6), we have

$$(3.6)\qquad N_t'^{[u\eta]} = \frac{1}{2}\int_0^t \frac{1}{m} \sum_{i,j=1}^{d} \frac{\partial}{\partial x_i}\Big(ma_{ij}\frac{\partial(u\eta)}{\partial x_j}\Big)(X_s)\,ds + \frac{1}{2}\int_0^t \sum_{i,j=1}^{d} a_{ij} f_i \frac{\partial(u\eta)}{\partial x_j}(X_s)\,ds.$$

Let $\chi_\varepsilon(x)$ be the mollifier and set $\rho_\varepsilon(x) = \rho * \chi_\varepsilon(x)$. Then, for any $v \in C_0^1(D)$,

$$(3.7)\qquad \lim_{\varepsilon\to 0} E(\rho_\varepsilon v, u\eta) = \lim_{\varepsilon\to 0} \frac{1}{2} \sum_{i,j=1}^{d} \int_D a_{ij} \frac{\partial(\rho_\varepsilon v)}{\partial x_i} \frac{\partial(u\eta)}{\partial x_j}\, m(x)\,dx.$$

On the other hand, by (3.5) and (3.6), we have

$$\begin{aligned}
(3.8)\qquad \lim_{\varepsilon\to 0} E(\rho_\varepsilon v, u\eta) &= \lim_{\varepsilon\to 0}\lim_{t\to 0} \frac{1}{t} E_{\rho_\varepsilon vmdx}[u\eta(X_0) - u\eta(X_t)] \\
&= \lim_{\varepsilon\to 0}\lim_{t\to 0} \frac{-1}{t} E_{\rho_\varepsilon vmdx}[N_t^{[u\eta]}] = \frac{-1}{2}\lim_{\varepsilon\to 0} \int \rho_\varepsilon v \sum_{i,j=1}^{d} \frac{\partial}{\partial x_i}\Big(ma_{ij}\frac{\partial(u\eta)}{\partial x_j}\Big)dx \\
&= \frac{-1}{2}\int \rho v \sum_{i,j=1}^{d} \frac{\partial}{\partial x_i}\Big(ma_{ij}\frac{\partial(u\eta)}{\partial x_j}\Big)dx \\
&= \lim_{t\to 0} \frac{-1}{t} E'_{\rho vmdx}\Big[\int_0^t \frac{1}{2} \sum_{i,j=1}^{d} \frac{1}{m}\frac{\partial}{\partial x_i}\Big(ma_{ij}\frac{\partial(u\eta)}{\partial x_j}\Big)(X_s)\,ds\Big] \\
&= \lim_{t\to 0} \frac{-1}{t} E'_{\rho vmdx}[N_t^{[u\eta]}] \\
&= \lim_{t\to 0} \frac{-1}{t} E'_{\rho vmdx}\Big[N_t'^{[u\eta]} - \int_0^t \sum_{i,j=1}^{d} a_{ij} f_i \frac{\partial(u\eta)}{\partial x_j}(X_s)\,ds\Big]
\end{aligned}$$

$$= \mathcal{E}'(v,u\eta) + \sum_{i,j=1}^{d} \int_D a_{ij}\, v\, f_i \frac{\partial(u\eta)}{\partial x_j}\, \rho m dx$$

$$= \frac{1}{2} \sum_{i,j=1}^{d} \int_D a_{ij} \frac{\partial v}{\partial x_i} \frac{\partial(u\eta)}{\partial x_j}\, \rho m dx + \sum_{i,j=1}^{d} \int_D a_{ij} v f_i \frac{\partial(u\eta)}{\partial x_j}\, \rho m dx.$$

Therefore

$$\lim_{\varepsilon\to 0} \sum_{i,j=1}^{d} \int_D a_{ij}\, v \frac{\partial \rho_\varepsilon}{\partial x_i} \frac{\partial(u\eta)}{\partial x_j}\, m dx = 2 \sum_{i,j=1}^{d} \int_D a_{ij} v f_i \frac{\partial(u\eta)}{\partial x_j}\, \rho m dx.$$

Taking η such that $\eta = 1$ on $\mathrm{Supp} v$, we have

$$\lim_{\varepsilon\to 0} \sum_{i,j=1}^{d} \int_D a_{ij}\, v \frac{\partial \rho_\varepsilon}{\partial x_i} \frac{\partial u}{\partial x_j}\, m dx = 2 \sum_{i,j=1}^{d} \int_D a_{ij}\, v\, f_i \frac{\partial u}{\partial x_j}\, \rho m dx.$$

Set $u = x_j$, then

$$\lim_{\varepsilon\to 0} \sum_{i=1}^{d} \int_D a_{ij}\, v \frac{\partial \rho_\varepsilon}{\partial x_i}\, m dx = 2 \sum_{i=1}^{d} \int_D a_{ij}\, v\, f_i\, \rho m dx.$$

For $\psi \in C_0^1(D)$, taking $v = (a^{-1})_{jk} \frac{1}{m} \psi$ and summing relative to j, we have

$$\lim_{\varepsilon\to 0} \int_D \psi \frac{\partial \rho_\varepsilon}{\partial x_k} dx = 2 \int f_k\, \psi\, \rho\, dx.$$

Hence

$$- \int_D \rho(x) \frac{\partial \psi}{\partial x_k} dx = 2 \int f_k\, \psi\, \rho\, dx$$

for all $\psi \in C_0^1(D)$. Which implies that ρ is differentiable and

$$\frac{\partial \psi}{\partial x_k} = 2\, f_k\, \rho. \tag{3.9}$$

Combining (3.9) with (2.5), we obtain the result.

<u>Remark.</u> By (2.16),

$$\sum_{i=1}^{d} \int_0^t f_i^2(X_s) a_{ii}(X_s) ds < \infty \qquad \text{a.e. } P'_x.$$

Hence $\sum_{i=1}^{d} f_i^2(x) a_{ii}(x) \rho(x) m(x) dx$ is a smooth measure of M'. This implies that the nest $\{K_n\}$ in Theorem 7 can be chosen as

$$f_i(x) = \frac{\partial}{\partial x_i} \log\rho \in L^2(\{K_n\}; \rho dx). \tag{3.10}$$

Remark. By a similar argument to Fukushima [3], we can show the following converse result of Theorem 7.

Let M and M' be the irreducible conservative diffusion processes associated with (3.2) and (3.3), respectively. Suppose that they satisfy ; (i) a sequence $\{K_n\}$ is M-nest iff M'-nest, and (ii) ρ is differentiable and $(\partial/\partial x_i)\log\rho \in L^2(\{K_n\};(1+\rho)dx)$. Then $P_x \sim P'_x$ for q.e. x (M).

Acknowledgements. This work was done during the stay at the Centre for Interdisciplinary Research (ZiF) of University of Bielefeld and the Bielefeld-Bochum Research Centre of Stochastic Processes (BiBoS). The author would like to thank Prof. S. Albeverio, Prof. Ph. Blanchard and Prof. L. Streit for the kind invitation and warm hospitality there.

References.

[1] M. Fukushima ; Dirichlet forms and Markov processes, Kodansha and North Holland, 1980.

[2] M. Fukushima ; On a representation of local martingale additive functionals of symmetric diffusions, Proc. of the LMS Durham Symp. on stochastic integrals, Lecture Notes in Math. 851, Springer, 1981, 112-118.

[3] M. Fukushima ; On absolute continuity of multidimensional symmetrizable diffusions, Proc. of the Symp. on Functional analysis in Markov processes, Lecture Notes in Math. 923, Springer, 1982, 146-176.

[4] H. Kunita and S. Watanabe ; On square integrable martingales, Nagoya Math. J., 30, 1967, 209-245.

[5] S. Orey ; Conditions for the absolute continuity of two diffusions, Trans. Amer. Math. Soc. 193, 1974, 413-426.

[6] S. Orey ; Radon-Nikodym derivatives of probability measures: Martingale methods, Tokyo Univ. of Education, 1974.

[7] Y. Oshima ; Some singular diffusion processes and their associated stochastic differential equations, Z. Wahr. verv. Geb. 59, 1982, 249-276.

[8] Y. Oshima and T. Yamada ; On some representation of continuous additive functionals locally of zero energy, J. Math. Soc. Japan 36, 1984, 315-339.

[9] M. L. Silverstein ; Symmetric Markov processes, Lecture Notes in Math. 426, Springer, 1974.

COLLECTIVE PHENOMENA IN STOCHASTIC PARTICLE SYSTEMS

Errico Presutti

Dipartimento Matematico, Universita' di Roma

ABSTRACT

Collective phenomena in stochastic particle systems are surveyed. Macroscopic field equations describe the evolution of some of these systems just like the hydrodynamical equations describe the dynamics of real physical fluids.

Shock wave phenomena and metastable behavior are also investigated.

Research partially supported by Nato grant n. 040.82 and by the " Consiglio Nazionale delle Ricerche".

1. Introduction

In this article I will discuss some collective phenomena appearing in the evolution of systems with many components. In particular I will consider features like the establishing of the hydrodynamical regime, as ruled by the hydrodynamical equations (Euler and Navier-Stokes equations), the formation and propagation of shock waves, the existence of metastable states and their transition to equilibrium.

All the above phenomena are experimentally observed in physical systems. They are therefore believed to be present in Hamiltonian models of interacting particle systems, but a mathematical derivation seems beyond the reach of the present techniques. I will consider the above questions in the frame of random interacting particle systems, where some progresses have been achieved in the last years. Such systems may exhibit a behavior analogous to that of real physical systems: hydrodynamical equations, fluctuation and dissipation theorems, shock waves, metastable states will emerge from our analysis. Precise mathematical properties are the counterpart of such phenomena ; the law of large numbers, the central limit theorem (invariance principle) and large deviation results.
Same or similar models as those which I will consider are introduced in such different frames as in genetics, chemistry, economics, statistics..... the terminology changes but the basic phenomena are the same.

Before describing models and results I will recall in Section 2 the scheme proposed by Morrey for deriving hydrodynamics. The crucial point is the notion of "local equilibrium", assuming local equilibrium one can derive the Euler equations. A critical discussion of such assumption is first given in Section 2 and will keep showing up in the other sections of this paper. I will also briefly recall the way the Hilbert and Chapman -Enskog expansion in the Boltzmann equation fits in the Morrey's approach to hydrodynamics.

In Section 3 I will describe stochastic models where the Morrey's scheme can be applied. Results refer to particular models except for the theory of space-time fluctuations in equilibrium (the Fluctuation-Dissipation theorem) which has been established for a large variety of models.

In Section 4 I will discuss models where shock waves are present, the shock wave front will be characterized by the spatial region where local equilibrium fails, "dynamical phase transition".

Finally, in Section 5, I will discuss some models which exhibit a metastable to stable transition and describe the so called "pathwise approach to metastability".

2. Local equilibrium and Euler equations.

We are in the setup of classical mechanics, we consider a system of identical point particles, having mass m and interacting via a smooth pair potential $V(|x|)$, which only depends on the distance

$|x|$ between particles. We assume that the initial state of the system is described by the probability

$$d\mu^{\varepsilon} = (Z^{\varepsilon})^{-1} \exp\Big(- \sum_{(q,v)} \beta(\varepsilon q) \big[\tfrac{1}{2} m (v-\bar{v}(\varepsilon q))^2 + \tfrac{1}{2} \sum_{q\neq q'} V(|q-q'|)$$

$$+ \lambda(\varepsilon q) \big] \Big) \prod dq\, dv \tag{2.1}$$

If $\beta(\cdot)$, $\bar{v}(\cdot)$ and $\lambda(\cdot)$ were constant, then $d\mu^{\varepsilon}$ would be the Gibbs measure at inverse temperature $\beta = (kT)^{-1}$, mean velocity $\bar{v}$ and chemical potential λ ; suitable assumption on the interaction V are required, like that V is superstable. To avoid ambiguities let us imagine that the particles are confined in some, very large, region.

If $\beta(\cdot)$, $\bar{v}(\cdot)$ and $\lambda(\cdot)$ are "slowly varying" functions, then $d\mu^{\varepsilon}$ describes approximate equilibria and it is called a local equilibrium distribution. Slowly varying of course means that $\beta(\cdot)$, $\bar{v}(\cdot)$ and $\lambda(\cdot)$ are essentially constant over distances of the order of the length of the correlation functions in the measure $d\mu^{\varepsilon}$. To make the statement sharp I fix once for all the functions $\beta(\cdot)$, $\bar{v}(\cdot)$ and $\lambda(\cdot)$ (assumed to be suitably smooth) and then I take the limit of small ε , the size of the region confining the particles being increased by ε^{-1}. When ε goes to zero the slowly varying condition is certainly fulfilled (phase transition should be neglected in this argument) so in the limit ε going to zero the"ideal" hydrodynamical situation is realized and we have

$$\lim \mu^{\varepsilon}(S(\varepsilon^{-1}x)f) = \mu_{\beta(x),\bar{v}(x),\lambda(x)}(f) \tag{2.2}$$

for all x. S(x) denotes space shift by x and f is any local bounded observable, $\mu_{\beta,\bar{v},\lambda}$ being the Gibbs measure with parameters β, v, λ.

We will assume that the validity of eq.(2.2) extends to positive times in the sense that the following limit holds:

$$\lim_{\varepsilon \to 0} \mu^{\varepsilon}(S(\varepsilon^{-1}x)\ T(\varepsilon^{-1}\tau)\ f) = \mu_{\beta(x,\tau),\bar{v}(x,\tau),\lambda(x,\tau)}(f) \tag{2.3}$$

There are two basic assumptions in eq.(2.3). One is a scaling property. Namely that states which are obtained one from the other by a scaling transformation [like the family μ^{ε} in eq.(2.2)] have "similar" evolutions when time is also suitably scaled. The scaling of space and time by the same parameter, ε^{-1}, as in the l.h.s. of eq.(2.3) foresees the space time invariance of the Euler equations, I do not think that at this stage it has any a priori justification.
The second assumption in eq.(2.3) is that stastical thermodynamic equilibrium describes the common limiting behavior of the states at each macroscopic space-time point (x,τ), in the limit of vanishingly small gradients [$\varepsilon \to 0$]. This is the so called "local equilibrium assumption", L E A.

The following theorem holds, cf. [M] and Theorem 0.1 of [DIPP].

2.1 Theorem

Assume that eqs.(2.2) and (2.3) hold. Assume also that $\beta(x,\tau)$, $\bar{v}(x,\tau)$ and $\lambda(x,\tau)$ are "smooth" functions and that for such values of the parameters the Gibbs state is unique [absence of phase transitions]. Then the corresponding mean velocity, energy and density satisfy the Euler equations.

2.2 Remarks

(a) The pressure appearing in the Euler equations is just the thermodynamic pressure defined by the Gibbs formula, as it can be proven under the assumptions of Theorem 2.1, [DIPP].

(b) The condition that $\beta(x,\tau), \bar{v}(x,\tau)$ and $\lambda(x,\tau)$ are smooth is not just a technical request, as we shall discuss at the end of the present section.

Before discussing meaning and validity of the local equilibrium assumption, L E A, let us see how the above procedure looks like in the frame of the Boltzmann equation.

I consider a rarefied gas and assume that its evolution can be described by the following Boltzmann equation

$$\frac{\partial}{\partial t} f_t + v \frac{\partial}{\partial x} f_t = Q(f_t, f_t) \tag{2.4}$$

where Q is the collision operator. I will assume that Q(f,f) vanishes at x if and only if f(x,v) is Maxwellian in v.

To describe quasi-homogeneous, slowly varying states, I fix an initial profile $\beta(x)$, $\bar{v}(x)$ and $\rho(x)$ [here I use the particles' density ρ rather than the chemical potential λ]. I then consider a family f_0^{ε} of initial states, where

$$f_0^{\varepsilon}(x,v) = \rho(\varepsilon x) \left[(2\pi)^{-1} \beta(\varepsilon x) m \right]^{\frac{1}{2}d} \exp\left[-\beta(\varepsilon x) \tfrac{1}{2} m (v - \bar{v}(\varepsilon x))^2 \right] \tag{2.5}$$

The behavior for small ε describes the small gradient, hydrodynamically ideal regime. Let $f_t^{\varepsilon}(x,v)$ denote the solution of eq.(2.4) with initial datum as in eq.(2.5). Then, according to eq.(2.3), set

$$F_t^{\varepsilon}(x,v) = f^{\varepsilon}_{\varepsilon^{-1}t}(\varepsilon^{-1}x,v)$$

i.e. x and t in F are the rescaled [macroscopic] space and time variables. It is easy to see that F_t^{ε} satisfies an equation like eq.(2.4) with $\varepsilon^{-1}Q$ in the place of Q [and, of course, with initial condition F_0^{ε}]. We are therefore in the setup of the Hilbert, Chapman - Enskog expansion. Namely for small ε one tries an asymptotic expansion of the form

$$F_t^{\varepsilon}(x,v) = \rho_t(x)\,[\,(2\pi)^{-1}\,\beta_t(x)m]^{\frac{1}{2}d}\,\exp[-\,\beta_t(x)\tfrac{1}{2}m(v-\bar{v}_t(x))^2]\,(1+\varepsilon(\cdot)+..) \tag{2.6}$$

The expansion is formally consistent if $\beta_t, \bar{v}_t$ and ρ_t are chosen to be the solution of the Euler equation [with initial value $\beta, \bar{v}, \rho$], and of the Navier-Stokes equation, for getting higher order terms in the expansion. We refer to [C] for a survey of the rigorous results obtained along this line.

Coming back to the local equilibrium assumption, L E A, let me first notice that L E A is a sufficient but certainly not a necessary condition for deriving the Euler equations. In the next section we shall see that a weaker condition, the so called "propagation of chaos", is in fact sufficient for deriving the [analogue of the] hy-

hydrodynamical equations for stochastic systems.

An attempt toward a derivation of the Euler equations with a condition weaker than L E A has been recently proposed by Sinai, [Si].

As we shall see in Section 4, we expect that L E A fails when shock waves phenomena are present. Since these are connected with singularities of the solutions of the Euler equations, it becomes clear that L E A and the hypothesis that the solution of the Euler equation is smooth, are in fact interrelated.

It is clear that L E A or any other similar assumption should put together "mixing like" properties of the time evolution and space-time scaling properties of the system. It is not known how much stronger the assumption should be than assuming good ergodic properties for all the extremal equilibrium states of the system. In the next section we will have a partial answer to the question, when we will consider the space time fluctuations in equilibrium.

We conclude the section by noticing that Morrey's procedure to derive the hydrodynamical behavior involves (1) a sequence of initial states like in eq.(2.2) and (2) a suitable space-time scaling when reading out the evolution from the different initial states, eq.(2.3). At no step we change the equations of motion, which are always the same. Other limits might be more appropriate in different physical situations; for instance long range forces, like the electro-magnetic ones, can be described by Vlasov equations, which involve a mean field limit procedure. I shall not consider such problems in this article so I simply refer to some recent excellent surveys, [MP] on the vortex and discrete approximations to the Euler and Navier-Stokes equations and [M] on the martingales methods for the macroscopic limit in mean field interacting particle systems.

3. Hydrodynamical behavior of random particle systems.

To be definite I will refer to a specific model and then briefly indicate which results extend to other systems. The model I consider is the symmetric zero range process on $\mathbb{Z}$ with jump intensity c(k) = 1 for $k \geq 1$. This is a system of particles moving on $\mathbb{Z}$. $\xi(x)$, $x \in \mathbb{Z}$, denotes the number of points at site x. The generator L of the process is such that for all bounded cylindrical functions f on $\mathbb{N}^{\mathbb{Z}}$

$$Lf = \frac{1}{2} \sum_{x \in \mathbb{Z}} \mathbb{1}(\xi(x) > 0) \left[f(\xi^{x,x+1}) + f(\xi^{x,x-1}) - 2f(\xi)\right] \tag{3.1}$$

where

$$\xi^{x,y}_x = \xi_x - 1 \qquad \xi^{x,y}_y = \xi_y + 1 \qquad \xi^{x,y}_z = \xi_z \quad z \neq x,y$$

It can be proven, cf. for instance [Li], that L defines a Feller semigroup T(t) and that the extremal equilibrium measures are probabilities μ_ρ, $\rho \geq 0$, which are product measures on $\mathbb{N}^{\mathbb{Z}}$ and give geometric distribution, with parameter ρ, to the $\xi(x)$ s, i.e.

$$\mu_\rho(\xi(x) = n) = (1+\rho)^{-n-1} \rho^n \qquad n \in \mathbb{N} \tag{3.2}$$

We have the following theorem:

3.1 <u>Theorem</u> ([FPV])

For $\varepsilon \in (0,1]$ *let* μ^{ε} *be the product probability on* $\mathbb{N}^{\mathbb{Z}}$ *which gives a geometric distribution to* $\xi(x)$ *with parameter* $\rho(\varepsilon x)$, *where* $\rho(r)$ *is a* C^{∞} *function on* $\mathbb{R}$ *such that*

$$0 < \rho' \leq \rho(r) \leq \rho'' < \infty \tag{3.3}$$

Then for every $r \in \mathbb{R}$, $\tau \geq 0$

$$\lim \quad \mu^{\varepsilon}(T(\varepsilon^{-2}\tau) S([\varepsilon^{-1}r]) f) = \mu_{\rho(r,\tau)}(f) \tag{3.4}$$

where $S(x)$ *denotes space shift by x [[a] denotes the integer part of a] and* $T(t)$ *is the semigroup with generator L given in eq.(3.1).* $\rho(r,\tau)$ *is the unique solution of the diffusion equation*

$$\frac{\partial}{\partial\tau}\rho(r,\tau) = \frac{1}{2}\frac{\partial}{\partial r}[D(\rho(r,\tau))\frac{\partial}{\partial r}\rho(r,\tau)] \qquad \rho(r,0) = \rho(r) \tag{3.5}$$

where the bulk diffusion coefficient D is given by

$$D(\rho) = (1+\rho)^{-2} \tag{3.6}$$

3.2 Remarks

(a) The above theorem is proven in [FPV], cf. Theorem 0.2 there. The assumption on the initial measure μ^{ε} can be considerably weakened.

(b) Notice that in Theorem 3.1 time is scaled like the square of the space: the absence of velocities [drifts] make the Euler limit as in eq.(2.3) trivial, hence the necessity of a diffusive scaling.

(c) Local equilibrium as in eq.(3.4) has been also proven for the following diffusive systems:

- the symmetric simple exclusion process, [GKMP],[GKS],[DIP], [DFIP],[DIPP]
- the voter model [PS]
- a stochastic evolution related to harmonic chains of oscillators [KMP]
- some branching processes [DS]
- a Spitzer model for traffic [F], [P Sc]
- stationary zero range models with external sources [DF]

In the Euler limit, systems with a drift, local equilibrium has been proven in the following cases:

- asymmetric simple exclusion processes [R], [BF]
- some zero range models [AK], [W]

Local equilibrium has been proven also for continuum systems, like in [R,2], where Wiener particles with hard core interactions are considered.

Also for some, degenerate, deterministic system local equilibrium has been proven, the gas of one dimensional hard rods, [BDS], and the chain of harmonic oscillators [DPST].

A corollary of Theorem 3.1 and in general of the validity of local equilibrium is the following :

3.3 Corollary

Under the same assumptions [and with the same notation] of Theorem 3.1 the following holds.

For $\varphi \in \mathcal{S}(\mathbb{R})$ *let*

$$X_\tau^\varepsilon(\varphi) = \varepsilon \sum_x \varphi(\varepsilon x)\, \xi(x, \varepsilon^{-2}\tau)$$

where $\xi(x,t)$ *denotes the occupation number at site x and time t. Let* $\mathbb{P}_{\mu^\varepsilon}$ *be the law on* $D(\mathbb{R}_+, \mathcal{S}'(\mathbb{R}))$ *induced by the* $X_\tau^\varepsilon(\varphi)$ *as variables in the zero range process with initial law* μ^ε. *Then* $\mathbb{P}_{\mu^\varepsilon}$ *converges weakly on the compacts to* $\mathbb{P}$ *, which is the law on* $D(\mathbb{R}_+, \mathcal{S}'(\mathbb{R}))$ *supported by the single trajectory* $\rho(r,\tau)$ *solution of eq.(3.5).*

3.4 Remarks

(a) The property proven in Corollary 3.3 is often taken as a basis for defining hydrodynamical behavior and hydrodynamical equations.

In this way one can avoid proving local equilibrium, as in Theorem 3.2.

(b) The fluctuations of $X^{\varepsilon}_{\tau}(\varphi)$ around its mean $\langle X^{\varepsilon}_{\tau}(\varphi)\rangle$ are given by

$$P_{\mu^{\varepsilon}}\left((X^{\varepsilon}_{\tau}(\varphi) - \langle X^{\varepsilon}_{\tau}(\varphi)\rangle)^2\right) = \varepsilon^2 \sum_{x,y} \varphi(\varepsilon x)\varphi(\varepsilon y)$$

$$P_{\mu^{\varepsilon}}\left((\xi(x,t) - \langle\cdot\rangle)(\xi(y,t) - \langle\cdot\rangle)\right) \tag{3.8}$$

where $t = \varepsilon^{-2}\tau$ and $P_{\mu^{\varepsilon}}$ is the law of the zero range process with initial measure μ^{ε}. The r.h.s. of eq.(3.8) vanishes when ε goes to zero if (1) $\mu^{\varepsilon}(\xi(x,t)^2)$ is uniformly bounded and (2) if $\xi(x,t)$ and $\xi(y,t)$ become mutually independent. This last request is however needed not for all $x \neq y$, but only when $|x-y|$ is larger than some value which can increase to infinity as ε vanishes. This is much less than local equilibrium which amounts to prove that $\xi(x)$ and $\xi(y)$ become independent for all $x \neq y$, and then that the limiting laws of the $\xi(x)$s are geometric.
The kind of independece needed for the r.h.s of eq.(3.8) to vanish [when ε goes to zero] is basically what is called the "propagation of chaos". This is directly connected to a result like that in Corollary 3.3 : we shall find examples where local equilibrium does not hold while propagation of chaos is present and convergence to a single trajectory occurs.

It is natural, after the density field, to study its fluctuations, namely the variables

$$Y_\tau^\varepsilon(\varphi) = \varepsilon^{\frac{1}{2}} \sum_{x} \varphi(\varepsilon x) \left[\xi(x, \varepsilon^{-2}\tau) - \langle \cdot \rangle \right] \tag{3.9}$$

where $\varphi \in S(\mathbb{R})$. An advantage of considering the fluctuations is that they have an interesting structure also at equilibrium, where the analysis is technically easier. The result in the following theorem refers to the zero range process considered so far, but its validity extends to a large class of models, as it will be discussed in the Remarks 3.6.

3.5 <u>Theorem</u> (Fluctuation-dissipation theorem)

Let P^ε be the law on $D(\mathbb{R}_+, S'(\mathbb{R}))$ induced by the $Y_\tau^\varepsilon(\varphi)$ as variables in the zero range process with initial measure μ_ρ.
Then P^ε converges weakly [when ε goes to zero] to a probability P supported by $C(\mathbb{R}_+, S'(\mathbb{R}))$. P is the law of a stationary Ornstein-Uhlenbeck generalized process, namely the Gaussian process with zero averages and covariance

$$P(Y_0(\varphi)\, Y_\tau(\psi)) = \iint dr\, dr'\; \varphi(r)\, \psi(r')\; \chi(\rho)\, (2\pi D(\rho)\tau)^{-\frac{1}{2}}$$

$$\exp[-(2D(\rho)\tau)^{-1}(r-r')^2] \tag{3.10}$$

$$\chi(\rho) = \sum_{x} \mu_\rho(\xi(0)[\xi(x) - \rho]) \tag{3.11}$$

where $D(\rho)$ *is the same as the bulk diffusion coefficient given in eq.(3.6) and* $Y_\tau(\varphi)$ *denotes the canonical process in* $C(\mathbb{R}_+, \mathcal{S}'(\mathbb{R}))$.

3.6 Remarks

(a) The above theorem is proven in [BR] where the result is obtained for a large class of symmetric zero range processes. It has then been extended to some exclusion with speed change system [DPSW] and recently to the physically more interesting case of interacting brownian particles [S1].

(b) The result might appear surprising, since it relates the covariance of the fluctuations in equilibrium to a parameter entering in the non equilibrium evolution, i.e. the diffusion coefficient in eq.(3.5). From a physical point of view this is very well understood and it goes under the name of " fluctuation dissipation theorem". It in fact relates non equilibrium dissipative effects to the equilibrium fluctuations.

We give below a heuristic argument to understand such connection, which is based on the observation that the kernel of the different timescovariance satisfies the linearized version of eq.(3.5).

We have

$$P(Y_0(\varphi)\, Y_\tau(\psi)) = \lim \mu_\rho\Big(\varepsilon^{\frac{1}{2}} \sum_x \varphi(\varepsilon x)(\xi(x)-\rho)\, \varepsilon^{\frac{1}{2}} \sum_y \psi(\varepsilon y) (\xi(y, \varepsilon^{-2}\tau) - \rho)\Big)$$

$$= \lim \mu_\rho \Big(\sum_x \varphi(\varepsilon x)(\xi(x)-\rho) \quad \varepsilon \sum_y \varphi(\varepsilon y) \ \xi(y, \varepsilon^{-2}\tau) \Big) \tag{3.12}$$

here μ_ρ denotes the law of the whole zero range process. The last term is the expected value of the density field $X^{\varepsilon}_{\tau}(\varphi)$ w.r.t. the signed measure $\mu_\rho \circ \sum_x \varphi(\varepsilon x)(\xi(x)-\rho)$. This is the first order term, in some small parameter λ, of the expansion of the probability $\mu_\rho \circ \exp[\lambda \sum_x \varphi(\varepsilon x)(\xi(x)-\rho)]$ times a normalization factor. The time zero density profile determined by such measure is

$$\rho(r,0) = \rho + \lambda \chi(\rho)\varphi(r) + o(\lambda) := \rho + \lambda \delta\rho(r,0) + o(\lambda) \tag{3.13}$$

At later times $\rho(r,0)$ will become $\rho(r,\tau)$, which is expected to be the solution of the diffusion equation (3.5), hence for small λ $\delta\rho(r,\tau)$ should satisfy the linearized diffusion equation

$$\frac{\partial}{\partial\tau}\delta\rho = \frac{1}{2} D(\rho) \frac{\partial^2}{\partial r^2}\delta\rho, \quad \delta\rho(r,0) = \chi(\rho)\varphi(r) \tag{3.14}$$

By inserting $\delta\rho(r,\tau)$ in eq.(3.12) we then get eq.(3.10).

(c) The proof of the theorem is based on the Holley and Stroock theory [HS1] and on a property of the stochastic system we are considering, which was first conjectured by Hermann Rost and then proven for the models considered in (a) above. After some manipulations the property can be reduced to the following one. Let f be a bounded cylindrical function on $\mathbb{N}^{\mathbb{Z}}$. The usual mixing condition is

$$\lim_{t\to\infty} \mu_\rho\big([f - \langle f \rangle] \, T(t) \, S(x) f \big) = 0 \tag{3.15}$$

where x is any fixed point in $\mathbb{Z}$. In the proof of the fluctuation dissipation theorem we have instead the expression

$$\mu_\rho([f- \langle f\rangle] T(t) \sum_{x\in\mathbb{Z}} S(x) f) \qquad (3.16)$$

and one needs to compute its limiting value when t diverges. Analogous expression for the Glauber dynamics in an Ising model would still go to zero, as noticed by Herbert Spohn, [S2]. In our case the presence of a conserved quantity, the particles'number, changes the asymptotics in eq.(3.16): to get zero in the limit one should subtract to f not only its average, $\langle f\rangle$, but also another term, namely $c_f(\xi(x) - \rho)$, where $c_f = \frac{d}{d\rho}\mu_\rho(f)$.
The physical interpretation is based on the fact the density fluctuations [referring to a conserved quantity] change in a slowlier time scale than the other fields, which therefore average out leaving only their component along the density fluctuation field, c_f. This provides an example of the kind of extra condition needed for proving hydrodynamical behavior, namely what conditions should be added to a "mixing condition" typically arising from the ergodic analysis of the equilibrium time evolution.

(d) Fluctuations can be studied also in non equilibrium. The results are very few, they refer to the zero range model we have been considering so far, [FPV], to simple exclusion [GKS],[DIPP] and to the voter model [HS4] and [PS]. Also in the non equilibrium case one finds that the different times covariance satisfies the linearized version of eq.(3.5).

4. Shock waves

The smoothing effect due to the diffusive term in the hydrodynamical equations considered in the previous section is lost in the Euler scaling. Singularities might develop and persist in time,and, as a consequence, the Euler equation might admit more than one solution.

At the physical level one has formation and propagation of shock waves. The onset of the phenomenon has been studied within the frame of the Boltzmann equation, cf. [C]. The phenomenon appears also for some discrete velocity models of Boltzmann equation, where computation can be carried to a deeper extent, [B], [Mo].

The analysis can be pushed through with great details in a few stochastic models. We shall see that the microscopic counterpart of the singularities present in the solutions of the Euler equations is the occurrence of a "dynamical phase transition ", in the terminology introduced by Dave Wick, [W]. Local equilibrium fails at the front wave, and a superposition of extremal states describes the micrsocopic state there, [cf. [T] for a survey on shock waves in stochastic systems].

Like before I will consider a specific model and I will only briefly recall which are the extensions. The model I will discuss is a slight modification of that considered in Section 3 : to have Euler scaling I consider the generator $\hat{L}$ in the place of L, where

$$\hat{L} f = \sum_{x} \mathbb{1}(\xi_x > 0) \; [\; f(\xi^{x,x+1}) - f(\xi)] \qquad (4.1)$$

Namely particles can only jump to the site on their right. The invariant measures for the corresponding Feller semigroup $\hat{T}(t)$ are the same as in the symmetric case.

4.1 Theorem ([AK],[W])

Let μ^{ε} be as in Theorem 3.1. Assume further that either (1) $\rho(r)$ is smooth and decreasing or (2) increasing and piecewise continuous. Let $\rho(r,\tau)$ be the entropic solution, [La], of

$$\frac{\partial \rho}{\partial \tau} = - v(\rho)\frac{\partial \rho}{\partial r} \qquad \rho(r,0) = \rho(r) \tag{4.2}$$

where

$$v(\rho) = (1+\rho)^{-2} \tag{4.3}$$

Then for any point (r,τ) of continuity for $\rho(r,\tau)$

$$\lim \mu^{\varepsilon}(\hat{T}(\varepsilon^{-1}\tau)\, S([\varepsilon^{-1}r])\, f) = \mu_{\rho(r,\tau)}(f) \tag{4.4}$$

for all bounded cylindrical functions f.

4.2 Remarks

(a) The theorem has been proven in [AK] assuming in the case (2) that $\rho(r)$ is piecewise constant. The extension to the above formulation is due to D. Wick, [W].

(b) The above theorem specifies which one, among the infinitely many possibilities, is the solution of eq.(4.2) which describes the evolution of the system in the Euler limit.
One might try to interpret the answer by saying that the small fluctuations, naturally present when one approximates the many particles system with a deterministic differential equation, can be described by adding to the r.h.s. of eq.(4.2) a noise term of the form $\lambda \frac{\partial^2 \rho}{\partial r^2}$, λ being some small number. The solution of the new equation, in the limit as λ goes to zero, becomes the entropic solution of eq.(4.2).

(c) Also the asymmetric simple exclusion process admits shock waves, the analogue of Theorem 4.1 has been proven in [R] [BF], cf. also [Li].

Theorem 4.1 does not say what happens at the singularities, at a microscopic level. The question is answered in the following

4.3 Theorem ([W])

Let μ be the product measure on $\mathbb{N}^{\mathbb{Z}}$ such that

$$\mu(\xi(x) = k) = \mu_{\rho(x)}(\xi(x) = k) \tag{4.5}$$

$$\rho(x) = 0 \text{ if } x < 0 \quad \text{and} \; = \rho_0 \text{ if } x \geq 0 \tag{4.6}$$

Then there is T(t) such that for any $\delta > 0$ and any bounded cylindrical function f on $N^{\mathbb{Z}}$ the following holds:

$$1 = \lim_{t\to\infty} \sup_x P_\mu\Big(T^{-1}\int_t^{t+T} (S(x)f)(\xi_s)\, ds \in \{[\mu_{\rho_0}(f) - \delta, \mu_{\rho_0}(f) + \delta] \cup [\mu_0(f) - \delta, \mu_0(f) + \delta]\}\Big) \tag{4.7}$$

where P_μ is the law of the zero range process with initial measure μ.

Furthermore let $\bar{\lambda}(r,t)$ be the solution of the equation

$$\frac{\partial\bar{\lambda}}{\partial t} = -v(\rho_0)\frac{\partial\bar{\lambda}}{\partial r} + \frac{1}{2}D(\rho_0)\frac{\partial^2}{\partial r^2}\bar{\lambda} \qquad \bar{\lambda}(r,0) = 0 \text{ if } r<0, = 1 \text{ if } r \geq 0 \tag{4.8}$$

where $v(\rho)$ is defined in eq.(4.3) and $D(\rho)$ in eq.(3.6).

Then given δ, $T(t)$ and f as above

$$\lim_{t\to\infty} \sup_x \Big| \bar{\lambda}(x,t) - P_\mu\Big(\Big|T^{-1}\int_t^{t+T} (S(x)f)(\xi_s)\, ds - \mu_{\rho_0}(f)\Big| < \delta\Big) \Big| = 0 \tag{4.9}$$

As a consequence of Theorem 4.3 we have that

$$\lim_{\varepsilon \to 0} T_{\varepsilon}^{-1} \int_{t\varepsilon^{-1}}^{t\varepsilon^{-1}+T_{\varepsilon}} ds\ T(s)\ S(\ v(\rho_0)t\ \varepsilon^{-1} + r\ \varepsilon^{-\frac{1}{2}})\ c\ \mu$$

$$= [\ 1 - \lambda(r,t)]\ \mu_0 + \lambda(r,t)\ \mu_{\rho_c} \tag{4.10}$$

$$[\ \lim T_{\varepsilon} = \infty\ ]$$

where $\lambda(r,t)$ satisfies the equation

$$\frac{\partial \lambda}{\partial t} = \frac{1}{2}\ D(\rho_0)\ \frac{\partial^2 \lambda}{\partial r^2} \tag{4.11}$$

with initial condition

$$\lambda\ (r,0) = 1 \quad \text{if} \quad r \geq 0 \quad \text{and} \quad = 0 \quad \text{if} \quad r < 0 \tag{4.12}$$

Namely

$$\bar{\lambda}\ (\ \varepsilon^{-1} v(\ \rho_0)t + \varepsilon^{-\frac{1}{2}} r,\ \varepsilon^{-1} t) = \lambda(r,t) \tag{4.13}$$

The interpretation of the result is therefore the following. Viewed on a scale of order $\varepsilon^{-\frac{1}{2}}$ the system is in a statistical mixture of pure states with parameter varying smoothly <u>on this scale</u> from zero to one.

There is another model where the shock wave region can be studied at a microscopic level, [GP], and I will describe it in the remaining of this section. I want to draw the attention on the model because I think there are still several interesting features to study there.

The model is the contact process. Particle are in $\mathbb{Z}^d$. They die independently of each other according to a Poisson time of intensity 1. They can also create new particles at neighboring sites, with intensity λ . There is a critical value λ_c above which there is an invariant state μ with positive density, besides the other extremal invariant state δ_ϕ supported by the configuration with no particles. Below λ_c only δ_ϕ is present, [Li],[D].

In [GP] the one-dimensional case with $\lambda > \lambda_c$ is considered. The initial measure μ_0 is chosen as the probability supported by the configuration with half line full and the other empty, namely $\eta(x) = 1$ for $x \leq 0$ and $= 0$ for $x > 0$ [$\eta(x) = 0,1$ is the occupation number at x]. The same kind of results as in Theorem 4.3 are proven in [GP].

The model is somewhat simpler than those previously considered, yet the phenomenology it exhibits is still very interesting. I think the next step should be the analysis of the two [or more] dimensional case. In this case μ_0 would be the measure supported by the configuration with the left half plane full and the right one empty. The propagation will have some distinct geometrical pattern and it would be nice to understand its structure.

I am too ignorant to say more about it, but, conforted by the many techniques developed so far, [Li], I hope to learn and to find the right tools for dealing with this problem.

5. Metastable states and transition to equilibrium.

In this section I will closely follow the approach proposed by M. Cassandro, A. Galves, E. Olivieri and M.E. Vares in [CGOV].

Let me introduce the main ideas discussing a model which we would like to see solved, but it has not, so far.
The model is the two dimensional Ising system with n.n. interaction. The system evolves in time according to Glauber dynamics and we will consider the case when the temperature is below the critical one. The system is in a large square box Λ with zero boundary conditions. The invariant measure is the Gibbs measure to which any other measure converges exponentially fast [the Markov evolution being Doblin]. In the infinite volume case there are two extremal invariant measures so we expect some consequence of that also at finite volume. More precisely let us take a realization ω of the process, $\omega \in D(\mathbb{R}_+, \{-1,1\}^\Lambda)$, and let us take the average on a suitable time interval T starting from any time t. Then we expect that the state

$$T^{-1} \int_t^{t+T} \delta_{\omega(s)} \, ds \tag{5.1}$$

[where $\delta_{\omega(s)}$ is the delta measure on $\{-1.1\}^\Lambda$ supported by $\omega(s)$] will be close to one of the extremal equilibria. More precisely this is what we would like to prove. Given any $\alpha > 0$ and a cylindrical function f, if the temperature, defining the Glauber dynamics, is small enough, then there is T such that, for Λ suitably large,

$$P(\{\omega : T^{-1} \int_{t}^{t+T} f(\omega(s))\, ds \in [\mu_-(f)-\alpha, \mu_-(f)+\alpha] \cup [\mu_+(f)-\alpha, \mu_+(f)+\alpha]\}) > 1 - \alpha \qquad (5.2)$$

where P is the law of the process with initial measure the Gibbs measure , and $\mu_\pm$ are the two extremal infinite volume equilibrium states. We also expect that if we follow a typical realization we will keep observing the same extremal state until a sudden transition to the other state occurs. This will happen in a very short time, much shorter than the persistence time of the previous equilibrium. The moment of transition should be unpredictable, that is, its distribution should be Poisson with characteristic time much longer than the length of the transition.

Some remarks on the above before going on. Firstly notice that the picture is in a sense similar to that in Section 3, when smooth hydrodynamical equations were derived. The system "likes" to stay close to the extremal equilibrium states, local equilibrium. In Section 3 a continuous family of equilibria was available, so that the state could change smoothly [in space and time] from one to the other. Here we have only two equilibrium states so that the transition is by necessity abrupt.

In facts the picture I have been drawing so far is not yet really definite. What is missing is the space time structure of the phenomenon: where the transition starts from and how it propagates. From this point of view the problem gets closer to that in Section 4.

Strictly speaking, if we want to study metastability we should consider the case when there is a small external magnetic field h, say $h > 0$. We would then start from an initial state close to μ_- [the stationary state when there is a magnetic field $-h$]. If h is very small we expect that the state μ_- persists for a long time before the transition to the "real" equilibrium μ_+ occurs. The transition back to μ_- will require a much longer wait and it becomes infinite when either the temperature goes to zero or the region Λ becomes infinite.

The above calls back to mind the Friedlin and Ventsel theory of the small random pertubations of deterministic evolutions. This is not fortuitous since the Friedlin and Ventsel theory plays here a basic role: we leave the small fluctuations of Section 3 to enter into the large deviation theory.

The simplest model to test the above ideas is the following. Consider the stochastic differential equation

$$dx = -\frac{\partial V}{\partial x}\, dt + \varepsilon\, dW \qquad x \in \mathbb{R} \tag{5.3}$$

where V is a double well potential with one minimum below the other. W is a standard Wiener process and ε is a small parameter, which eventually goes to zero.

The states μ_- and μ_+ are replaced by states corresponding to the two minima of V.

The scheme outlined in the beginning of

this section for the Ising model has been carried through in [CGOV] for a discrete version of eq.(5.3). The model has then been reconsidered in [KN] . An extension to more dimensions has been obtained in [GOV]. A model similar in a sense to the above ones is that considered by Dawson in [D]. He has in fact many Wiener particles which move in an external double well potential and interact with each other via a mean field type of interaction. It turns out that for suitable values of the parameters the particles prefer to stay close together either in one or the other of the two wells. Dawson studies the critical fluctuations for this system.

The above models miss the space structure essential in the analysis of metastability in real physical systems. A first step in this direction has been established in [CGOV], where the authors study [besides the already mentioned discrete version of eq.(5.3)] the metastable behavior present in the contact process.

Very recently in [COP] metastability is studied for an infinite dimensional version of eq.(5.3). They consider a stochastic equation of the form

$$d\phi(x,t) = -V'(\phi(x,t))dt + \frac{\partial^2\phi}{\partial x^2}\, dt + \varepsilon\, dW \tag{5.4}$$

where V is a duouble well potential with two minima and dW is a white noise in space and time; x ranges in a finite interval [-L,L].

One should think of the above equation as representing a vibrating string in an external field and in the presence of stochastic perturbations. The heigth of the string at (x,t) is $\phi(x,t) \in \mathbb{R}$. The elastic interaction with its neighborhoods is taken into account by the term $\frac{\partial^2 \phi}{\partial x^2}$. The external potential V favors two positions of the string, corresponding to the minima of V. The term $\varepsilon\, dW$ is the noise which simulates interactions with a thermal bath.

The case of symmetric V $[= \phi^4 - \phi^2]$ has been studied in [FI] where the authors characterize the tunneling effect which determines the transition from one minima to the other. The model has been further studied in [COP] where also the asymmetric case [metastability] has been considered. The way the transition appears is studied as a function of the boundary conditions for the string. According to the way they are chosen one can select the space region where the transition starts from and the way it propagates. In [COP] the string becomes infinite and at the same time the noise goes suitably to zero. In this way the model can be interpreted as a Kac Lebowitz and Penrose limit starting from an Ising model.

There are more open, intersting, problems than solved in this line. One is, for instance, an analysis of a model in an infinite region, which, ultimately leads to the study of the infinite two dimensional Ising model.

Another problem I consider very interesting is that of deriving eq.(5.4) as a limit of some particle system with short range interactions. A possible model is that proposed by Joel Lebowitz. He considers a system of particles in $\mathbb{Z}$ (the same can be extended to more dimensions). Each site is occupied by one particle and its spin has value ±1. Time evolution is determined by two effects. There is a spin flip process which is the usual Glauber dynamics with nearest neigbor ferromagnetic interactions. Besides this there is a stirring simple exclusion process for the particles. Namely there is a Poisson process [each one independent of the others] at each pair of n.n. sites. When the event occurs the particles which are there exchange their positions, one with the other. Denote by L_G the Glauber generator and by L_e the simple exclusion stirring generator. The generator of the whole process is $L_G + \lambda L_e$, where λ is a suitable parameter, to be scaled with space.

To simulate eq.(5.4) one takes $\lambda = \varepsilon^{-2}$ and considers profiles which vary on a space scale ε^{-1}. In [DFL] it is proven that the spin density field

$$X_t^{\varepsilon}(\varphi) = \varepsilon \sum_x \varphi(\varepsilon x)\sigma(x,t) \quad , \varphi \in \mathcal{S}(\mathbb{R}) \tag{5.5}$$

induces [under suitable assumptions on the initial measure for the process (μ^{ε})] a law on $D(\mathbb{R}_+, \mathcal{S}'(\mathbb{R}))$ which converges weakly to the law supported by the trajectory m(r,t) which satisfies the equation

$$m(r,t) = - V'(m(r,t)\, dt + \frac{1}{2}\frac{\partial^2}{(\partial r)^2} m \tag{5.6}$$

$$V(m) = \alpha m^4 - \beta m^2 \tag{5.7}$$

where α and β are determined by the parameters which enter in L_G.

In [DFL] the flucuation field around m(r,t) [cf. Section 3] is also studied: convergence to a generalized Ornstein Uhlenbeck process is proven, like in the non equilibrium version of Theorem 3.5.

The next natural question [at present completely open] is to study the large deviations and the metastable behavior in this model.

6. Concluding remarks

Here I briefly report some of the points emerged during the discussions after the lecture I gave at the II Bibos symposium in Bielefeld, my lecture being based on the present article.

Michiel Hazewinkel pointed out that in connections with the topics discussed in Section 5 the works of Zeev Schuss and Johan Grasmann on exit times of stochastically perturbed differential equations might be relevant.

A generalization of eq.(5.4) to two dimensions, not mentioned in Section 5, leads to the study of Euclidean field theory in one space dimension. In [IM] the case of ϕ^4 potentials is examined.

Gianna Nappo and Enza Orlandi are presently studying a model of particles which in an appropriate limit should lead to an equation of the form (5.6). In this case the analysis of the fluctuations and of the large deviations might be easier than in [DFL].

As Geoffrey Sewell pointed out the main fact still missing in constructing the hydrodynamics of Stochastic Models is the analysis of the Navier-Stokes limit in a system with non trivial Euler equation. Some ideas in this direction can be drawn from [DS], cf. also Section 3 of [DIPP]. The question can be posed also in an equilibrium frame, by studying equilibrium fluctuations.
Such analysis has been carried through for one dimensional hard rods systems by Herbert Spohn, [S3].
The non equilibrium case, in the same model, is presently under investigation by Carlo Boldrighini and Dave Wick, the analysis starts from the scheme established in [BDS].

The limit ε going to zero is related to the limit Ω going to infinity, in the Van Kampen's approach.
One might view the Markov process introduced by Van Kampen after partitioning the phase of the system as the basic stochastic process which I consider in Section 3. In this respect the particles in the models considered in Section 3 are interpreted as individual excitations of the system [the qualitative feature to be preserved in the transcription is the conservation law, which is responsible for the long time tails, characteristics of the hydrodynamical regime]. As an example the simple exclusion process can be interpreted in terms of a local thermalization process in an Ising model. cf. [GKMP] and also [KMP].

References

[AK] E.D. Andjel, C. Kipnis. Derivation of the hydrodynamical equations for the zero range interaction process.
Annals of Probability 12,325-334 (1984)

[B] J. Bradwell Shock structure in a simple discrete velocity gas.
Phys. of Fluids 8, 1243 (1964)

[BDS] C. Boldrighini, R.L. Dobrushin, Yu.M. Sukhov. One dimensional hard core caricature of hydrodynamics.
J. Stat. Phys. 31, 577 (1983)

[BF] A. Benassi, J.P. Fouque. Loi de grands nombres pour l'exclusion simple non symetrique.
Preprint 1984

[BR] T. Brox, H. Rost. Equilibrium fluctuations of stochastic particle systems: the role of conserved quantities.
Annals of Probability 12, 744 (1984)

[C] R.E. Caflish. Fluid dynamics and the Boltzmann equation.
Studies in Statistical Mechanics, Vol. 10, "Non equilibrium phenomena" (1983). North Holland.

[CGOV] M. Cassandro, A. Galves,, E. Olivieri, M.E. Vares.
Metastable behavior of stochastic dynamics: a pathwise approach.
J. Stat. Phys. 35, 603-634 (1984)

[COP] M. Cassandro, E. Olivieri, P. Picco.
In preparation.

[D] R. Durrett Oriented percolation in two dimensions.
Annals of Probability 12,999-1040 (1984)

[Da] D. Dawson. Critical dynamics and fluctuations for a mean field model of cooperative behaviour.
J. Stat. Phys. 31, 1 (1983)

[DF] A. De Masi, P. Ferrari. A remark on the hydrodynamics of the zero range process.
J. Stat. Phys. 36,81-87 (1984)

[DFIP] A. De Masi, P. Ferrari, N. Ianiro, E. Presutti.
Small deviations from local equilibrium for a process which exhibits hydrodynamical behavior. II
J. Stat. Phys. 29, 81 (1982)

[DFL] A. De Masi, P. Ferrari, J.L. Lebowitz. In preparation.

[DIP] A. De Masi, N. Ianiro, E. Presutti. Small deviations

from local equilibrium for a process which exhibits hydrodynamical behavior. I.
J. Stat. Phys. 29,57 (1982)

[DIPP] A. De Masi, N. Ianiro, A. Pellegrinotti, E. Presutti.
A survey of the hydrodynamical behavior of many particle systems.
Stdies in Statistical Mechanics, Vol. 11 "Non equilibrium phenomena II, from stochastics to hydrodynamics".
North Holland (1984)

[DPST] R.L. Dobrushin, A. Pellegrinotti, Yu.M. Sukhov, L. Triolo.
One dimensional harmonic lattice caricature of hydrodynamics.
Preprint 1985

[DPSW] A. De Masi, E. Presutti, H. Spohn, D. Wick.
Asymptotic equivalence of fluctuation fields for reversible exclusion with speed change processes.
Preprint 1984.

[DS] R.L. Dobrushin, R. Siegmund-Schultze. The hydrodynamic limit for systems of particles with independent evolution.
Math. Nacr. 105, 199 (1982)

[F] J. Fritz. On the asymptotic behavior of Spitzer's model for evolution of one dimensional point systems.
J. Stat. Phys.

[FI] W. Faris, G. Jona Lasinio. Large fluctuations for a non linear heat equation with noise.
Journ. Phys. A15, 3025 (1982)

[FPV] P. Ferrari, E. Presutti, M.E. Vares. Hydrodynamics of a zero range model.
Submitted to Annals of Probability.

[GKMP] A. Galves, C. Kipnis, C. Marchioro, E. Presutti.
Non equilibrium measures which exhibit a temperature gradient: study of a model.
Commun Math. Phys. 81, 127 (1982)

[GKS] A. Galves, C. Kipnis, H. Spohn. Fluctuations theory for the symmetric simple exclusion process.
Unpublished.

[GOV] A. Galves, E. Olivieri, M.E. Vares. Metastability for a class of dynamical systems subject to small random perturbations.
Preprint IHES/P/84/55. October 1984.

[GP] A. Galves, E. Presutti. In preparation.

[HS1] R. Holley, D.W. Strock. Generalized Ornstein-Uhlenbeck processes and infinite branching Brownian motions.
Kyoto Univ. Res. Inst. Math. Sci. Publ. A14, 741 (1978).

[HS2] R. Holley, D.W. Stroock. Central limit phenomena for various interacting systems.
Ann. Math. 110, 333 (1979)

[KMP] C. Kipnis, C. Marchioro, E. Presutti. Heat flow in an exactly solvable model.
J. Stat. Phys. 22, 67 (1982)

[KN] C. Kipnis, C. Newman. Metastable behavior of infrequently observed, weakly random, one dimensional diffusion processes.
Preprint 1984.

[IM] G. Jona-Lasinio, P.K. Mitter. On the stochastic quantization of field theory.
Preprint PAR LPTHE 84/36 (1985)

[La] P. Lax. Formation and decay of shock waves.
Am. Math. Monthly, March 1972.

[Li] T. Liggett Interacting particle systems.
Springer Verlag 1985.

[Me] M. Metivier. Quelques problemes lies aux systemes infinis de particules et leur limites.
Preprint, October 1984.

[M] C. Morrey. On the derivation of the equations

of hydrodynamics from statistical mechanics.
Comm. Pure Applied Math. 8, 279 (1955)

[Mo] R. Monaco. Shock wave propagation in gas mixtures by means of a discrete velocity model of the Boltzmann equation.
Acta Mechanica (1985)

[MP] C. Marchioro, M. Pulvirenti. Vortex methods in two dimensional fluid dynamics.
Lecture Notes in Physics 203 (1984), Springer-Verlag.

[PS] E. Presutti, H. Spohn. Hydrodynamics of the voter model.
Annals of Probability 11, 867 (1983).

[PSc] E. Presutti, E. Scacciatelli. Time evolution of a one dimensional point system: a note on Fritz's paper.
J. Stat. Phys.

[R] H. Rost. Non equilibrium behavior of a many particle system: density profile and local equilibrium.
Z W 58, 41 (1981)

[R,2] H. Rost. Diffusion de spheres dures en R: comportament macroscopique et equilibre local.
Lecture Notes in Mathematics 1059,127 (1984), Springer-Verlag.

[S1] H. Spohn. Equilibrium fluctuations for interacting brownian particles.
Submitted to Commun. Math. Phys.

[S2] H. Spohn. Large scale behavior of equilibrium time correlation functions for some stochastic Ising model.
Lecture Notes in Physiscs 173, 304 (1982), Springer and Verlag.

[S3] H. Spohn. Hydrodynamical theory for equilibrium time correlation functions of hard rods.
Ann. Phys. 141, 353 (1982).

[Si] Ya. G. Sinai. Lectures given in Budapest, January 1985.

[T] L. Triolo. Particle models for macroscopic equations. Proceedings German-Italian Symposium " Application of Mathematics in Technology". 1984. B.G. Teubner, Stuttgart.

[W] D. Wick. A dynamical phase transition in an infinite particle system.
J. Stat. Phys. 38, 1015 (1985)

BOUNDARY PROBLEMS FOR STOCHASTIC PARTIAL DIFFERENTIAL EQUATIONS

Yu.A. ROZANOV
Steklov Mathematical Institute
Moscow, USSR
and
Research Center Bielefeld-Bochum-Stochastics
University of Bielefeld

§1. Introduction. Boundary Values Problem.

Quite a few well-known stochastic models can be represented by means of a partial differential equation

$$L\xi(t) = \eta(t), \quad t \in T \quad , \tag{1}$$

with a linear differential operator

$$L = \Sigma a_k(t)D^k$$

and a stochastic source η in a domain $T \subseteq R^d$ of a white noise type, say, given as a generalized random field

$$\eta = (\varphi,\eta) \; , \quad \varphi \in C_0^\infty(T) \; ,$$

over a standard Schwartz space $C_0^\infty(T)$ of infinitely differentiable functions $\varphi = \varphi(t)$, $t \in T$, with compact supports $\operatorname{supp} \varphi \subseteq T$. Of course the differential equation (1) has to be treated as a generalized equation

$$(L^*\varphi,\xi) = (\varphi,\eta), \; \varphi \in C_0^\infty(T) \; , \tag{2}$$

for the random function ξ in the domain T_k, where

$$L^* = \Sigma(-1)^k \, D^k a_k(t)$$

is a formal conjugate differential operator (we assume that all coefficients of L are infinitely differentiable).

As an example of ξ given by means of the stochastic differential equation (1) one can take

Brownian Motion

$$\frac{d}{dt}\xi(t) = \eta(t), \quad t \in R^1_+ ,$$

Brownian Sheet

$$\frac{\partial^2}{\partial t_1 \partial t_2}\xi(t) = \eta(t), \quad t \in R^2_+ ,$$

in the domain $T = R^2_+ = \{t : t_1, t_2 > 0\}$ of a variable $t = (t_1, t_2)$ on the plane,

Levy Brownian Motion

$$\Delta^{\ell}\xi(t) = \eta(t), \quad t \in R^d \setminus \{0\} ,$$

in R^d with an odd dimension $d = 2\ell + 1$ where Δ is the Laplace operator,

Generalized Stationary Field

$$L\xi(t) = \eta(t), \quad t \in R^d ,$$

with the operator $L = \Sigma a_k D^k$ which corresponds to a spectral density

$$f(\lambda) = \frac{1}{|\Sigma a_k (i\lambda)^k|^2}, \quad \lambda \in R^d .$$

We would like to note that all these random fields represented by the corresponding equation (1) enjoy the so-called *Markov property* which despite of different approaches has in common that the random field in a domain $S \subseteq T$, is conditionally independent of its part outside of S under fixed boundary conditions on a boundary $\Gamma = \partial S$ given by means of all events from a σ-algebra

$$A(\Gamma) = \cap A(\Gamma^{\varepsilon})$$

where the σ-algebra $A(\Gamma^{\varepsilon})$ represents all events generated by the random field in the ε-neighbourhood of Γ.

What kind of events associated with a boundary behavior of ξ do form the σ-algebra $A(\Gamma)$? One can look for some explicit form of these events as they are in case of Brownian motion $\xi(t)$, $t \in R^1_+$, when for a "future" domain $S = (t, \infty)$ with a current time-point $t > 0$ the corresponding boundary conditions can be set in a form of a fixed boundary value $\xi(t)$ at $\Gamma = \{t\}$. A problem suggested seems to be the most interesting in a case when ξ essentially is the generalized field and there are no values of

$$\xi = (\varphi,\xi), \ \varphi \in C_0^\infty(T) ,$$

which can be treated somehow as boundary values on the boundary Γ .

The generalized equations (1)-(2) give us nothing but a linear functional

$$\xi = (x,\xi), \ x = L^*\varphi ,$$

which is continuous with respect to a norm $\|\varphi\|$ in a standard $L_2(T)$-space because of the very equation (2). One can get an idea to treat $x = L^*\varphi$ as the Schwartz distribution and to extend ξ over all distributions $x = L^*f$ with $f \in L_2(T)$. It turns out that this way leads to some remarkable results.

Namely, assuming the fact that L^* is the non-degenerate operator which for any non-zero $f \in L_2(T)$ gives us non-zero distribution

$$L^*f = (\varphi,L^*f) = (L\varphi,f), \ \varphi \in C_0^\infty(T) ,$$

let us introduce a Hilbert space

$$X = L^* \ L_2(T) \tag{3}$$

with an inner product

$$(L^*f,L^*g)_X = (f,g)_{L_2} , f,g \in L_2(T) .$$

The continuous (in mean square sense) linear functional

$$\xi = (x,\xi) \overset{\text{def}}{=\!=} (\varphi,\eta), \ x = L^*\varphi ,$$

given for all $x \in L^*C_0^\infty(T)$ can be extended by continuity over all $x \in X$ because $C_0^\infty(T)$ is dense in $L_2(T)$-space and

$$L^* : \ L_2(T) \to X$$

is a *unitary* operator so we can set

$$(x,\xi) = \lim(L^*\varphi,\xi) \tag{4}$$

for $x = \lim L^*\varphi$ represented by the limit of the corresponding $L^*\varphi, \ \varphi \in C_0^\infty(T)$.

The corresponding values

$$(x,\xi), \quad \text{supp } x \subseteq \Gamma \tag{5}$$

over all distributions $x \in X$ with supports on the boundary $\Gamma = \partial S$ of the domain $S \subseteq T$ have proved to be the natural boundary values of the random field ξ; in particular they generate the very σ-algebra $A(\Gamma)$ which was above the starting point of our interest. These boundary values can also be used as boundary data in a FORECAST PROBLEM which requires to find the best linear extrapolation of the field ξ over the domain S by means of its values outside of S. The result is that the corresponding FORECAST $\hat{\xi} = \hat{\xi}(t), t \in S$, can be made by means of a partial differential equation

$$L^*L\hat{\xi}(t) = 0, \quad t \in S , \tag{6}$$

with the boundary conditions

$$(x,\hat{\xi}) = (x,\xi), \text{ supp } x \subseteq \Gamma ; \tag{7}$$

as a matter of fact the stochastic boundary problem (6)-(7) for the linear differential operator $L^*L \geq 0$ is of the Dirichlet-Sobolev type and it has the unique solution $\hat{\xi} = \hat{\xi}(t), t \in S$.

Our approach to the stochastic model (1)-(2) is based on the proper continuity of the corresponding generalized source

$$\eta = (\varphi,\eta), \; \varphi \in C_0^\infty(T) ,$$

which has been treated as the white noise in $L_2(T)$-space. This approach can also be applied to another case, for example, to the well-known

Markov Free Field

$$(1-\Delta)\xi(t) = \eta(t), \quad t \in R^d ,$$

with the white noise $\eta = (\varphi,\eta), \varphi \in C_0^\infty(T)$ in the Sobolev space $W = W_2^1(R^d)$ which is a closure of the test functions space $C_0^\infty(R^d)$ with respect to the inner product

$$(u,v)_W = (u,Lv), \; u,v \in C_0^\infty(R^d),$$

associated with the operator $L = 1-\Delta$. To apply now our approach we have to substitute L_2-space by Sobolev's W-space and to introduce the corresponding space of our test distributions $x \in X$

$$X = L^{*}W .$$

This change with respect to (3) is reflected for example in the FORECAST PROBLEM which solution $\hat{\xi} = \hat{\xi}(t)$, $t \quad S$, in the domains $S \subseteq R^{d}$ can be given by means of the corresponding differential equation

$$L\hat{\xi}(t) = 0, \; t \in S ,$$

with the boundary conditions

$$(x,\hat{\xi}) = (x,\xi), \quad \text{supp}\; x \subseteq \Gamma ,$$

on the boundary $\Gamma = \partial S$ (cf. (6)-(7)).

Let us consider the Equation (1) in some domain $T_0 \subseteq T$,

$$L\xi(t) = \eta(t), \; t \in T_0 , \tag{1'}$$

with the random source η,

$$\eta = (\varphi,\eta), \; \varphi \in C_0^{\infty}(T_0) ,$$

continuous in the standard wide sense with respect to $\|\varphi\|_{L_2}$. The Equation (1') has a variety of solutions ξ which are well-defined over all <<test distributions>> $x \in X$ with the corresponding values

$$(x,\xi), \quad \text{supp}\; x \subseteq \bar{T}_0 ,$$

in the closure $\bar{T}_0$ of the domain T_0 continuous with respect to $\|x\|$. The boundary values

$$(x,\xi), \quad \text{supp}\; x \subseteq \Gamma_0 ,$$

on the boundary $\Gamma_0 = \partial T_0$ <u>can be partly determined</u> by the very equation (1') which gives all values

$$(x,\xi) = (f,\eta), \; x = L^{*}f, \; f \in L_2(T_0) ,$$

and in particular the boundary values

$$(x,\xi), \; x \in X^{-}(\Gamma_0) = X(\Gamma_0) \cap L^{*}L_2(T_0)$$

where

$$X(\Gamma_0) = \{x \in X, \quad \text{supp } x \subseteq \Gamma_0\}$$

represents a generalized boundary formed with the generalized variable $x \in X$ on the boundary Γ_0 of the domain T_0 .

It is remarkable that on another part $X^+(\Gamma)$ of the generalized boundary

$$X(\Gamma) = X^-(\Gamma) \oplus X^+(\Gamma)$$

in the direct sum indicated the boundary values

$$(x,\xi) = (x,\xi_{\Gamma_0}), \; x \in X^+(\Gamma_0) \tag{2'}$$

can be given arbitrarily by means of any random function ξ_{Γ_0} of a type considered.

Theorem. The equation (1') with the arbitrarily given boundary conditions (2') has the unique solution ξ which can be represented over all test distributions in the closure of the domain T_0 as

$$(x,\xi) = (Gx,\eta) + (\pi x,\xi_{\Gamma_0})$$

where π is the projection onto the boundary subspace $X^+(\Gamma)$ in X parallel to $L^* L_2(T_0)$ and G is the «Green operator» giving the solution $g = Gx \in L_2(T_0)$ of the conjugate equation

$$L^* g = x - \pi x \; .$$

It turns out that the solution of the stochastic boundary problem (1') - (2') enjoys the Markov property in the case of the boundary conditions (2') independent of the random source η inside of the domain T_0 represented by a generalized random field with independent values. Namely, the following result holds true.

Theorem. Consider the random function ξ in any set S' conditionally independent of its part outside of S' conditioned with its boundary values

$$(x,\xi), \; \text{supp } x \subseteq \Gamma$$

on the boundary $\Gamma = \partial S'$ of S'. The best linear FORECAST of ξ in S by means of all boundary data

$$(x,\xi), \; \text{supp } x \subseteq \Gamma \; , \tag{3'}$$

can be given as the unique solution of the equation

$$L^*L\,u(t) = 0,\ t \in S_0$$

in the interior domain S_0 of S with the boundary conditions (3').

These results can be extended on a very general case in a framework suggested in [1], [2].

§2. General Boundary Conditions for Linear Differential Equations

■ We are going to consider here a general linear differential operator

$$L = \Sigma\, a_k(t) D^k$$

in $L_2(T)$-space and the equation

$$L\,u(t) = f(t),\ t \in S\,, \tag{8}$$

in the domain $S \subseteq T$ with respect to $u \in W$ from a certain functional class W where the operator L will be determined in a very direct way. We do not assume anything about coefficients of this operator (they could be even Schwartz distributions) so from the beginning L is well-defined on the corresponding Schwartz space $C_0^\infty(T)$ of infinitely differentiable test functions $\varphi \in C_0^\infty(T)$,

$$L\varphi = \Sigma\, a_k D^k \varphi \in L_2(T)\ .$$

One can take $W \supseteq C_0^\infty(T)$ in a way that L be continuous with respect to the corresponding norm $\|\varphi\|_W, \varphi \in C_0^\infty(T)$, so it can be extended to the closure $\overline{C_0^\infty(T)}$ in W; we take

$$W = \overline{C_0^\infty(T)}$$

as a completion of $C_0^\infty(T)$ with respect to a seminorm

$$\|\varphi\|_W = \|L\varphi\|_{L_2}$$

where any $u \in W$ can be treated as a limit

$$u = \lim \varphi$$

of the corresponding $\varphi \in C_0^\infty(T)$ and we set

$$Lu = \lim \varphi\ .$$

Of course

$$F = LW = \overline{LC_0^\infty(T)}$$

is a closure of $LC_0^\infty(T)$ in $L_2(T)$ and W is the Hilbert space with the inner product

$$(u,v)_W = (Lu,Lv)_{L_2} .$$

Considering $u \in W$ in the domain $S \subseteq T$ we are dealing with a fairly rich functional class and there is a variety of solutions $u \in W$ of the Equation (8) for any $f \in F$.

It turns out that the solution $u \in W$ is determined uniquely by means of the corresponding boundary conditions

$$(u,x) = (u_\Gamma,x),\ x \in X^+(\Gamma) , \tag{9}$$

which prescribe the <u>boundary values</u> (u,x) of $u \in W$ over certain Schwartz distributions x with supports $\text{supp } x \subset \Gamma$ on the boundary $\Gamma = \partial S$. Namely, let X be a space of all Schwartz distributions

$$x = (\varphi,x),\ \varphi \in C_0^\infty(T) ,$$

continuous with respect to $\|L\varphi\|_{L_2}$. All of them are well-defined on the space W by the corresponding limit

$$(u,x) = \lim\ (\varphi,x)$$

for any $u = \lim \varphi$. We suggest to treat $u \in W$ as a function

$$u = (u,x),\ x \in X ,$$

of a generalized variable X with values (u,x) which are the result of application of the corresponding Schwartz distribution $x \in X$ to $u \in W$. Let us call X the test distributions space. ■

□ To avoid a confusion with complex conjugate values let us assume that we deal with the <u>real</u> $L_2(T)$-space and the <u>conjugate operator</u> L^* is introduced as follows: L^*f for any $f \in L_2(T)$ is a Schwartz distribution

$$L^*f = (\varphi,L\ f) = (L\varphi,f)_{L_2} ,\ \varphi \in C_0^\infty(T) ,$$

given by means of a standard inner product in L_2-space; of course

$$L^*f = \Sigma - (1)^k D^k [a_k f]$$

in a case of L with infinitely differentiable coefficients, say.

Lemma 1. The test distributions space can be represented as

$$X = L^*F = L^*L_2(T) .$$

Proof. Any $x \in X$ gives us a linear functional (φ, x) which is continuous with respect to $L\varphi$ in the Hilbert space

$$F = LW = \overline{LC_0^\infty(T)} \subseteq L_2(T)$$

and it can be represented by means of some $f \in F$ in a form

$$(\varphi, x) = (L\varphi, f)_{L_2} = (\varphi, L^* f), \ \varphi \in C_0^\infty(T) ,$$

so this x coincides with the distribution L^*f ; moreover, for any $f \in F$ the distribution

$$x = (\varphi, L^* f) = (L\varphi, f)_{L_2} , \ \varphi \in C_0^\infty(T) ,$$

is continuous with respect to $L\varphi \in L_2$, so it belongs to our space X. For any $f \in L_2(T)$ the distribution L^*f coincides with L^*Pf where P is the projection operator onto the closure $F = \overline{LC_0^\infty(T)}$ of $LC_0^\infty(T)$ in $L_2(T)$-space because

$$(\varphi, L^*f) = (L\varphi, f)_{L_2} = (L\varphi, Pf)_{L_2} =$$

$$= (\varphi, L^*Pf), \ \varphi \in C_0^\infty(T) .$$

The proof is finished.

Thus we deal now with the very same class of the Schwartz distributions $x \in X$ as it was suggested in (3). The only change must be done with respect to the inner product in X which we set now as

$$(L^*f, L^*g)_X = (Pf, Pg)_{L_2} , \ f, g \in L_2(T) .$$

By the very definition of our Hilbert space $W = \overline{C_0^\infty(T)}$ any test distribution $x = (u, x)$, $u \in C_0^\infty(T)$, can be treated as a linear continuous functional on W. On the hand, any linear continuous functional $x = (u,x)$ on W is continuous with respect to $\|u\|_W = \|Lu\|_{L_2}$, $u \in C_0(T)$, so it gives us some test distribution $x \in X$. Thus X is a dual space to W. Moreover, taking into account the representation

$$(u,x) = (u,L^*g) = (Lu,g)_{L_2} \tag{10}$$

of the bilinear form

$$(u,x),\ u \in W,\ x \in X\ ,$$

we see that any $u \in W$ is a linear functional

$$u = (u,x) \overset{\text{def}}{=\!=} (x,u),\ x \in X$$

On the Hilbert space X with

$$\|x\|_X = \|g\|_{L_2}\ ,\quad x = L^*g, g \in F\ .$$

Obviously any linear continuous functional $u = (x,u)$ on $x = L^*g$, $g \in F$ can be represented as

$$(g(Lu)_{L_2} = (Lu,g)_{L_2}\ ,\ g \in F\ ,$$

and it can be identified with the corresponding $u \in W$ for we have one-to-one correspondence in the relationship $LW = F$ by means of the unitary operator $L: W \to F$. One can easily verify that W and X are the dual Hilbert spaces conjugate to each other with the bilinear form (10),

$$\|u\|_W = \sup_{\|x\|_X = 1} |(u,x)|$$

and

$$\|x\|_X = \sup_{\|u\|_W = 1} |(u,x)|\ .$$

We see that any function $u \in W$ is uniquely determined by its value (u,x) over all distributions $x \in X$. It will be beneficial to treat $u \in W$ as the linear continuous functional $u = (x,u)$, $x \in X$, with values

$$(x,u) \overset{\text{def}}{=\!=} (u,x).$$ ■

■ Going back to the Equation (8) in the domain S let us appeal to the corresponding boundary values

$$(u,x), \ \text{supp}\, x \subseteq \Gamma,$$

over our test distributions $x \in X$ with supports in the boundary $\Gamma = \partial S$. Obviously a collection of all these distributions is a certain subspace $X(\Gamma) \subseteq X$; let us call $X(\Gamma)$ boundary space.

One has to expect that some of the boundary values are determined uniquely by the very equation (8) for any solution $u \in W$. Namely, the equation (8) gives us the function Lu in the domain S by means of the corresponding $f \in F$ and according to (10) we have

$$(u,x) = (u,L^*g) = (Lu,g)_{L_2(T)} = (f,g)_{L_2(S)}$$

for all $x = L^*g$ with $g \in L_2(s)$ so the values (u,x) over all test distributions $x \in L^*L_2(S)$ from a closure of $L^*L_2(S)$ in X are determined by the very equation (8) and it holds true in particular for all generalized boundary points $x \in X^-(\Gamma)$ from the boundary subspace

$$X^-(\Gamma) = X(\Gamma) \cap L^*L_2(S) .$$

Thus all boundary values

$$(u,x), \ x \in X^-(\Gamma) ,$$

are uniquely determined by the very equation (8).

It is remarkable that the other part of the boundary values can be arbitrarily described for the solution $u \in W$ of the equation (8); to be more precise, this is true for the boundary values

$$(u,x), \ x \in X^+(\Gamma) ,$$

at all generalized boundary points from any *boundary subspace* $X^+(\Gamma) \subseteq X(\Gamma)$ which forms a direct sum

$$X(\Gamma) = X^-(\Gamma) \otimes X^+(\Gamma) . \tag{11}$$

This fact is reflected in the boundary conditions (9) given by means of the corresponding values

$$(u_\Gamma,x) \overset{\text{def}}{=\!=} (x,u_\Gamma), \ x \in X^+(\Gamma) ,$$

of an arbitrary linear continuous functional

$$u_\Gamma = (x, u_\Gamma),\ x \in X^+(\Gamma)$$

on the boundary subspace $X^+(\Gamma)$.

Let us introduce the space $X(\bar{S})$ of all test distributions $x \in X$ with supports $\operatorname{supp} x \subseteq \bar{S}$ in a closure $\bar{S} = S \cup \Gamma$ of the domain S. Let

$$F(S^c)^\perp = F \ominus F(S^c)$$

be an orthogonal complement to a subspace

$$F(S^c) = F \cap L_2(S^c)$$

of all $g \in F$ with supports $\operatorname{supp} g \subseteq S^c$ in a complementary $S^c = T \setminus S$.

The following result [1] holds true.

Theorem. The differential equation (8) with any source $f \in F$ has the unique solution $u \in W$ in $\bar{S}$ with the arbitrarily given boundary conditions (9) and it can be represented over all test distributions $x \in X(\bar{S})$ as

$$(x, u) = (G_\Gamma^+ x, f) + (\pi_\Gamma^+ x, u_\Gamma) \tag{12}$$

where Π_Γ^+ is a projection operator onto the boundary subspace $X^+(\Gamma)$ parallel to the subspace $L^* F(S^c)^\perp$ in the direct sum

$$X(\bar{S}) = L^* F(S^c)^\perp \oplus X^+(\Gamma)$$

and G_Γ^+ is an operator giving us $g = G_\Gamma^+ x \in F(S^c)^\perp$ by means of the conjugate differential equation

$$L^* g = x - \pi_\Gamma^+ x$$

with $x - \pi_\Gamma^+ x \in L^* F(S^c)^\perp$.

The linear continuous form

$$(g, f),\ g \in F(S^c)^\perp ,$$

which appears as the first term in the formula (12) can be determined by means of $f \in F$ in the very domain S because according to Lemma 2 the projection $PL_2(S)$ is dense in $F(S^c)^\perp$ and we have

$$(Pg,f) = (Pg,f)_{L_2(T)} = (g,f)_{L_2(T)} =$$

$$(g,f)_{L_2(S)}\ ,\ g \in L_2(S)\ ;$$

as a matter of fact the first term in the formula (12) represents the values

$$(x,u_1) = (G_\Gamma^+ x,f) = (f,g)$$

of the solution $u = u_1$ with zero boundary conditions so the operator G_Γ^+ plays here a role of a generalized Green's function; correspondingly the second term in (12) gives us the values

$$(x,u_2) = (\pi_\Gamma^+ x,u_\Gamma)$$

of the solution $u = u_2$ of the equation (8) with zero source $f = 0$.

Proof. As has been mentioned, the equation (8) gives us all values (u,x) over

$$x \in L^* L_2(S) = L^* P L_2(S)$$

remember, P is the orthogonal projection onto $F = \overline{LC_0^\infty(T)}$ in $L_2(T)$. The orthogonal complement

$$F^\perp = L_2(T) \ominus F$$

is formed by all L^*-harmonic functions $g \in L_2(T)$, i.e.

$$L^* g(t) = 0,\ t \in T\ ,$$

for all $x = L^* g$ with $g \in F^\perp$ are zero distributions

$$L^* g = (\varphi, L^* g) = (L\varphi, g) = 0,\ \ \varphi \in C_0^\infty(T).$$

Lemma 2. The subspace

$$F(S^c)^\perp = F \ominus F(S^c)$$

coincides with the closure $\overline{PL_2(S)}$ in $L_2(T)$ and

$$L^* L(S^c)^\perp = L^* L_2(S)\ .$$

Proof. The intersection

$$F(S^c) = F \cap L_2(S^c)$$

has its orthogonal complement in $L_2(T)$-space generated by the sum of the corresponding orthogonal complements $F^\perp$ and $L_2(S) = L_2(S^c)^\perp$, namely

$$L_2(T) \ominus F(S^c) = \overline{F^\perp + L_2(S)} .$$

Taking into account that $F(S^c) \subseteq F$ we obtain

$$L_2(T) \ominus F(S^c) = [F^\perp \oplus F] \ominus F(S^c) =$$

$$= F^\perp \oplus [F \ominus F(S^c)] = F^\perp \oplus F(S^c)^\perp$$

and

$$F(S^c)^\perp = \overline{[L_2(S) + F^\perp]} \ominus F^\perp =$$

$$= \overline{P[L_2(S) + F^\perp]} = \overline{PL_2(S)} .$$

For the unitary operator $L^*: F \to X$ we have

$$\overline{L^*L_2(S)} = \overline{L^*PL_2(S)} =$$

$$= L^*\overline{PL_2(S)} = L^*F(S^c) \quad .$$

That is all.

To prove our theorem, let us appeal to the space X of our test distributions $x = L^*g$ with $g \in F$ and to $L^* : F \to X$ as the unitary operator. We have

$$(\varphi,x) = (L\varphi,g)_{L_2} = (L^*L\varphi,L^*g)_X =$$

$$= (L^*L\varphi,x)_X , \varphi \in C_0^\infty(T) .$$

Let us consider the subspace $X(\bar{S}) \subset X$ of all $x \in X$ with supports in $\bar{S}$ and the boundary space $X(\Gamma) \subseteq X(\bar{S})$. Obviously the subspace $X(\Gamma) \subseteq X(\bar{S})$ is formed by all $x \in X(\bar{S})$ which vanish in the domain S ,

$$(\varphi,x) = (L^*L\varphi,x)_X = 0, \varphi \in C_0^\infty(S) .$$

Thus we have the orthogonal sum

$$X(\bar{S}) = \overline{L^*LC_0^\infty(S)} \oplus X(\Gamma)$$

where

$$\overline{LC_o^\infty(S)} \subseteq \overline{PL_2(S)} = F(S^c)^\perp$$

and

$$L^* F(S^c)^\perp = \overline{L^* L_2(S)} \subseteq X(S) \ .$$

Remember in the direct sum (11),

$$X(\Gamma) = X^-(\Gamma) \oplus X^+(\Gamma) \ ,$$

we have

$$X^-(\Gamma) = X(\Gamma) \cap \overline{L^* F_2(S)}$$

so obviously the following direct decomposition holds true:

$$\begin{aligned} X(\bar{S}) &= [\overline{L^* LC_o^\infty(S)} + X^-(\Gamma)] + X(\Gamma) = \\ &= L^* F(S^c)^\perp + X^+(\Gamma) \ . \end{aligned} \tag{13}$$

As has already been emphasized, the equation (8) with $f \in F$ treated in the very direct way gives us nothing but values

$$\begin{aligned} (u, L^* g) &= (Lu, g)_{L_2} = (f, g)_{L_2} = \\ &= (f, Pg)_{L_2} = (Pg, f)_{L_2} \ , \ g \in L_2(S) \ , \end{aligned}$$

of the bilinear form (10) so by the very equation (8) we are given all values

$$(x, u) \stackrel{\text{def}}{=\!=} (u, x) = (f, g)_{L_2} = (g, f)_{L_2} \ , \ g \in F(S^c)^\perp \ ,$$

which determine its solution $u \in W$ over all test distributions

$$x = L^* g \in X^-(S) \stackrel{\text{def}}{=\!=} L^* F(S^c)^\perp \ .$$

With the arbitrarily given boundary conditions (9) we can determine a linear continuous functional

$$u = (x, u) \stackrel{\text{def}}{=\!=} (u, x), \ x \in X \ ,$$

with the prescribed values

$$(u,x) = (u_\Gamma,x),\ x \in X^+(\Gamma)\ ,$$

and

$$(u,x) = (f,g)_{L_2}\ ,\quad x = L^*g \in X^-(S)$$

which are uniquely determined on the direct sum

$$X(\bar{S}) = X^-(S) \oplus X^+(\Gamma)\ ,$$

for according to the direct decomposition (13) we have

$$x = (x - \pi_\Gamma^+ x) + \pi_\Gamma^+ x,\quad \pi_\Gamma^+ x \in X^+(\Gamma),\ x - \pi_\Gamma^+ x \in X^-(S)\ .$$

This ends the proof.

One can treat (8) as the <u>generalized equation</u>

$$(L^*\varphi,u) = (\varphi,f),\ \varphi \in C_0^\infty(S)$$

for the corresponding $u \in W$ as the function $u = (x,u)$ over the test distributions $x \in X$. It is easy to verify that in this way we get the very same result for the boundary problem (8), (9) as was given by our theorem.

■ Considering the generalized differential equation

$$(L^*\varphi,u) = (\varphi,f),\ \varphi \in C_0^\infty(S)\ ,$$

we can in our framework treat the source $f \in F$ as a generalized function

$$f = (\varphi,f),\ \varphi \in C_0^\infty(S)\ , \tag{14}$$

which is continuous with respect to

$$\|P\varphi\|_{L_2} = \|L^*\varphi\|_X\ . \tag{15}$$

By some reason we would like to indicate the case when the equation (8) does have the solution $u \in W$ for any generalized function $f = (\varphi,f)$ which is continuous with respect to the norm $\|\varphi\|_{L_2}$ in L_2-space. Obviously it holds true if and only if the orthogonal projection operator

$$P : L_2(S) \to F = \overline{LC_0^\infty(T)}$$

on the <u>subspace</u> $L_2(S)$ in $L_2(T)$ is invertible, i.e. P^{-1} does exist and it is bounded. This can be characterized as the following

<u>COMMON CASE</u>: either we have $F = L_2(T)$ or the orthogonal complement $F^{\perp} = L_2(T) \ominus F$ forms a non-zero angle with $L_2(S)$ in $L_2(T)$-space.

(Roughly speaking, if any L^*-harmonic function $g \in H$ belongs to the subspace $L_2(S)$ over $S \subset T$ then being zero in the neighborhood $T\setminus\bar{S}$ of the boundary of the domain T it is the solution of the differential equation $L^*g(t) = 0$, $t \in T$, with <u>zero</u> boundary conditions so $g = 0$; thus

$$H \cap L_2(S) = 0$$

and one can verify his feeling about our "COMMON CASE" being common case by an example of the Laplace operator $L = \Delta$.)

In the COMMON CASE, our theorem on the boundary problem (8) - (9) can be improved in the following way. There is no necessity to introduce the bilinear form

$$(Pg,f) = (g,f), \; g \in L_2(S) ,$$

on $Pg \in F(S^c)^{\perp}$ which allows us to define the first term in the formula (12), for in the COMMON CASE we have one-to-one correspondence in the relationship $PL_2(S) = F(S^c)^{\perp}$, so in the formula (12) we can substitute $g = G_\Gamma^+ x \in F(S^c)^{\perp}$ by the corresponding $P^{-1}g \in L_2(S)$, $(P^{-1}g,f) = (g,f)$ and introduce <u>another</u> operator

$$G_\Gamma^+ : X(\bar{S}) \to L_2(S)$$

which for any $x \in X(\bar{S})$ gives us $g = G_\Gamma^+ x$ by means of the <u>conjugate equation</u>

$$L^*g = x - \pi_\Gamma^+ x \qquad (16)$$

with the <u>unique</u> solution $g \in L_2(S)$. ∎

∎ One can be interested in the equation (8) with the source f being a <u>vector</u> function in a Hilbert space, say. To deal with this case, let us introduce $f \in \underline{F}$ as a class of all generalized <u>vector</u> functions in the Hilbert space considered which are continuous with respect to $\|P\varphi\|_{L_2}$ as was indicated in (14-(15). For any $f \in \underline{F}$ we define

$$(P\varphi,f) \overset{\text{def}}{=\!=} (\varphi,f), \; \varphi \in C_0^\infty(S) ,$$

and by continuity it can be extended to the vector linear function

$$(g,f),\ g \in F(S^c)^{\perp}\ ,$$

on the subspace

$$F(S^c)^{\perp} = \overline{P\ L_2(S)} = \overline{P\ C_0^{\infty}(S)}\ ,$$

- see Lemma 2. We have to introduce $u \in \underline{W}$ as a class of vector functions which are well-defined over all test distributions $x \in X(\bar{S})$ in the closure $\bar{S}$ of the domain S with the corresponding vector values

$$(x,u) \equiv (u,x)$$

which form a linear continuous vector function on the generalized variable $x \in X(\bar{S})$. In a similar way the boundary conditions can be represented by means of the arbitrarily given linear continuous vector function

$$u_{\Gamma} = (x,u_{\Gamma}) \equiv (u_{\Gamma},x)$$

on the boundary generalized variable $x \in X^{+}(\Gamma)$. It is easy to see that our approach to the boundary problem (8)-(9) can be applied to the vector functions and the following result hodls true.

Generalized Theorem. For every vector source $f \in \underline{F}$ the generalized equation (8) has the unique solution $u \in \underline{W}$ with any arbitrarily given boundary conditions (9); this $u \in \underline{W}$ can be represented over all test distributions $x \in X(\bar{S})$ by the corresponding formula (12).

(Of course in the COMMON CASE the first term in the formula (12) can be obtained by means of the conjugate equation (16) for $G_{\Gamma}^{+}x = g \in L_2(S)$.) ■

■ We would like to note that our approach to the boundary problem of the (8)-(9) type can be applied to the operator L in any functional Hilbert space which is local with respect to L, i.e., any elements $L\varphi$, u, f are orthogonal in this space if $\varphi \in C_0^{\infty}(T)$ and f have disjoint supports (see [2]). As an example, our Hilbert space X of the test distributions is local with respect to the operator $L^{*}L$,

$$(L^{*}L\varphi,x) = (\varphi,x) = 0$$

if $\varphi \in C_0^{\infty}(T)$, $x \in X$ are with disjoint supports and the following result holds true:

For any generalized vector function

$$y = (\varphi,y),\ \varphi \in C_0^{\infty}(S)\ ,$$

which is continuous with respect to

$$\|\varphi\|_W = \|L\varphi\|_{L_2} = \|L^*L\varphi\|_X$$

the generalized equation

$$L^*Lu(t) = y(t),\ t \in S\ ,$$

has the unique solution $u \in \underline{W}$ with arbitrarily given boundary conditions

$$(x,u) = (x,u_\Gamma),\ x \in X(\Gamma)\ ; \tag{18}$$

this $u \in \underline{W}$ can be represented over all test distributions $x \in X(\bar{S})$ as

$$(x,u) = (G_\Gamma x,y) + (\pi_\Gamma x,u_\Gamma) \tag{19}$$

where π_Γ is the orthogonal projection operator onto the boundary space $X(\Gamma) \subseteq X$ and $G_\Gamma x = v \in \overline{C_0^\infty(S)}$ is the unique solution of the equation

$$L^*Lv = x - \pi_\Gamma x$$

in $\overline{C_0^\infty(\bar{S})} \subseteq W$. The first term in the formula (19) represents the values of the solution $u \in W$ with zero boundary conditions (18) by means of the linear continuous function

$$(v,y),\ v \in \overline{C_0^\infty(S)}$$

over the closure of $C_0^\infty(S_1)$ in our functional space $W = \overline{C_0^\infty(T)}$ defined from the very beginning; the second term represents the values over the test distributions $x \in X(\bar{S})$ of the solution of the equation (17) with zero source $y = 0$. (One can exercise that the solution $u \in \underline{W}$ of the boundary problem (8)-(9) in the same time is the solution of the equation (17) with the source $y = L^*f$ and the corresponding boundary conditions (18).) ■

■ Exercise. Let us consider $L = d/dt$ in the finite interval $T \subseteq R^1$. By our approach we get the corresponding Sobolev spaces

$$W = W_2^1\ ,\quad X = W_2^{-1}\ .$$

The space $F = LW$ is formed by all functions $f \in L_2(T)$, $\int_T f(t)dt = 0$. The corresponding $H = L_2(T) \ominus F$ formed by all L^*-harmonic functions $g(t) \equiv \text{const.},\ t \in T$, gives a trivial illustration to our COMMON CASE when the subspace H has a non-zero angle with $L_2(S)$ in $L_2(T)$-space. The generalized equation (8):

$$(-\frac{d}{dt}\varphi, u) = (\varphi, f),\ \varphi \in C_0^\infty(S)\ ,$$

gives us by the limit $\varphi \to 1_{(t,b)}$ to the (t,b)-indicator function $g = 1_{(t,b)}$ in $L_2(S)$ the values

$$u(b) - u(t) = \lim_{\varphi \to g} (\varphi, f),\ a < t < b\ .$$

The boundary values of $u \in W$ are linear combinations

$$(u,x) = c_1 u(a) + c_2 u(b)$$

which are given by means of the delta-distributions

$$x = c_1\delta(t-a) + c_2\delta(t-b)$$

on the boundary $\Gamma = \{a,b\}$; the corresponding $X^-(\Gamma)$ is represented by these boundary distributions with $c_2 = -c_1$ and they give us

$$(u,x) = c\int_a^b f(t)dt = c[u(b)-u(a)]$$

with $c = c_2 = -c_1$. Taking arbitrarily the boundary value $c_1u(a) + c_2u(b)$ at any generalized point $x = c_1\delta(t-a) + c_2\delta(t-b)$ with $c_2 \neq -c_1$ we reconstruct the unique solution

$$u = u(t),\ a \le t \le b\ .$$

(One can exercise another case of the equation (8) which is not so trivial considering the harmonic operator $L = \Delta$ in the certain domains $S \subseteq T \subseteq R$; $X^+(\Gamma)$ in the boundary conditions (9) one can get the well-known Dirichlet/Neumann problems in the Sobolev spaces

$$W = W_2^2\ ,\quad X = W_2^{-2}\ .)$$ ■

§ 3 . Stochastic Boundary Problems

■ We are going to investigate some properties of a random field ξ associated with the linear differential operator L in our general framework by means of the partial differential equation

$$L\xi(t) = \eta(t),\ t \in S\ , \tag{20}$$

with the stochastic source η in the domain S and the boundary conditions on the boundary $\Gamma = \partial S$ given as

$$(x,\xi) = (x,\xi_\Gamma),\ x \in X^+(\Gamma)\ , \tag{21}$$

with linear random functional

$$\xi_\Gamma = (x,\xi_\Gamma),\ x \in X^+(\Gamma)\ ,$$

on the corresponding boundary subspace $X^+(\Gamma)$ of the test distributions $x \in X$.

One can here apply the Generalized Theorem of § 2 considering vector functions in the standard Hilbert space H of random variables h, $E|h|^2 < \infty$, with the inner product of its elements h_1,h_2 given as the expectation $Eh_1\bar{h}_2$. The random field ξ considered is a vector function in the closure $\bar{S}$ of the domain S with values in H well-defined over all test distributions $x \in X(\bar{S})$ with supports in $\bar{S}$ and it represents the unique solution $\xi \in \underline{W}$ of the <u>stochastic partial differential equation</u> (20) with arbitrarily given boundary conditions (21) according to our Generalized Theorem.

This stochastic model seems to be most interesting in a case when the stochastic source

$$\eta = (\varphi,\eta),\ \varphi \in C_0^\infty(S)\ ,$$

is of the standard white noise type and actually this was a main reason for us to introduce the COMMON CASE which allows us to deal in the equation (20) with any source $\eta = (\varphi,\eta)$ continuous with respect to $\|\varphi\|_{L_2}$. We assume that this very COMMON CASE holds true although it is not necessary for our further results which could be treated in the general framework. (We would like to note that all stochastic models of our § 1 represent the COMMON CASE of equation (20) in the corresponding domain $S = T$ without any boundary conditions (21) which are not required for the boundary ∂T of the very domain T, for all boundary test distributions with supports on ∂T are <u>zero</u> elements $x \in X$ in our Hilbert space X, so actually in our general framework we have zero boundary values for any solution $\xi \in \underline{W}$ of equation (20) in the domain $S = T$.)

Remember, according to the Generalized Theorem of § 2 the solution $\xi \in \underline{W}$ of the arbitrary boundary problem (20)-(21) can be described over all test distributions as

$$(x,\xi) = (G_\Gamma^+ x, \;) + (\pi_\Gamma^+ x, \xi_\Gamma), \; x \in X(\bar{S}), \tag{22}$$

where $G_\Gamma^+ x = g \in L_2(S)$ is the unique solution of the conjugate equation (16) and π_Γ^+ is the projection onto the corresponding boundary subspace $X^+(\Gamma)$ parallel to the subspace $F(S^c)^\perp = L^* L_2(S)$ in the test distributions Hilbert space X . ■

■ It is worthy to note that for any $h \in H$ the inner product

$$u = (x,u) = E(x,\xi)\bar{h} \; , \; x \in X(\bar{S}),$$

represents over the test distributions the corresponding values of some function $u \in W$ from our functional space $W = \overline{C_0^\infty(T)}$; obviously this scalar function $u \in W$ in the closure $\bar{S}$ of the domain S is the unique solution of the differential equation (8) with the generalized source

$$f = (\varphi,f) = E(\varphi,\eta)\bar{h} \; , \; \varphi \in C_0^\infty(\,S) \; . \tag{23}$$

and with the boundary conditions (9) given by means of

$$u_\Gamma = (x,u_\Gamma) = E(x,\xi_\Gamma)\bar{h} \; , \; x \in X^+(\Gamma) \; .$$

For $h = 1$ we deal here with mean values of ξ; let us assume that $E(x,\xi) \equiv 0$. Then for any $h \in H$ the function $u = E(\cdot,\xi)\bar{h} \in W$ describes a correlation of random variable h with our random field ξ. Let us determine the corresponding $u \in W$ for a value

$$h = (x,\xi)$$

of the field ξ itself at the generalized point $x \in X(\bar{S})$.

Suppose the boundary data ξ_Γ in the boundary conditions (21) on $\Gamma = \partial S$ do not stochastically depend on the random source η of the equation (20) in the domain S.

Of course the correlation of the values of our random field ξ depends on the correlation operator B of the random source η in the equation (20),

$$(\varphi,B\psi)_{L_2} = \overline{E(\varphi,\eta)(\psi,\eta)}; \quad \varphi,\psi \in L_2(S) \; .$$

Let us represent $x \in X(\bar{S})$ as it was done in the formula (22)

$$x = G_\Gamma^+ x + \pi_\Gamma^+ x \; ;$$

remember, π_Γ^+ is the corresponding projection onto the boundary subspace $X^+(\Gamma)$ and

$g = G_\Gamma^+ x \in L_2(S)$ is the solution of the conjugate equation (16),

$$L^* g = x - \pi_\Gamma^+ x \ .$$

then

$$E(\varphi,\eta)\overline{(\pi_\Gamma^+ x,\xi)} = E(\varphi,\eta)\overline{(\pi_\Gamma^+ x,\xi_\Gamma)} = 0$$

and

$$E(\varphi,\eta)\overline{(x,\xi)} = E(\varphi,\eta)\overline{(L^* g,\xi)} =$$

$$= E(\varphi,\eta)\overline{(g,\eta)} = (\varphi,Bg), \ \varphi \in C_0^\infty(S) \ ;$$

thus for the random variable $h = (x,\xi)$ the corresponding function f indicated in (23) is

$$f = Bg \ .$$

At any generalized boundary point $y \in X^+(\Gamma)$ we have

$$E(x,\xi)\overline{(L^* g,\xi)} = E(y,\xi_\Gamma)\overline{(g,\eta)} = 0$$

and

$$E(y,\xi)(x,\xi) = E(y,\xi_\Gamma)(\pi_\Gamma^+ x,\xi) =$$

$$= E(y,\xi_\Gamma)(\pi_\Gamma^+ x,\xi_\Gamma) \ .$$

We see that the following result holds true.

Theorem. The correlation function

$$u = E(\cdot,\xi)\overline{(x,\xi)} \in W$$

of the random field ξ,

$$(y,u) = E(y,\xi)\overline{(x,\xi)}$$

over the test distributions $y \in X(\bar{S})$, can be obtained for any

$$x = L^* g + \pi_\Gamma^+ x \in X(\bar{S})$$

as the unique solution of the differential equation

$$Lu(t) = Bg(t),\ t \in S,$$

with the boundary conditions

$$(y,u) = E(y,\xi_\Gamma)(\pi_\Gamma^+ x,\xi_\Gamma),\ y \in X^+(\Gamma).$$

(Of course we have here the boundary problem of the (8)-(9) type.) ■

■ Let us now consider BOUNDARY FORECAST PROBLEMS!

One can be interested in the FORECAST $\hat{\xi} = (x,\hat{\xi})$ of our random field $\xi = (x,\xi)$ over the test distributions $x \in X(\bar{S})$ by means of the corresponding boundary data

$$(x,\xi) = (x,\xi_\Gamma)\ ,\quad x \in X^+(\Gamma),$$

in the boundary conditions (21). Let us hold the assumption made earlier that this boundary conditions on the <u>boundary</u> $\Gamma = \partial S$ of the domain S do <u>not</u> stochastically depend on the random source η <u>inside</u> of S. Then according to the formula (22) we obviously have

$$\hat{\xi} = (x,\hat{\xi}) = (\pi_\Gamma^+ x,\xi),\quad x \in X(\bar{S})\ .$$

and this result can be given as follows.

<u>Theorem</u>. The unique solution $u \in \underline{W}$ of the differential equation

$$Lu(t) = 0,\quad t \in S\ , \tag{24}$$

in the domain S with the boundary conditions

$$(x,u) = (x,\xi_\Gamma),\quad x \in X^+(\Gamma), \tag{25}$$

represents the best FORECAST $u = \xi$ for the random field ξ by means of the boundary data ξ_Γ .

Suppose now we are given <u>all</u> boundary values

$$(x,\xi),\quad x \in X(\Gamma);$$

how can the best FORECAST be made in this case? We are going to answer this important question assuming that the stochastic source η in the equation (20) is represented as the generalized Gaussian field

$$\eta = (\varphi,\eta),\quad \varphi \in L_2(S),$$

with <u>independent values</u> which have zero means and a correlation

$$E(\varphi,\eta)(\psi,\eta) = \int \sigma^2(t)\varphi(t)\psi(t)dt, \quad \varphi,\psi \in L_2(S),$$

where a coefficient $\sigma^2(t) \neq 0$ is non-degenerate and the operator L in our scheme can be substituted by $\sigma^{-1}L$ so we can assume that a transformation

$$\eta \to \sigma^{-1}\eta \ , \quad L \to \sigma^{-1}L$$

is made and our random field ξ is represented by the corresponding equation (20) with the stochastic source η of the white noise type.

<u>Theorem</u>. The unique solution $u \in \underline{W}$ of the differential equation

$$L^*L\, u(t) = 0, \quad t \in S\ , \tag{26}$$

with the boundary conditions

$$(x,u) = (x,\xi),\ x \in X(\Gamma), \tag{27}$$

represents the best linear FORECAST

$$(x,u) = (x,\hat{\xi}),\ x \in X(\bar{S}),$$

of the random field ξ in the domain S by means of its boundary values on the boundary $\Gamma = \partial S$.

Remember, the boundary subspace $X^+(\Gamma)$ which appears in the boundary conditions (21) is a complementary to the boundary subspace $X^-(\Gamma)$ - see (11).

<u>Proof</u>. As we know by § 2 there is the orthogonal decomposition

$$L^*L_2(S) = \overline{L^*LC_0^\infty(S)} \oplus X^-(\Gamma)$$

of the subspace

$$L^*L_2(S) = L^*PL_2(S) = L^*F(S^c)^\perp$$

of the test distributions $x \in X(\bar{S})$ where the values $(x,\xi) = (G_\Gamma^+x,\eta)$ are determined by the very equation (20) apart from the boundary values (21) which all together give us the formula (22). Remember, the operator L^* is unitary with respect to the inner product $(Pf,Pg)_{L_2}$, $f,g \in L_2(T)$, where P is the orthogonal projector onto $F = \overline{LC_0^\infty(T)}$ in $L_2(T)$-space, so for all $g = G_\Gamma^+y \in L_2(S)$ with $y = L^*g \in X^-(\Gamma)$ and $f \in \overline{LC_0^\infty(S)} \subseteq F$ we have

$$(Pf,Pg)_{L_2} = (f,g)_{L_2} = 0.$$

In the case of the Gaussian white noise source η it means that the values

$$(L^*f,\xi) = (f,\eta), \quad x = L^*f, \quad f \in \overline{LC_0^\infty(S)}$$

and

$$(x,\xi) = (L^*g,\xi) = (g,\xi), \quad x = L^*g \in X^-(\Gamma),$$

are stochastically independent.

Taking into account that according to our assumption the boundary conditions in (21) do not stochastically depend on the source η in the equation (20), we see that all boundary values

$$(x,\xi), \quad x \in X(\Gamma),$$

do not depend on the random variables (f,η), $f \in \overline{LC_0^\infty(S)}$ generated by a random source

$$\eta^* = (\varphi,\eta^*) \overset{\text{def}}{=\!=} (L\varphi,\eta), \; \varphi \in C_0^\infty(S),$$

and our random field ξ fits to the differential equations

$$L^*L\xi(t) = \eta^*(t), \; t \in S, \tag{28}$$

with the boundary conditions

$$(x,\xi) = (x,\xi_\Gamma), \quad x \in X(\Gamma), \tag{29}$$

given by the corresponding boundary values which -- let us repeat -- do not stochastically depend on the random source η^* in the equation (28). For any random field ξ of this type according to the general formula (19) we have

$$(x,\xi) = (G_\Gamma x,\eta^*) + (\pi_\Gamma x,\xi_\Gamma), \quad x \in X(\bar{S}), \tag{30}$$

where π_Γ is the orthogonal projection onto the whole boundary space $X(\Gamma)$ in our test distributions Hilbert space and $G_\Gamma x = f \in LC_0^\infty(S)$ is the solution of the conjugate equation $L^*f = x - \pi_\Gamma x$. The first term in the formula (30) is stochastically independent of all boundary values of the field ξ on the boundary $\Gamma = \partial S$ so the best FORECAST we are looking for must be given by the second term in the formula (30)

$$(x,\hat{\xi}) = (\pi_\Gamma x,\xi), \quad x \in X(\bar{S}), \tag{31}$$

which represents as we know already the unique solution $u = \hat{\xi} \in \underline{W}$ of the boundary problem (26)-(27).

Exercise. Let us consider the operator $L = d/dt$ in the infinite interval $T = R^1_+$. In this case

$$F = \overline{LC_0^\infty(T)} = L_2(T)$$

and the delta-distributions

$$x = \delta(s-t) = L^* 1_{(0,t)}$$

form a complete system in the test distributions space $X = L^* L_2(T)$ because the indicator functions $f = 1_{(0,t)}$ of intervals $(0,t)$, $t \in T$, form a complete system in $L_2(T)$-space; one can set

$$\xi(t) = (x,\xi), \quad x = \delta(s-t),$$

and the corresponding function $\xi = \xi(t)$, $t \in T$, represents

$$\xi = (x,\xi), \quad x \in X,$$

originally determined by means of our test distributions. For any linear combination $f = \Sigma c_k 1_{(0,t_k)}$ we have

$$\|\Sigma c_k 1_{(0,t_k)}\|^2_{L_2} = \Sigma c_k c_j \min(t_k,t_j)$$

and a random function $\xi = \xi(t)$, $t \in T$, represents $\xi \in \underline{W}$ if and only if

$$E|\Sigma c_k \xi(t_k)|^2 =$$

$$= E|(L^* f,\xi)|^2 \le C\|f\|^2_{L_2} =$$

$$= C \sum_{k,j} c_k c_j \min(t_k,t_j);$$

in particular

$$\xi(0) = 0, \quad E|\xi(t)|^2 \le Ct$$

for any $\xi \in \underline{W}$. Let us set

$$\int f(t)\eta(t)dt = (f,\eta), \quad f \in L_2(T)$$

for a generalized random source η in $L_2(T)$-space.

The equation

$$\frac{d}{dt}\xi(t) = \eta(t), \quad a < t < b ,$$

in the interval $S = (a,b)$ with $a = 0$ does not require any boundary conditions for we have $\xi(0) = 0$, so it has the unique solution

$$\xi(T) = \int_0^t \eta(s)ds, \quad a \leq t \leq b ,$$

with the boundary value $\xi(b) = \int_0^b \eta(s)ds$ which is determined by the very equation itself. For the interval $S = (a,b)$ with $a > 0$ and $b = \infty$ the boundary value

$$\xi(a) = (x,\xi), \quad x = \delta(s-a),$$

can be taken arbitrarily; with zero source $\eta = 0$ the corresponding equation

$$\frac{d}{dt}u(t) = 0, \quad t > a ,$$

with the boundary condition

$$u(a) = \xi(a)$$

has the unique solution $u \in \underline{W}$

$$u(t) \equiv \xi(a), \qquad t \geq a$$

(one can observe that a linear function $u(t) = \xi(a) + c(t-a)$ belongs to $\underline{W}$ if and only if $c = 0$); this solution $u(t) \equiv \xi(a)$ according to our general theorem gives us the best FORECAST $u(t) = \xi(t)$, $t \geq a$, for the $\xi(t)$, $t \geq a$, in the case of the Gaussian white noise source η (of course, in this case $\xi = \xi(t)$ is the Brownian Motion). The Brownian Motion ξ can be treated as the solution of the equation

$$-\frac{d^2}{dt^2}\xi(t) = \eta^*(t), \quad a < t < b,$$

with the random source

$$\eta^* = -L^*\eta (L^* = -\frac{d}{dt})$$

and corresponding boundary conditions

$$\xi(a) = \xi_\Gamma(a), \quad \xi(b) = \xi_\Gamma(b)$$

on the boundary $\Gamma = \{a,b\}$ which do not depend on the random source $\eta^*(t)$ in the open interval $S = (a,b)$. A particular solution of this differential equation with zero boundary conditions represents the well-known Brownian Bridge. One can set a similar Brownian Bridge $\xi(t)$, $a \le t \le b$, on arbitrarily taken random "supports" $\xi(a) = \xi_\Gamma(a)$, $\xi(b) = \xi_\Gamma(b)$ at the end points $t = a,b$ which are stochastically independent of the random source $\eta^*(t)$, $a < t < b$; the best FORECAST for the random function $\xi = \xi(t)$, $a \le t \le b$, by means of the boundary values $\xi(a)$, $\xi(b)$ can be given as the unique solution $\hat{\xi} = u \in \underline{W}$

$$\hat{\xi}(t) = \xi(a) + \frac{\xi(b) - \xi(a)}{b - a}(t-a), \quad a \le t \le b,$$

of the differential equation

$$-\frac{d^2}{dt^2}u(t) = 0, \quad a < t < b,$$

with the boundary conditions

$$u(a) = \xi(a), \; u(b) = \xi(b).$$

In a case of the random source $\eta = \sigma\eta_0$ which differes from the white noise η_0 with the coefficient $\sigma(t) \neq 0$ the best FORECAST for the corresponding ξ is the unique solution $u = \hat{\xi} \in \underline{W}$,

$$\hat{\xi}(t) = \xi(a) + \frac{\xi(b) - \xi(a)}{\int_b^a \sigma(s)^2 ds} \int \sigma(s)^2 ds, \; a \le t \le b,$$

of the equation

$$-\frac{d}{dt}\sigma(t)^2\frac{d}{dt}u(t) = 0, \quad a < t < b,$$

$$u(a) = \xi(a), \quad u(b) = \xi(b).$$

§ 4. Forecast Problem and Markov Property

One can imagine that the that the random field given in the domain S by means of the stochastic model (20)-(21) represents a part of some random field ξ of the same type in the domain $S_0 \supseteq S$, i.e. ξ is generated by the (corresponding) differential equation

$$L\xi(t) = \eta(t), \; t \in S_0 \tag{32}$$

with the boundary conditions

$$(x,\xi) = (x,\xi_{\Gamma_0}), \; x \in X^+(\Gamma_0), \tag{33}$$

on the boundary $\Gamma_0 = \partial S_0$.

Suppose the field ξ is observable outside of the certain domain $S \subseteq S_0$ and we are to give the best FORECAST of ξ in the domain S by means of all data

$$(y,\xi), \; \text{supp } y \subseteq \bar{S}_0 \backslash S ,$$

collected over all $y \in X(\bar{S}_0\backslash S)$ from the subspace of our test distributions with supports outside of S. Considering this FORECAST PROBLEM we assume that the source η in the equation (32) is the random field with independent values in the domain S_0 and it does not stochastically depend on the boundary conditions (33) on the boundary $\Gamma_0 = \partial S_0$.

The best FORECAST for the values

$$(x,\xi), \; x \in X(\bar{S}),$$

we are looking for is determined by means of the corresponding conditonal expectation

$$(x,\hat{\xi}) = E \{ (x,\xi) | (y,\xi), \; y \in X(\bar{S}_0\backslash S) \}$$

and we treat this

$$\hat{\xi} = (x,\hat{\xi}), \; x \in X(\bar{S}),$$

as the certain random field $\hat{\xi} \in \underline{W}$.

Theorem. Let the source η be the Gaussian white noise. Then $\hat{\xi} = u \in \underline{W}$ can be given by means of the unique solution of the stochastic boundary problem (26)-(27).

Proof. As we already know, our field ξ coincides in the domain $S \subseteq S_0$ with the unique solution of the corresponding boundary problem (28)-(29) with the boundary conditions on the boundary $\Gamma = S$ given by means of the boundary values (x,ξ), $x \in X(\Gamma)$, of the very field ξ outside of S. We are going to show that the first term

$$(G_\Gamma x,\eta) = (L^* f,\xi), \; x \in X(\bar{S}),$$

in the formula (30) with $L^* f = x - \pi_\Gamma x$ and $f \in \overline{LC_0^\infty(S)}$ is independent of the field

ξ outside of S ; then obviously the second term

$$(\pi_\Gamma x,\xi),\ x \in X(S),$$

given by means of the boundary values (x,ξ), $x \in X(\Gamma)$, of the field ξ outside of the domain S represents the best FORECAST we are looking for and this ends the proof.

One can easily verify that very similarly to the direct decomposition (13) we have

$$X(\bar{S}_0) = [\overline{L^* L C_0^\infty(S)} \oplus X^-(S_0 \setminus S)] + X^+(\Gamma_0) \tag{34}$$

where $X^-(S_0\setminus S)$ is the orthogonal complement to $\overline{L^* L C_0^\infty(S)}$ in the subspace

$$L^* L_2(S_0) = L^* P L_2(S) = L^* F(S_0^c)^\perp$$

and $X^-(S_0\setminus S)$ is formed by all test distributions $x = L^* g$, $g \in L_2(S_0)$, with supports $\operatorname{supp}\ x \subseteq \bar{S}_0\setminus S$ <u>outside</u> of the domain $S \subseteq S_0$. Taking any of these $x = L^* g$ we get

$$(f,g) = \lim (L\varphi,g) = \lim (\varphi,L^* g) =$$

$$= \lim (\varphi,x) = 0$$

for all

$$f = \lim \varphi \in \overline{LC_0^\infty(S)},\quad \varphi \in C_0^\infty(S),$$

because $\operatorname{supp}\ x$ is outside of S. Thus, in the case of Gaussian white noise source η the values

$$(L^* f,\xi) = (f,\eta),\ f \in \overline{LC_0^\infty(S)},$$

are stochastically independent of the values

$$(x,\xi),\ x \in X^-(\bar{S}_0\setminus S),$$

as well as on the boundary values

$$(x,\xi) = (x,\xi_{\Gamma_0}),\ x \in X^+(\Gamma_0),$$

according to our assumption that the boundary conditions (33) are independent of the source in the differential equation (32). Going back to the formula (30), we see that its first term is stochastically independent of the random field ξ outside of the domain S, to be more precise, it is independent of a family of all random variables

$$(x,\xi),\ x \in X(\bar{S}_0\backslash S) = X^-(S_0\backslash S) + X^+(\Gamma_0),$$

and in particular it is independent of the second term

$$(\pi_\Gamma x,\xi), \pi_\Gamma\ x \in X(\Gamma) \subseteq X(\bar{S}_0\backslash S)$$

of the formula (30).

Thus we have gotten what we were looking for and apart from that we have actually proved the Markov property of the field considered which can be formulated as follows.

The random field ξ in any domain $S \subseteq S_0$ is conditionally independent of its part outside of S conditioned with respect to all its boundary balues on the boundary $\Gamma = \partial S$, or, to be more precise, the values (x,ξ), supp $x \subseteq \bar{S}$, of the field in $\bar{S}$ conditioned with respect to the boundary data (x,ξ), supp $x \subseteq \Gamma$, are independent of the values (x,ξ), supp $\subseteq \bar{S}_0\backslash S$ of ξ outside of the domain S. We have actually proved that the random function ξ in any set S, i.e. $\xi = (x,\Gamma)$, supp $x \subseteq S$, is conditionally independent of its part out of S conditioned with its boundary values (x,ξ) supp $x \subseteq \Gamma$, on the boundary Γ of S and one can verify it with application of our approach to the corresponding closure $\bar{S} = \dot{S} \cup \Gamma$ of S with the interior domain $\dot{S}$. (A point of view on the Markov property with respect to any set known as a global Markov property is borrowed here from [3].)

We would like to emphasize that this Markov property holds true due to the fact that a σ-algebra A_1 of events generated in $\bar{S}$ by the random variables

$$(x,\xi),\ x \in L^* L_2(S),$$

is independent of a σ-algebra A_2 generated in $\bar{S}_0\backslash S$ by the random variables

$$(x,\xi),\ x \in \left[L^* L_2(S_0) \ominus L^* L_2(S)\right],$$

and

$$(x,\xi_{\Gamma_0}),\ x \in X^+(\Gamma_0).$$

Indeed, A_1 contains a σ-algebra $\mathcal{L}_1$ generated by the boundary values (x,ξ), $x \in X^-(\Gamma)$ and A_2 contains a σ-algebra $\mathcal{L}_2$ generated by the boundary values (x,ξ), $x \in X^+(\Gamma)$, so a σ-algebra $\mathcal{L} = \mathcal{L}_1 \cup \mathcal{L}_2$ generated by all boundary values

(x,ξ), $x \in X(\Gamma)$, splits A_1 and A_2 as well as $A_1 \cup \mathcal{L}$ and $A_2 \cup \mathcal{L}$ which correspondingly represent all events generated by the field ξ in $\bar{S}$ and outside of S thanks to the direct decomposition (34). One can see that we don't here involve the assumption about ξ to be Gaussian which has been used to prove independence A_1 and A_2.

Suppose now the stochastic source η is represented by any random field with independent values and apart from that

$$F \stackrel{\text{def}}{=\!=} \overline{LC_0^\infty(T)} = L_2(T) \tag{35}$$

(it holds true for all stochastic models presented in § 1, say). Then we have $L^*: L_2(T) \to X$ as the unitary operator and

$$L^* L_2(S_0) \ominus L^* L_2(S) = L^* L_2(S_0 \setminus S)$$

so the σ-algebra A_1 generated actually by the random variables

$$(g,\eta),\ g \in L_2(S),$$

is independent of σ-algebra A_2 generated correspondingly by the random variables

$$(g,\eta),\ g \in L_2(S_0 \setminus S),$$

and

$$(x,\xi_{\Gamma_0}),\ x \in X^+(\Gamma_0).$$

We are getting by the same argument as before that $\mathcal{L}$ <u>splits</u> $A_1 \cup \mathcal{L}$ and $A_2 \cup \mathcal{L}$, i.e. the σ-algebra $A_1 \cup \mathcal{L}$ generated by the random variables (x,ξ), supp $x \subseteq \bar{S}$, is conditionally independent of the σ-algebra $A_2 \cup \mathcal{L}$ generated by the random variables (x,ξ), supp $\subseteq \bar{S}_0 \setminus S$, conditioned with respect to the boundary σ-algebra $\mathcal{L}$ generated by the (x,ξ), supp $\subseteq \Gamma = \partial S$. Thus, the random field enjoys the Markov property.

■ Let us go back to our FORECAST PROBLEM. According to the formula (30) the best FORECAST

$$\hat{\xi} = (x,\hat{\xi}),\ x \in X(\bar{S}),$$

for the corresponding $\xi = (x,\xi)$ differs from the best <u>linear</u> FORECAST

$$u = (x,U) = (\pi_\Gamma x,\xi)$$

by the conditional expectation

$$(x,\hat{\xi}) - (x,u) = E\{(G_\Gamma x,\eta)|_{A}\} =$$

$$= E\{(G_\Gamma x,\eta)|_{\mathcal{L}}\} = E\{(G_\Gamma x,\eta)|_{\mathcal{L}_1}\} ,$$

where, as you remember, A_2 is the σ-algebra generated by the field outside of the domain $S \subseteq S_0$ and $\mathcal{L} = \mathcal{L}_1 \cup \mathcal{L}_2$ represents the boundary σ-algebra with components $\mathcal{L}_1$, $\mathcal{L}_2$ generated by the boundary values (x,ξ) with the corresponding $y \in X^-(\Gamma)$, $X^+(\Gamma)$. In the formula suggested we can finally take the condition with respect to $\mathcal{L}_1$ because the random variables $(G_\Gamma x,\xi)$ and σ-algebra $\mathcal{L}_1 \subseteq A_1$ are all together independent of the σ-algebra $\mathcal{L}_2 \subseteq A_2$.

Let us set our results as follows.

Theorem. The stochastic model (32)-(33) with the source η as the random field with independent values represents in the domain S_0 the Markov field ξ. The best FORECAST for ξ in the domain $S \subseteq S_0$ by means of the data outside of S can be given by the formula

$$(x,\hat{\xi}) = (x,u) + E\{(G_\Gamma x,\eta)|(y,\xi), y \in X^-(\Gamma)\} , \quad x \in X(S),$$

where $u = (x,u)$ is the solution of the boundary problem (26)-(27) giving the best linear FORECAST. ■

■ One can be interested to know when the generalized function

$$\xi = (x,\xi) \in \underline{W}$$

over all our test distributions $x \in X$ in the domain T can be identified as some "regular" function $\xi = \xi(t)$, $t \in T$, or as the generalized function

$$\xi = (\varphi,\xi), \ \varphi \in C_0^\infty(T),$$

over the standard test functions $\varphi = \varphi(t)$, $t \in T$, say. Of course, the best one holds true when we have a dense imbedding $C_0^\infty(T) \subseteq X$, so

$$X = \overline{C_0^\infty(T)} \ .$$

For example, we have $X = C_0^\infty(T)$ when the differential operator L is non-degenerate in a way that

$$\| \varphi \|_{L_2} \geq c \left(\int_{T_{loc}} |\varphi|^2 dt \right)^{1/2} \quad \varphi \in C_0^\infty(T),$$

for any bounded domain T_{loc} with a closure $\bar{T}_{loc} \subseteq T$ because then

$$|(\varphi,x)| \leq C \|L\varphi\|_{L_2}$$

for any Schwartz distribution $x \in C_0^\infty(T)$.

In a case of $X = \overline{C_0^\infty(T)}$ any test distribution $x \in X$ can be approximated by the test functions $\varphi \in C_0^\infty(T)$,

$$x = \lim \varphi ;$$

let us call the imbedding $C_0^\infty(T) \subseteq X$ is <u>local</u> with respect to $\Gamma = \text{supp } x$ if for any of its neighborhood Γ^ε we can take the corresponding limit $\lim \varphi = x$ with $\varphi \in C_0^\infty(\Gamma^\varepsilon)$. For example, we have this property with respect to any compact Γ when a <u>multiplicator</u> with a function $\omega \in C_0^\infty$ is well-defined on the test distributions space X and it represents a <u>continuous</u> operator in X which for the differential operator L means that

$$\|L(\omega\varphi)\|_{L_2} \leq C\|L\varphi\|_{L_2} , \ \varphi \in C_0^\infty(T) .$$

As a matter of fact, the local imbedding $C_0^\infty(T) \subseteq X$ in $X = \overline{C_0^\infty(T)}$ with respect to any compact Γ can be observed for various classes of the differential operators L (in particular for the operators L which are represented in the stochastic models of our § 1) and it allows to conclude that the <u>Markov property</u> of the corresponding ξ as it has been formulated in § 1 for generalized random field over standard test functions $\varphi \in C_0^\infty(T)$ is <u>equivalent</u> to our Markov property of ξ considered as generalized random field over the test distributions $x \in X$ because we have the boundary space $X(\Gamma)$ on the boundary $\Gamma = \partial S$ of the domain $S \subseteq T$ as

$$X(\Gamma) = \cap \overline{C_0^\infty(\Gamma^\varepsilon)}$$

and for the <u>Gaussian</u> ξ the boundary σ-algebra $\mathcal{L}$ generated by all random variables (x,ξ), $x \in X(\Gamma)$, coincides with the σ-algebra

$$A(\Gamma) = \cap A(\Gamma^\varepsilon)$$

which happened to be a starting point of our interest in the boundary problems.

References

[1] Yu. A. Rozanov, The Generalized Dirichlet Problem, Dokl. Acad. Sci. USSR, 266, N5, 1982 (English trans. in Soviet Math. Dokl. 26, N2, 1982).

[2] Yu. A. Rozanov, General Boundary Problems for Linear Differential Operators and the Conjugate Equations Method, Proceedings of the Steklov Math. Institute 166, 1984.

[3] M. Röckner, Generalized Markov fields and Dirichlet forms, Acta Applic. Mathematicae 3, 285-311 (1985).

GENERALIZED ONE-SIDED STABLE DISTRIBUTIONS

W.R. Schneider
Brown Boveri Research Center
CH-5405 Baden, Switzerland

1. Introduction

Stable distributions play an eminent role in the theory of addition of random variables [1-3]; a close connection to the renormalization group idea has been exhibited in [4]. A particular case are the one-sided (i.e. support in R_+) stable distributions. In the sequel two distribution functions F and G are considered equivalent if for some $\lambda > 0$

$$G(x) = F(\lambda x) \tag{1.1}$$

holds. Up to equivalence the one-sided stable distributions form a one-parameter family F_α, $0 < \alpha < 1$, characterized by its Laplace-Stieltjes transform

$$\int_0^\infty e^{-px}\, dF_\alpha(x) = e^{-p^\alpha} \quad . \tag{1.2}$$

It follows [3] that F_α has a density which will be denoted by f_α, whose Laplace transform is given by

$$\int_0^\infty e^{-px} f_\alpha(x)dx = e^{-p^\alpha} \quad . \tag{1.3}$$

A direct characterization in terms of "known" functions has been lacking so far. This fact has found repeatedly particular attention, e.g. in [3], [5], [6]. This unsatisfactory situation will be remedied as a byproduct of the study of generalized one-sided stable distributions, $F_{m,\alpha}$, with $0 < \alpha < 1$ as above and m running through the positive integers. For m = 1 the ordinary one-sided stable distribution is recovered (up to equivalence). In Section 2 a homogeneous linear integral equation containing the two parameters m and α is introduced and investigated by means of the Mellin transform. A solution, denoted by $g_{m,\alpha}$, is found and shown to be expressible in terms of Fox functions [7-9]. Series expansion and asymptotic behaviour of $g_{m,\alpha}$ are given using the general results on Fox functions in [8]. Furthermore, it turns out that $g_{m,\alpha}$ is a probability density on R_+. Hence,

$$f_{m,\alpha}(x) = x^{-2}\, g_{m,\alpha}(x^{-1}) \tag{1.4}$$

is also a probability density on R_+; by definition, $f_{m,\alpha}$ is the density of the generalized one-sided stable distribution $F_{m,\alpha}$. The Laplace transform $\phi_{m,\alpha}$ of $f_{m,\alpha}$ is studied in Section 3. Via Mellin transform it is shown that also $\phi_{m,\alpha}$ can be expressed in terms of Fox functions, reducing to an exponential function of the form $\exp(-\lambda p^{\alpha})$ for $m = 1$. Thus, F_{α} is equivalent to $F_{1,\alpha}$. Particular attention is also paid to the case $m = 2$ which plays a role in one-dimensional random lattice systems [10-13]. In a sense to be made precise there exists a generalized domain of attraction for $m = 2$ (whereas the usual notion of domain of attraction is tied to $m = 1$). It is conjectured that analogous results hold also for $m \geq 3$. Section 4 contains the Fox function representation of the general (two-sided) stable distributions. The definition of the Fox functions together with a collection of a few basic properties may be found in Section 5.

2. Integral Equation and Solution

The integral equation to be solved is

$$g(x) = x^{m+\alpha-2} \int_x^{\infty} dy\; y^{\alpha}(y-x)^{-\alpha}\; g(y) \tag{2.1}$$

where

$$m = 1,2,3,\ldots \quad , \qquad 0 < \alpha < 1 \tag{2.2}$$

and $0 \leq x < \infty$. The right hand side of (2.1) makes sense e.g. for g continuous and falling off faster than any inverse power of x as x tends to infinity, the resulting function having the same properties. Hence, the Mellin transform $\hat{g}$ of g, defined by

$$\hat{g}(s) = \int_0^{\infty} dx\; x^{s-1}\; g(x) \quad , \tag{2.3}$$

exists for $0 < \mathrm{Re}\, s < \infty$ (and is regular in this half-plane). Inserting (2.1) into (2.3) and interchanging the order of integration (which is justified by Fubini's theorem in the version of [14], e.g.) yields

$$\hat{g}(s) = \hat{g}(s+m+\alpha-1)\,\frac{\Gamma(1-\alpha)\Gamma(s+m+\alpha-2)}{\Gamma(s+m-1)} \quad , \tag{2.4}$$

i.e. a difference equation for $\hat{g}$. By using repeatedly

$$\Gamma(z+1) = z\,\Gamma(z) \tag{2.5}$$

one verifies that

$$\hat{g}_{m,\alpha}(s) = Ab^{s}\,\frac{1}{\Gamma(s-1)}\prod_{k=1}^{m}\Gamma\left(\frac{s+k-2}{a}\right) \tag{2.6}$$

with

$$a = m+\alpha-1 \quad , \quad b = \left(\frac{a^{m}}{\Gamma(1-\alpha)}\right)^{1/a} \tag{2.7}$$

solves (2.4). The prefactor A is arbitrary and will be fixed in the following by the requirement

$$\int_{0}^{\infty} dy\, g(y) = \hat{g}(1) = 1 \quad . \tag{2.8}$$

This leads to

$$Aab\prod_{k=2}^{m}\Gamma\left(\frac{k-1}{a}\right) = 1 \quad . \tag{2.9}$$

Inversion of the Mellin transform yields

$$g(x) = \frac{1}{2\pi i}\int ds\, x^{-s}\,\hat{g}(s) \tag{2.10}$$

where the path of integration runs parallel to the imaginary axis from $c-i\infty$ to $c+i\infty$ with $c > 0$ arbitrary. As is easily verified this path may be deformed into one running from $-\infty-ic$ to $-\infty+ic$ counterclockwise around the negative real axis. A change of s into -s and comparison with [8] or (5.1)-(5.5) yields

$$g_{m,\alpha}(x) = A\, H^{m0}_{1m}\left(\frac{x}{b} \,\middle|\, \begin{matrix} (-1,1) \\ (\frac{k-2}{a}, \frac{1}{a})_{k=1,\dots,m} \end{matrix} \right) \tag{2.11}$$

where H^{mn}_{pq} denotes the general Fox function (in the notation of [9] which seems to have become standard, but was neither in use when Fox [7] introduced these functions nor when Braaksma [8] investigated in great detail their properties). Analyticity properties, series expansion and asymptotic behaviour of Fox functions may be found in [8]. A brief outline may be found in Section 5. The conditions under which these results hold are easily checked for the particular case (2.11). Thus, g(x) has an analytic continuation into the complex plane cut along the negative real axis. It has the following absolutely convergent series expansion

$$g_{m,\alpha}(x) = Aa \sum_{k=1}^{m} \sum_{n=1}^{\infty} \frac{c_{k,n}}{\Gamma(1-k-na)} \frac{(-1)^n}{n!} \left(\frac{x}{b}\right)^{k-2+na} \tag{2.12}$$

with

$$c_{k,n} = \prod_{j=1}^{m}{}' \,\Gamma\left(\frac{j-k}{a} - n\right) \tag{2.13}$$

where the prime indicates omission of $j = k$. The leading term of (2.12) for small x is the one with $k = 1$, $n = 1$. After some algebra and use of (2.9) it is given by

$$g_{m,\alpha}(x) \sim x^{m+\alpha-2} \quad . \tag{2.14}$$

Formally, this result is obtained by setting $x = 0$ in the integral of (2.1) and using (2.8).

The asymptotic behaviour of g(x) for x large is given by

$$g_{m,\alpha}(x) \sim Bx^{\sigma} e^{-\kappa x^{\tau}} \tag{2.15}$$

where

$$\tau = \frac{m+\alpha-1}{1-\alpha} \quad , \quad \kappa = \frac{1-\alpha}{m+\alpha-1} \Gamma(1-\alpha)^{1/(1-\alpha)} \tag{2.16}$$

and

$$B = (2\pi)^{(m-2)/2} (1-\alpha)^{-1/2} \Gamma(1-\alpha)^{\varepsilon} (m+\alpha-1)^{-\delta} \prod_{k=2}^{m} \Gamma\left(\frac{k-1}{m+\alpha-1}\right)^{-1} \quad ,$$

$$\sigma = m-2+\frac{m\alpha}{2(1-\alpha)} \quad , \quad \varepsilon = \frac{\sigma+1}{m+\alpha-1} \quad , \quad \delta = \frac{(1-\alpha)(m-1)}{2(m+\alpha-1)} \quad . \tag{2.17}$$

From (2.12) and (2.15) it follows that $g_{m,\alpha}(x)$ has only finitely many, say N, positive zeros. Assume $N > 0$. Inserting in (2.1) for x the largest zero leads to a contradiction, as the left hand side vanishes whereas the right hand side is positive. Hence, $g_{m,\alpha}(x)$ is positive for x positive, which, together with (2.8) implies, that $g_{m,\alpha}$ is a probability density.
The probability densities $f_{m,\alpha}$ on R_+ are defined by

$$f_{m,\alpha}(x) = x^{-2} g_{m,\alpha}(x^{-1}) \tag{2.18}$$

for x positive. They satisfy the homogeneous linear integral equation

$$x^m f(x) = \int_0^x dy (x-y)^{-\alpha} f(y) \tag{2.19}$$

as is seen by transforming (2.1) accordingly.
From (2.12) the series expansion of $f_{m,\alpha}$ in negative (non-integer) powers of x is obtained; in particular, (2.14) yields the asymptotic behaviour

$$f_{m,\alpha}(x) \sim x^{-m-\alpha} \tag{2.20}$$

as x tends to infinity.
The asymptotic behaviour for small x is given by

$$f_{m,\alpha}(x) \sim B\, x^{-\sigma-2}\, e^{-\kappa x^{-\tau}} \tag{2.21}$$

with B, σ, κ, τ defined in (2.16) and (2.17).

3. Laplace Transform of $f_{m,\alpha}$

The Laplace transform of the probability density $f_{m,\alpha}$, defined by (2.18), will be denoted by $\phi_{m,\alpha}$; it is given by

$$\phi_{m,\alpha}(p) = \int_0^\infty dx\, e^{-px}\, f_{m,\alpha}(x) \quad . \tag{3.1}$$

To obtain an explicit expression for $\phi_{m,\alpha}$ its Mellin transform $\hat{\phi}_{m,\alpha}$ with

$$\hat{\phi}_{m,\alpha}(s) = \int_0^\infty dp\, p^{s-1}\, \phi_{m,\alpha}(p) \tag{3.2}$$

is introduced. Insertion of (3.1) and interchange of the order of integration (by Fubini's theorem [14]) yields

$$\hat{\phi}_{m,\alpha}(s) = \Gamma(s)\, \hat{f}_{m,\alpha}(1-s) \quad . \tag{3.3}$$

Now, the Mellin transform $\hat{f}_{m,\alpha}$ of $f_{m,\alpha}$ is related to the Mellin transform $\hat{g}_{m,\alpha}$ of $g_{m,\alpha}$ by

$$\hat{f}_{m,\alpha}(s) = \hat{g}_{m,\alpha}(2-s) \quad . \tag{3.4}$$

Hence, by (2.6)

$$\hat{f}_{m,\alpha}(s) = Ab^{2-s} \frac{1}{\Gamma(1-s)} \prod_{k=1}^{m} \Gamma\left(\frac{k-s}{a}\right) \quad . \tag{3.5}$$

This yields with (3.3)

$$\hat{\phi}_{m,\alpha}(s) = Ab^{1+s} \prod_{k=1}^{m} \Gamma\left(\frac{s+k-1}{a}\right) \quad , \tag{3.6}$$

or finally with arguments analogous to the ones leading from (2.6) to (2.11)

$$\phi_{m,\alpha}(p) = Ab\, H^{m0}_{0m}\left(\frac{p}{b} \,\Big|\, \overline{} \atop \left(\frac{k-1}{a}, \frac{1}{a}\right)_{k=1,\dots,m}\right) \quad . \tag{3.7}$$

Its series expansion is

$$\phi_{m,\alpha}(p) = Aab \sum_{k=1}^{m} \sum_{n=0}^{\infty} c_{k,n} \frac{(-1)^n}{n!} \left(\frac{p}{b}\right)^{k-1+na} \tag{3.8}$$

with the coefficients $c_{k,n}$ defined in (2.13). For $p = 0$ (3.8) leads to

$$\phi_{m,\alpha}(0) = Aabc_{1,0} = 1 \tag{3.9}$$

in accordance with (2.8).

For p large the asymptotic behaviour of $\phi_{m,\alpha}(p)$ is

$$\phi_{m,\alpha}(p) \sim Cp^{\beta} \exp(-\lambda p^{\gamma}) \tag{3.10}$$

where

$$\gamma = \frac{m+\alpha-1}{m} \quad , \quad \lambda = \frac{m}{m+\alpha-1} \quad \Gamma(1-\alpha)^{1/m} \tag{3.11}$$

and

$$\beta = \frac{(m-1)(1-\alpha)}{2m}$$

$$C = (2\pi)^{(m-1)/2} \, Aa \, m^{-1/2} \, b^{1-\beta} \quad . \tag{3.12}$$

The cases m = 1 and m = 2 will now be treated in more detail.

(I) <u>m = 1</u>

$\phi_{1,\alpha}$ may be expressed in terms of an exponential function. This follows from (3.8), (2.13) and (2.9) when m = 1 is inserted:

$$\phi_{1,\alpha}(p) = \sum_{n=0}^{\infty} \frac{(-1)^n}{n!} \left(\frac{p}{b}\right)^{na} = \exp\left(-\left(\frac{p}{b}\right)^a\right) \tag{3.13}$$

where

$$a = \alpha \quad , \quad b = \left(\frac{a}{\Gamma(1-\alpha)}\right)^{1/a} \quad . \tag{3.14}$$

Consequently, the stable density f_α and $f_{1,\alpha}$ are equivalent,

$$f_\alpha(x) = b^{-1} f_{1,\alpha}(b^{-1}x) \quad , \tag{3.15}$$

which leads to the following Fox function representation of f_α:

$$f_\alpha(x) = \alpha^{-1} x^{-2} H_{11}^{10} \left(x^{-1} \,\middle|\, \begin{matrix}(-1,1)\\(-\alpha^{-1},\alpha^{-1})\end{matrix}\right) \quad . \tag{3.16}$$

Its series expansion is

$$f_\alpha(x) = \sum_{n=1}^{\infty} \frac{1}{\Gamma(-n\alpha)} \frac{(-1)^n}{n!} x^{-1-n\alpha} \tag{3.17}$$

or, using $\Gamma(z)\Gamma(1-z)\sin\pi z = \pi$,

$$f_\alpha(x) = \frac{1}{\pi x} \sum_{n=1}^{\infty} \frac{\Gamma(n\alpha+1)}{n!} (-x^{-\alpha})^n \sin(-\pi n\alpha) \tag{3.18}$$

in accordance with [3].
The asymptotic behaviour for x small is given by

$$f_\alpha(x) \sim D\, x^{-\mu} \exp(-\omega x^{-\tau}) \tag{3.19}$$

with

$$\tau = \frac{\alpha}{1-\alpha} \quad , \quad \mu = \frac{2-\alpha}{2(1-\alpha)} \tag{3.20}$$

and

$$\omega = (1-\alpha)\alpha^{\alpha/(1-\alpha)}$$

$$D = \{[2\pi(1-\alpha)]^{-1} \alpha^{1/(1-\alpha)}\}^{1/2} \quad , \tag{3.21}$$

in accordance with [15] (note the different prefactor of p^{α} which here is one, see (1.3)).

(II) m = 2

$\phi_{2,\alpha}$ may be expressed in terms of a modified Bessel function (second kind):

$$\phi_{2,\alpha}(p) = \frac{2}{\Gamma(\beta)} \left(\frac{p}{b}\right)^{1/2} K_{\beta}\left(2\left(\frac{p}{b}\right)^{1/2\beta}\right) \tag{3.22}$$

with β related to α by

$$\beta = 1/(1+\alpha) \quad . \tag{3.23}$$

This may be seen by using

$$K_{\beta}(z) = \frac{\pi}{2\sin\pi\beta} \left(I_{-\beta}(z) - I_{\beta}(z)\right) \tag{3.24}$$

where I_{σ} denotes the modified Bessel function (first kind) with the series expansion

$$I_{\sigma}(z) = (z/2)^{\sigma} \sum_{n=0}^{\infty} \frac{(z/2)^{2n}}{n!\ \Gamma(n+\sigma+1)} \quad . \tag{3.25}$$

Combining (3.22) - (3.25) and taking

$$\Gamma(c-n)\Gamma(n+1-c) = (-1)^{n}\ \pi/\sin\pi c \tag{3.26}$$

into account yields a series expansion coinciding with the one obtained from (3.13) for m = 2. As an application of the representation (3.22) the double integral

$$C_{\alpha} = \int_{0}^{\infty}\int_{0}^{\infty} dxdy\ (x+y)^{-1}\ g_{2,\alpha}(x) g_{2,\alpha}(y) \tag{3.27}$$

is evaluated analytically (in [10] $C_{o}^{(\alpha)} = C_{1-\alpha}$ was computed numerically). The substitution $x \to 1/x$, $y \to 1/y$ yields

$$C_\alpha = \int_0^\infty \int_0^\infty dxdy\ xy(x+y)^{-1}\ f_{2,\alpha}(x)\ f_{2,\alpha}(y) \quad . \tag{3.28}$$

Insertion of

$$(x+y)^{-1} = \int_0^\infty dp\ e^{-p(x+y)} \tag{3.29}$$

leads to

$$C_\alpha = \int_0^\infty dp[\frac{d}{dp}\ \phi_{2,\alpha}(p)]^2 \quad . \tag{3.30}$$

With a partial integration, the explicit expression (3.22) and

$$\int_0^\infty dz\ z\ K_\beta(z)^2 = \frac{\pi\beta}{2\sin\pi\beta} \tag{3.31}$$

[16] the result

$$C_\alpha = \frac{\pi\beta}{\sin\pi\beta}\ \frac{(\beta^2\Gamma(1-\alpha))^\beta}{\Gamma(\beta)^2} \quad , \qquad \beta = 1/(1+\alpha) \tag{3.32}$$

is obtained.

An alternative way to treat the Laplace transform $\phi_{m,\alpha}$ of $f_{m,\alpha}$ starts from the integral equation (2.19). Inserting (2.19) into (3.1) yields

$$(-\frac{d}{dp})^m\ \phi_{m,\alpha}(p) = \Gamma(1-\alpha)p^{\alpha-1}\ \phi_{m,\alpha}(p) \quad . \tag{3.33}$$

For m = 1 the general solution of (3.33) is

$$\phi_{m,\alpha}(p) = \Lambda\ \exp\ \{-\ \frac{\Gamma(1-\alpha)}{\alpha}\ p^\alpha\} \tag{3.34}$$

with arbitrary Λ, which however is fixed to $\Lambda = 1$ by the requirement $\phi_{m,\alpha}(0) = 1$. For m = 2 the general solution of (3.33) is

$$\phi_{m,\alpha}(p) = p^{1/2}\{\Lambda_1 K_\beta(z) + \Lambda_2 I_\beta(z)\} \tag{3.35}$$

with arbitrary $\Lambda_{1,2}$ and

$$z = 2\beta\Gamma(1-\alpha)^{1/2} p^{(1+\alpha)/2} \quad , \quad \beta = \frac{1}{1+\alpha} \tag{3.36}$$

as is easily verified using the differential equation

$$\frac{d^2f}{dz^2} + \frac{1}{z}\frac{df}{dz} - \frac{\beta^2}{z^2} f - f = 0 \tag{3.37}$$

for the modified Bessel functions.
As a Laplace transform $\phi_{m,\alpha}$ is bounded which implies $\Lambda_2 = 0$ whereas Λ_1, is fixed to

$$\Lambda_1 = \frac{2}{\Gamma(\beta)} (\beta^2 \Gamma(1-\alpha))^{\beta/2} \tag{3.38}$$

by the requirement $\phi_{m,\alpha}(0) = 1$.
In conclusion, for $m = 1$ and $m = 2$ the integral equation (2.19) has a unique solution which is simultaneously a probability density. For $m \geq 3$ the question of uniqueness remains open.

As is well known [1-3] a distribution function F whose support is in R_+ and satisfies

$$1-F(x) \sim (\frac{c}{x})^\alpha \quad , \quad 0 < \alpha < 1 \quad , \tag{3.39}$$

for large x belongs to the domain of attraction of the stable distribution F_α, i.e.

$$F^{*n}(c'n^{1/\alpha}x) \underset{n\to\infty}{\to} F_\alpha(x) \quad , \quad c' = c\Gamma(1-\alpha)^{1/\alpha} \tag{3.40}$$

Here, * denotes convolution. Remarkably, also $F_{2,\alpha}$ occurs as limit distribution albeit in a more complex way: Let F be as above and G_s, $s > 0$, a family of distribution functions on R_+ satisfying

$$G_s(x) = \iint_{C_{s,x}} dF(y)\, dG_s(z) \tag{3.41}$$

with

$$C_{s,x} = \{(y,z) \in R_+^2 \mid c^{-1}s^{1/\alpha}y + z/(1+sz) < x\} \tag{3.42}$$

Existence, uniqueness and weak continuity in s follow from the results in [12] whereas in [13] it has been shown that

$$G_s \underset{s \to 0}{\to} F_{2,\alpha} \quad . \tag{3.43}$$

It is conjectured that also $F_{m,\alpha}$, $m \geq 3$, are limit distributions in a suitable sense.

4. Two-Sided Stable Distributions

The general stable distribution $F_{\alpha,\beta}$ is characterized by its Fourier-Stieltjes transform [3]

$$\int e^{ikx}\, dF_{\alpha,\beta}(x) = \exp \psi_{\alpha,\beta}(k) \tag{4.1}$$

with

$$\psi_{\alpha,\beta}(k) = - |k|^{\alpha} \exp (i \tfrac{\pi}{2} \beta), \qquad k \geq 0 \tag{4.2}$$

and

$$\psi_{\alpha,\beta}(-k) = \overline{\psi_{\alpha,\beta}(k)} \quad . \tag{4.3}$$

The ranges of the parameters α and β are as follows

$$0 < \alpha < 1 \quad , \qquad |\beta| \leq \alpha \tag{4.4}$$

$$1 < \alpha \leq 2 \quad , \qquad |\beta| \leq 2-\alpha \tag{4.5}$$

For $\beta = \alpha$ $(-\alpha)$ the support of $F_{\alpha,\beta}$ is R_- (R_+) whereas in all other cases the support is R.
It is remarkable that also these two-sided stable distributions may be expressed in terms of Fox functions. A short sketch is given below; details are presented in [17] where also special values of α and β are considered which allow representations in terms of the more familiar hypergeometric functions (correcting thereby errors in the literature).

From (4.1)-(4.3) it follows [3] that $F_{\alpha,\beta}$ has a density $f_{\alpha,\beta}$ which is obtained by inverse Fourier transform

$$f_{\alpha,\beta}(x) = \mathrm{Re}\,\frac{1}{\pi}\int_0^\infty dk\; e^{-ikx}\,\exp\psi_{\alpha,\beta}(k) \qquad . \tag{4.6}$$

Obviously, the relation

$$f_{\alpha,\beta}(-x) = f_{\alpha,-\beta}(x) \tag{4.7}$$

holds. Hence, it is sufficient to consider $f_{\alpha,\beta}(x)$ for $x \geq 0$.
In particular, $f_{\alpha,\beta}$ is characterized by its Mellin transform

$$\hat{f}_{\alpha,\beta}(s) = \int_0^\infty dx\; x^{s-1}\, f_{\alpha,\beta}(x) \tag{4.8}$$

which after some manipulation [17] is obtained from (4.6), (4.8),

$$\hat{f}_{\alpha,\beta}(s) = \frac{\Gamma(s-1)\Gamma(1+\varepsilon-\varepsilon s)}{\Gamma(1+\gamma-\gamma s)\Gamma(-\gamma+\gamma s)} \qquad , \tag{4.9}$$

where

$$\varepsilon = \alpha^{-1} \qquad , \qquad \gamma = \frac{\alpha-\beta}{2\alpha} \qquad . \tag{4.10}$$

For $\alpha > 1$ (4.9) leads to

$$f_{\alpha,\beta}(x) = H^{11}_{22}\left(x \,\middle|\, \begin{matrix} (-\varepsilon,\varepsilon),\ (-\gamma,\gamma) \\ (-1,1),\ (-\gamma,\gamma) \end{matrix}\right) \quad ; \tag{4.11}$$

note that condition (5.9) reads here

$$(1+\gamma) - (\varepsilon+\gamma) = 1-\varepsilon \quad > 0 \tag{4.12}$$

which is equivalent to $\alpha > 1$. For $\alpha < 1$ consider

$$g_{\alpha,\beta}(x) = x^{-2}\, f_{\alpha,\beta}(x^{-1}) \tag{4.13}$$

with Mellin transform

$$\hat{g}_{\alpha,\beta}(s) = \hat{f}_{\alpha,\beta}(2-s) \quad . \tag{4.14}$$

For $|\beta| < \alpha$ this leads to

$$f_{\alpha,\beta}(x) = x^{-2}\, H_{22}^{11}\left(x^{-1} \,\middle|\, \begin{matrix} (0,1)\ , & (1-\gamma,\gamma) \\ (1-\varepsilon,\varepsilon), & (1-\gamma,\gamma) \end{matrix}\right) \tag{4.15}$$

with condition (5.9) reading

$$(\varepsilon+\gamma) - (1+\gamma) \quad > 0 \quad , \tag{4.16}$$

i.e. $\alpha < 1$. For $\beta = -\alpha$ (4.9) yields

$$\hat{f}_{\alpha,-\alpha}\,(2-s) = \frac{\Gamma(1-\varepsilon+\varepsilon s)}{\Gamma(s)} = \varepsilon\,\frac{\Gamma(-\varepsilon+\varepsilon s)}{\Gamma(s-1)} \tag{4.17}$$

leading to

$$f_{\alpha,-\alpha}(x) = \varepsilon\, x^{-2}\, H_{11}^{10}\left(x^{-1} \,\middle|\, \begin{matrix} (-1,1) \\ (-\varepsilon,\varepsilon) \end{matrix}\right) \quad . \tag{4.18}$$

Comparison with (3.16) shows that

$$f_{\alpha,-\alpha}(x) = f_{\alpha}(x) \quad , \tag{4.19}$$

i.e. the one-sided stable distribution with support R_+.

5. Fox Functions

The Fox function [7-9]

$$H^{mn}_{pq}(z) = H^{mn}_{pq}\left(z \;\middle|\; \begin{matrix} (a_j,\alpha_j)_{j=1,\ldots,p} \\ (b_j,\beta_j)_{j=1,\ldots,q} \end{matrix}\right) \tag{5.1}$$

is defined by the contour integral

$$H^{mn}_{pq}(z) = \frac{1}{2\pi i} \int_L K^{mn}_{pq}(s)\, z^s \, ds \tag{5.2}$$

with

$$K^{mn}_{pq}(s) = \frac{A(s)B(s)}{C(s)D(s)} \tag{5.3}$$

where

$$A(s) = \prod_{j=1}^{m} \Gamma(b_j-\beta_j s)$$

$$B(s) = \prod_{j=1}^{n} \Gamma(1-a_j+\alpha_j s)$$

$$C(s) = \prod_{j=m+1}^{q} \Gamma(1-b_j+\beta_j s)$$

$$D(s) = \prod_{j=n+1}^{p} \Gamma(a_j-\alpha_j s) \tag{5.4}$$

Here, m,n,p,q are integers satisfying

$$0 \le n \le p \quad , \quad 1 \le m \le q \tag{5.5}$$

In the cases $n = 0$, $m = q$, $n = p$ (5.4) has to be interpreted as $B(s) = 1$, $C(s) = 1$, $D(s) = 1$, respectively.
The parameters $a_j (j=1,\ldots,p)$ and $b_j (j=1,\ldots,q)$ are complex whereas $\alpha_j (j=1,\ldots,p)$ and $\beta_j (j=1,\ldots,q)$ are positive. They are restricted by the condition

$$P(A) \cap P(B) = \emptyset \tag{5.6}$$

where

$$\begin{aligned} P(A) &= \{s = (b_j+k)/\beta_j \quad | \quad j = 1,\dots,m \quad ; \quad k = 0,1,2,\dots\} \\ P(B) &= \{s = (a_j-1-k)/\alpha_j \quad | \quad j = 1,\dots,n \quad ; \quad k = 0,1,2,\dots\} \end{aligned} \tag{5.7}$$

are the sets of the poles of A and B, respectively. The contour L in (5.2) runs from $s = \infty-ic$ to $\infty+ic$ with

$$c > |\mathrm{Im}\, b_j|/\beta_j \qquad (j = 1,\dots,m) \tag{5.8}$$

such that P(A) lies to the left, P(B) to the right of L.

The following additional condition is assumed to hold throughout

$$\mu = \sum_{j=1}^{q} \beta_j - \sum_{j=1}^{p} \alpha_j > 0 \tag{5.9}$$

(in [8] also the case $\mu = 0$ is treated). Under these conditions $H_{pq}^{mn}(z)$ is an analytic function for $z \neq 0$, in general multiple-valued (one-valued on the Riemann surface of log z). It is given by

$$H_{pq}^{mn}(z) = -\sum_{s \varepsilon P(A)} \mathrm{res}\left(\frac{A(s)B(s)}{C(s)D(s)} z^s\right) \quad , \tag{5.10}$$

res standing for residuum. If all poles of A are simple, i.e.

$$(b_j + k)/\beta_j \neq (b_{j'} + k')/\beta_{j'} \tag{5.11}$$

for $j \neq j'$ with $j,j' = 1,\dots, m$ and $k,k' = 0,1,2,\dots$, then (5.10) yields

$$H_{pq}^{mn}(z) = \sum_{j=1}^{m} \sum_{k=0}^{\infty} c_{j,k} \frac{(-1)^k}{k!\beta_j} z^{(b_j+k)/\beta_j} \tag{5.12}$$

with

$$c_{j,k} = \frac{A_j(s_{j,k})B(s_{j,k})}{C(s_{j,k})D(s_{j,k})} \quad , \qquad s_{j,k} = (b_j+k)/\beta_j \tag{5.13}$$

and A_j defined by

$$A(s) = A_j(s)\,\Gamma(b_j-\beta_j s) \quad . \tag{5.14}$$

Let δ be given by

$$\delta = \left(\sum_{j=1}^{m} \beta_j - \sum_{J=n+1}^{p} \alpha_j\right)\pi \tag{5.15}$$

and assume

$$\delta > \frac{\pi}{2}\mu \quad . \tag{5.16}$$

Then, asymptotically

$$H_{pq}^{mn}(z) \sim \sum_{s\varepsilon P(B)} \text{res}\left(\frac{A(s)B(s)}{C(s)D(s)}\, z^s\right) \tag{5.17}$$

as $|z| \to \infty$ uniformly on every closed subsector of

$$|\arg z| < \delta - \frac{\pi}{2}\mu \quad . \tag{5.18}$$

In the case where all poles of B are simple (5.17) may be written in a form analogous to (5.12).

Obviously, for n = 0 the above statement becomes void. For this case exponentially small asymptotic behaviour is derived in [8]. In particular, for m = q (which implies $\delta = \mu\pi$) the asymptotic behaviour for $|z| \to \infty$ is given by

$$H_{pq}^{q0}(z) \sim (2\pi)^{q-p}\, e^{i\pi(\alpha-1/2)}\, E(ze^{i\pi\mu}) \quad , \tag{5.19}$$

uniformly on every closed sector (vertex in 0) contained in $|\arg z| < \mu\pi/2$, where

$$E(z) = \frac{1}{2\pi i\mu} \sum_{k=0}^{\infty} A_k \, (\beta\mu^{\mu} z)^{(1-\alpha-k)/\mu} \exp(\beta\mu^{\mu} z)^{1/\mu} \quad . \tag{5.20}$$

The constants α and β are given by

$$\alpha = \sum_{j=1}^{p} a_j - \sum_{j=1}^{q} b_j + (q-p+1)/2 \tag{5.21}$$

and

$$\beta = \prod_{j=1}^{p} \alpha_j^{\alpha_j} \prod_{j=1}^{q} \beta_j^{-\beta_j} \quad , \tag{5.22}$$

respectively. The coefficients $A_k (k = 0,1,2,\ldots)$ are determined by

$$\frac{A(s)B(s)}{C(s)D(s)} (\beta\mu^{\mu})^{-s} \sim \sum_{k=0}^{\infty} \frac{A_k}{\Gamma(\mu s+\alpha+k)} \quad . \tag{5.23}$$

In particular

$$A_o = (2\pi)^{(p-q+1)/2} \mu^{\alpha-1/2} \prod_{j=1}^{p} \alpha_j^{1/2-a_j} \prod_{j=1}^{q} \beta_j^{b_j-1/2} \tag{5.24}$$

Fox functions have found applications in other parts of probability theory [18]. Their connection with Lévy distributions however seems to have been unnoticed so far, to the best of the author's knowledge.

<u>References</u>

[1] Lévy, P. : Théorie de l'addition des variables aléatoires. Paris: Gauthier-Villars 1954.

[2] Gnedenko, B.V., Kolmogorov, A.N. : Limit distributions for sums of independent random variables. Reading: Addison Wesley 1954.

[3] Feller, W. : An introduction to probability theory and its applications, Vol. II. New York: John Wiley 1971.

[4] Jona-Lasinio, G. : The renormalization group: A probabilistic view. Nuovo Cimento 26B, 99-119 (1975).

[5] Mandelbrot, B.B. : The fractal geometry of nature. New York: W.H. Freeman 1983.

[6] Montroll, E.W., Shlesinger, M.F. : On the wonderfull world of random walks. In: Nonequilibrium phenomena II (Studies in statistical mechanics, Vol. 11). Lebowitz, J.L., Montroll, E.W., (eds.). Amsterdam: North Holland 1984.

[7] Fox, C. : The G and H Functions as symmetrical Fourier kernels. Trans. Amer. Math. Soc. 98, 395-429 (1961).

[8] Braaksma, B.L.J. : Asymptotic expansions and analytic continuations for a class of Barnes-integrals. Compos. Math. 15, 239-341 (1963).

[9] Gupta, K.G., Jain, U.C. : The H-function-II. Proc. Nat. Acad. Sci. India A36, 594-602 (1966).

[10] Bernasconi, J., Schneider, W.R., Wyss, W. : Diffusion and hopping conductivity in disordered one-dimensional lattice systems. Z. Physik B37, 175-184 (1980).

[11] Alexander, S., Bernasconi, J., Schneider, W.R., Orbach, R. : Excitation dynamics in random one-dimensional systems. Rev. Mod. Phys. 53, 175-198 (1981).

[12] Schneider, W.R. : Existence and uniqueness for random one-dimensional lattice systems. Commun. Math. Phys. 87, 303-313 (1982).

[13] Schneider, W.R. : Rigorous scaling laws for Dyson measures. In: Stochastic Processes - Mathematics and Physics. Proceedings of the first BiBoS-Symposium. Albeverio, S., Blanchard, Ph., Streit, L., (eds.). Lecture notes in mathematics. Berlin: Springer (1985).

[14] Reed, M., Simon, B. : Methods of modern mathematical physics I: Functional analysis. New York: Academic Press 1972.

[15] Skorohod, A.V. : Asymptotic formulas for stable distribution laws. Selected translations in mathematical statistics and probability, Vol. 1, 157-161 (1961).

[16] Gradshteyn, I.S., Ryzhik, I.M. : Tables of integrals, series, and products. New York: Academic Press (1965).

[17] Schneider, W.R. : Stable distributions: Fox function representation and generalization. First Ascona-Como international conference (1985): Stochastic processes in classical and quantum systems. To appear in: Lecture notes in physics. Berlin: Springer (1986).

[18] Srivastava, H.M., Kashyap, B.R.K. : Special functions in queuing theory and related stochastic processes. New York: Academic Press (1982).

QUANTUM FIELDS, GRAVITATION AND THERMODYNAMICS

Geoffrey L. Sewell

Department of Physics, Queen Mary College, London E1 4NS

ABSTRACT

The thermalisation of quantum fields by gravitational ones associated with certain event horizons, as in the Hawking-Unruh effect, is shown to be a general, model-independent consequence of the basic axioms of quantum theory, general relativity and statistical thermodynamics, closely connected with the PCT theorem.

1. INTRODUCTION

The interplay between the developments of Quantum Field theory and Statistical Thermodynamics has led to striking advances in both these areas of Physics (cf. Ref. [1]). A connection between these developments and those of General Relativity was initiated by Hawking's [2] argument that Black Holes emit thermal radiation as a result of the action of their gravitational fields on ambient quantum fields. Most significantly, the interplay between Quantum Theory, General Relativity and Statistical Thermodynamics was crucial to his argument since, according to a purely classical picture, Black Holes emit nothing. From the statistical mechanical standpoint, his result is remarkable, since it implies that a quantum field can be thermalised by the action of certain ***secular*** forces (gravitational ones), whereas the traditional view is that heat is generated by stochastic forces. Similar results have subsequently been obtained by Unruh [3] and Davies [4] for the thermalisation of quantum fields by gravitational ones corresponding to uniform accelerations in flat space-time. However, the arguments in the pioneering works [2-4] are limited to the exacting solvable models of free scalar quantum fields and, moreover, are somewhat lacking in rigour.

For these reasons, I made a general rigorous approach to the subject [5], based on axiomatic field theory and statistical mechanics, and this provided a derivation of the Hawking-Unruh thermalisation effect that is appreciable to interacting, as well as free, quantum fields. Thus, apart from the gain in rigour, this approach has the advantage of demonstrating, in a model-independent way, how the basic principles of quantum theory, general relativity and statistical mechanics conspire to achieve the result that certain gravitational fields, including those of Refs. [2-4], act so as to thermalise ambient quantum fields. The stochasticity required for this thermalisation stems, in fact, from the quantum fluctuations of the fields. In particular, in the case where the gravitational forces correspond, via the Principle of Equivalence, to uniform acceleration in flat space-time, the thermalisation is just a consequence of a relativity of temperature with respect to acceleration: specifically, the fluctuations that, for an inertial observer, are those of the vacuum, are seen by accelerated observers to be thermal.

The purpose of this talk is to demonstrate the structure of the argument leading from the axioms of quantum field theory and statistical mechanics to the Hawking-Unruh thermalisation effect. I shall start, in Section 2, by sketching the general formulation of statistical mechanics that is applicable both to finite and infinite systems. This formulation is needed here because a quantum field has an infinite number of degrees of freedom, and so the traditional quantum statistics of finite systems is inapplicable to it (cf. Ref [6,7]). In Section 3, I shall make some rather elementary observations concerning the geometry of both Minkowski space-time and a certain class of curved space-times that include those of Black Holes. In Section 4, I shall sketch the axiomatic approach to quantum field theory and state two key theorems that ensue from it. Finally, in Section 5, I shall show how the Hawking-Unruh thermalisation effect arises as a simple consequence of standard results of axiomatic quantum field theory and statistical mechanics, specified in the previous Sections, and I shall briefly discuss its observational consequences. For simplicity, the explicit treatment of the thermalisation effect will be confined to fields in flat space-time: details of the more complex theory of fields on curved manifolds are provided in Ref. [5].

2. QUANTUM STATISTICAL PRELIMINARIES

In the general formulation of quantum theory (cf. [6,7]), applicable both to finite and infinite systems, the observables of a system correspond to the self-adjoint elements of a *-algebra $\mathcal{A}$, and the states are represented by the linear functionals, ρ, on $\mathcal{A}$ that satisfy the conditions of positivity ($\rho(A^*A) \geq 0$) and normalisation ($\rho(I) = 1$). $\rho(A)$ is then interpreted as the expectation value of the observable A for the state ρ. The states of the system thus form a convex set, and the pure states are taken to be its extremal elements. This characterisation of pure states reduces to the standard one in the conventional formulation of finite systems.

The dynamics of a system is taken to correspond to a one-parameter group $\{\alpha_t | t \in \mathbb{R}\}$ of automorphisms of $\mathcal{A}$, the time-translate of an observable A being

$$A_t = \alpha_t A \tag{1}$$

To pass from the abstract algebraic picture to a concrete Hilbert space description, one invokes the classic Gelfand-Naimark-Segal (GNS) theorem, which tells us that each state ρ induces a representation of the algebra $\mathcal{A}$ in a Hilbert space $\mathcal{H}$, with cyclic vector† Ψ, such that

$$\rho(A) \equiv (\Psi, A\Psi), \tag{2}$$

where the symbol A denotes both an element of $\mathcal{A}$ and its representation in $\mathcal{H}$. Note that [6] in the case of an infinite system, unlike that of a finite one, there are inequivalent irreducible representations of A, in fact an infinity of them! For such a system $\mathcal{A}$ is generally taken to consist of observables in bounded spatial regions, and thus the family of states ρ_σ corresponding to the density matrices σ in the GNS space $\mathcal{H}$ of ρ ($\rho_\sigma(A) \equiv Tr(\sigma A)$) are essentially localised modifications of ρ. Furthermore, the GNS representation is irreducible if and only if the state ρ is pure.

Suppose now that ρ is a stationary state, i.e. $\rho(A_t) \equiv \rho(A)$. Then, in this case, the time-translational automorphisms $\{\alpha_t\}$ are implemented in H by a one-parameter unitary group $\{U_t\}$ that leave the cyclical vector Ψ invariant, i.e.

$$A_t = U_t A U_t^{-1} \quad \text{and} \quad U_t\Psi = \Psi. \tag{3}$$

In fact, U_t is defined for such states by

$$U_t A\Psi = A_t\Psi. \tag{4}$$

Under suitable continuity conditions for α_t (e.g. that $\rho(AB_t)$ is continuous in t), the unitary group U_t is continuous and so, by Stone's theorem, it defines a Hamiltonian H by the standard formula

$$U_t = \exp(iHt/h). \tag{5}$$

Thermal equilibrium states may be characterised, for reasons we shall presently discuss, by the Kubo-Martin-Schwinger (KMS) condition (cf. [8]). For finite temperature $T = (k\beta)^{-1}$, this is the condition on the state ρ that, for any pair of elements A, B of A, there is a function $F_{AB}(z)$ of the complex variable z, that is analytic in the strip $\text{Im}z \in (o, \hbar\beta)$ and continuous on its boundaries, such that

$$F_{AB}(t) = \rho(BA_t) \quad \text{and} \quad F_{AB}(t + ih\beta) = \rho(A_t B)\ \forall\, t \in \mathbb{R}. \tag{6}$$

---

†A vector Ψ is cyclic with respect to an algebra $\mathcal{A}$ in $\mathcal{H}$ if the action of $\mathcal{A}$ on Ψ generates the space $\mathcal{H}$.

In the Hilbert space representation induced by ρ, this relation may be inferred from equations (2) - (6) to be equivalent to the condition that there is a conjugation† J of the Hilbert space $\mathcal{H}$, such that

$$JA^{*}\Psi = \exp(-\tfrac{1}{2}H)A\Psi \quad \forall\ A \text{ in } \mathcal{A}. \tag{7}$$

The KMS condition has been proved to have the following properties, which represent the principal reasons for assuming that it characterises equilibrium states (cf. [9]).
(a) For a finite system, it reduces to the condition for a canonical equilibrium state.
(b) For an infinite system, it represents the condition for behaviour as a thermal reservoir that drives finite systems to which it is weakly coupling into thermal equilibrium at the same temperature T. [10]
(c) It also represents the condition for the fulfillment of various dynamical and thermodynamical requirements of equilibrium states.

Finally we note that since, by (4) and (5), any stationary state ρ satisfies the condition $H\Psi = 0$, the condition for a *ground, i.e. zero temperature, state* is simply that the Hamiltonian operator H is positive. Since, by equations (2), (4) and (5),

$$\rho(BA_t) = (B^{*}\Psi, \exp(iHt/h)A\Psi), \tag{8}$$

the positivity of H implies that $\rho(BA_t)$ is the boundary value, on Imz = 0, of a function that is analytic in the upper half plane, namely

$$F(z) = (B^{*}\Psi, \exp(iHz/h)A\Psi). \tag{9}$$

3. SPACE-TIME GEOMETRY

We shall now note some elementary properties of Minkowski space, and then point out their analogues for a class of curved space-times that includes those of Black Holes.

Minowski Space. In a standard way, we denote the points of Minkowski space-time X $(=\mathbb{R}^4)$ by $(x^{(0)} = ct, x^{(1)}, x^{(2)}, x^{(3)})$, t being the time-coordinate and $x^{(1)}, x^{(2)}, x^{(3)}$ the spatial ones. The Minkowski metric is given by the formula

$$ds^2 = (dx^{(0)})^2 - (dx^{(1)})^2 - (dx^{(2)})^2 - (dx^{(3)})^2. \tag{10}$$

The Rindler wedges, $X^{(\pm)}$, are the submanifolds of X given by (cf. Fig. 1)

$$\left.\begin{aligned} &x^{(1)} > |x^{(0)}| \quad \text{for } X^{(+)} \\ \text{and}\quad &x^{(1)} < -|x^{(0)}| \quad \text{for } X^{(-)} \end{aligned}\right\} \tag{11}$$

† A conjugation J is an antilinear transformation of $\mathcal{H}$, such that $J^2 = I$ and $(Jf,Jg) = (g,f)$.

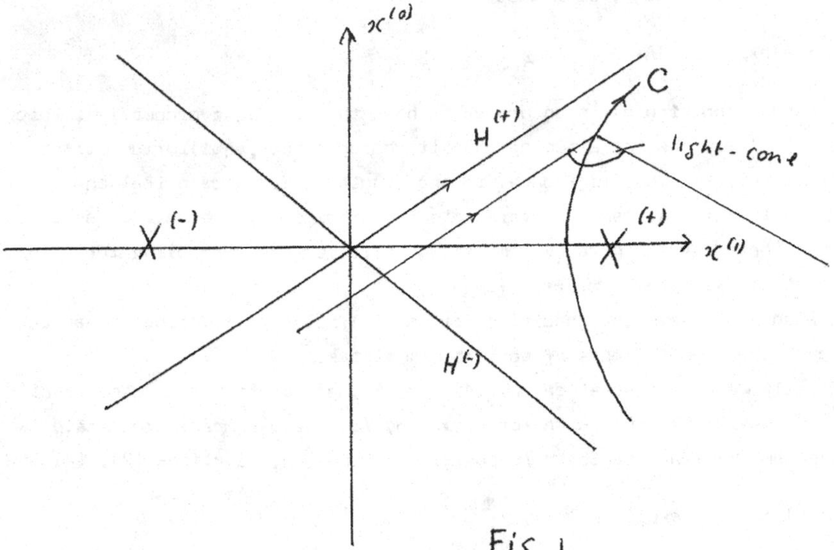

Fig 1

We denote the boundaries $x^{(1)} = \pm x^{(0)}$ of these wedges by $H^{(\pm)}$ (cf. Fig. 1). The wedge $X^{(+)}$ can be parametrised in terms of Rindler coordinates $(\xi, \tau, x^{(2)}, x^{(3)})$

$$x^{(1)} = \xi \cosh \tau, \ x^{(0)} = \xi \sinh \tau \tag{12}$$

with ξ, τ running through $\mathbb{R}_+$ and $\mathbb{R}$, respectively. Thus, by equations (10) and (12), the metric in $X^{(+)}$ is given by

$$(ds^2)_{X^{(+)}} = \xi^2 d\tau^2 - d\xi^2 - (dx^{(2)})^2 - (dx^{(3)})^2 \ . \tag{13}$$

τ is therefore a time coordinate for $X^{(+)}$. Furthermore, the trajectories of uniform acceleration along $Ox^{(1)}$ in $X^{(+)}$ are the curves, C, on which ξ, $x^{(2)}$ and $x^{(3)}$ are constant (cf. Ref. [11], Ch. 6). From (12) and (13), it follows that the proper time and the acceleration on C is given by

$$\tau_{prop} = \tau\xi/c \tag{14}$$

and $$\alpha = c^2/\xi \tag{15}$$

respectively.

We note here that the boundaries $H^{(\pm)}$ of $X^{(+)}$ are horizons in the following sense.

The past light cone of any point P in $X^{(+)}$ does not cross the boundary $H^{(+)}$ (cf. Fig. 1), and therefore no signal can reach $X^{(+)}$ from anywhere across $H^{(+)}$. Likewise, no signals emitted inside $X^{(+)}$ can cross the boundary $H^{(-)}$. For these reasons, $H^{(\pm)}$ are termed *event horizons*.

We now make the important remark that ***time-translations for a uniformly accelerated observer in*** $X^{(+)}$ ***correspond to Lorentz boosts for an inertial one.*** To prove this, one simply notes that time-translations for the former observer correspond to the transformations $\tau \to \tau + s$. By equations (12), these induce the space-time transformations

$$x^{(1)} \to x^{(1)} \cosh s + x^{(0)} \sinh s \;;\; x^{(0)} \to x^{(0)} \cosh s + x^{(1)} \sinh s, \tag{16}$$

which are just the Lorentz boosts for velocity $c \tanh s$ along $Ox^{(1)}$.

Curved Space-times. Consider now a class of space-times of the form $\bar{X} = \mathbb{R}^2 \times Y$, pointwise $\bar{x} = (u, v; y)$, with metric given by

$$ds^2 = f(u^2-v^2)(dv^2-du^2) - g(u^2-v^2)d\sigma_y^2 \tag{17}$$

with f and g smooth positive-valued functions on $\mathbb{R}$. Thus, v is a time coordinate and u, y are spatial ones. We define $\bar{X}^{(\pm)}$ to be the open submanifolds of $\bar{X}$ given by

$$\left.\begin{array}{ll} u > |v| & \text{for } X^{(+)} \\ \text{and} \quad u < -|v| & \text{for } X^{(-)} . \end{array}\right\} \tag{18}$$

Thus $\bar{X}^{(+)}$ may be parametrised by (ξ, τ, y), according to the prescription

$$u = \xi \cosh \tau, \quad v = \xi \sinh \tau \tag{19}$$

with ξ, τ running through $\mathbb{R}_+$ and $\mathbb{R}$, respectively; and hence, by (17) and (19), the metric in $\bar{X}^{(+)}$ is given by

$$ds^2 = f(\xi^2)\,(\xi^2 d\tau^2 - d\xi^2) - g(\xi^2)\, d\sigma_y^2 \tag{20}$$

τ is therefore a time coordinate for $X^{(+)}$.

We see from the formulae (17) - (20) that the geometry of $\bar{X}$ and its submanifolds $\bar{X}^{(\pm)}$ may be formulated analogously with that of Minkowski space-time and the Rindler wedges. In particular, the boundaries $(u = \pm v)$ of $\bar{X}^{(\pm)}$ are again event horizons, and time-translations $\tau \to \tau + s$ in $\bar{X}^{(+)}$ correspond to generalised Lorentz transformations

$$u \to u \cosh s + v \sinh s, \; v \to v \cosh s + u \sinh s. \tag{21}$$

This formulation of $\bar{X}^{(+)}$, $\bar{X}$ may easily be seen to cover the case where the former manifold is the exterior Schwarzschild space-time, and the latter one is its Kruskal extension (cf. Ref. [11], Ch. 31). In this case, the event horizons, $u = \pm v$, form the surface of the Black Hole of $\bar{X}^{(+)}$, and time-translations in that manifold correspond to generalised Lorentz transformations of the Kruskal space. Hence, in this case, the parallel between the manifolds $\bar{X}$, $\bar{X}^{(+)}$, on the one hand, and the Minkowski and Rindler spaces, X, $X^{(+)}$, on the other, may be thought of in the form

Schwarzschild/Kruskal = Rindler/Minkowski.

4. QUANTUM FIELD THEORY

We shall restrict ourselves here to quantum fields in Minkowski space-time: for a generalisation to the curved spacetimes just described by the formulae (17) - (21), see Ref. [5].

We describe quantum fields in Minkowski space X according to Wightman's axiomatic scheme, which represents the minimal requirements of quantum theory and relativity [12]. For notational simplicity, we confine the formulation to real scalar fields. Other fields can be treated analogously, with the same results.

Thus, we take the ingredients of a quantum field theory to be the following.

$\mathcal{H}$, a Hilbert space.
U, a continuous unitary representation in $\mathcal{H}$ of the proper, Poincare group G.
$\phi(x)$, a Hermitian operator-valued distribution, representing the quantum field.
Ψ, a vector in $\mathcal{H}$, representing the vacuum state, that is cyclic with respect to the algebra of observables[†] $\mathcal{A}$, generated by the 'smeared field' $\int \phi(x) f(x) dx$, with f running through the Schwarz space $\mathcal{S}$.

The theory is centred on the Wightman functions

$$W(x_1, \ldots, x_n) = (\Psi, \phi(x_1) \ldots \phi(x_n)\Psi), \tag{22}$$

which are taken to be tempered distributions, so as to cover the requirement that the canonical commutation relations involve δ-functions. Since Ψ is cyclic with respect to the algebra $\mathcal{A}$ of the smeared fields, it follows that the distributions W carry all the information concerning the states given by the vectors and density matrices in $\mathcal{H}$.

The Wightman axioms concerning $(\mathcal{H}, U, \phi, \Psi)$, and thus W, are the following.

(I) $\phi(x)$ and $\phi(x')$ intercommute if x and x' have spacelike separation. This is the requirement of relativistic causality.

(II) $$U(g)\,\phi(x)\,U(g^{-1}) = \phi(gx) \tag{23}$$

and $$U(g)\Psi = \Psi, \tag{24}$$

which are the requirements of a relativistically covariant field and invariant vacuum, respectively. It follows from the latter condition that the Hamiltonian H, defined as $(-i\hbar)$ times the generator of the time-translational subgroup of $U(G)$, annihilates the

---

† One could regard this algebra as a representation in $\mathcal{H}$ of an abstract one, as discussed in Section 2.

state Ψ, i.e.

$$H\Psi = 0. \tag{25}$$

(III) The Hamiltonian H, which we have just defined, is a positive operator. In view of (25), this is the condition of stability of the vacuum against creation of particles.

The essential power of these axioms stems from the fact that they imply that the Wightman distributions W are boundary values of analytic functions of complex variables $W(z_1, \ldots, z_n)$, over certain domains in $\mathbb{C}^{4n}$ (cf. Ref. [12]). Here the analyticity properties arise from the positivity of H in the same way they do for the correlation functions of ground states, discussed at the end of Section 2. Furthermore, the axioms (I) and (II), of local commutativity and relativistic covariance, serve to extend the domains of analyticity of the functions W. In particular, they permit an extension of the one-parameter subgroup $\{V(s)/s\in\mathbb{R}\}$ of U(G), representing the Lorentz boosts (16), to imaginary values of s, where the resultant transformations then correspond to rotations in the plane $Ox^{(0)}x^{(1)}$ and, in the case $s = i\pi$, to space-time inversions $x^{(0)} \to -x^{(0)}$, $x^{(1)} \to -x^{(1)}$.

The following two key theorems have been derived from the Wightman axioms by methods involving the analytic extensions of the distributions W, and in particular of the Lorentz transformations V.

<u>PCT Theorem 12. There is a conjugation J_{PCT} of H such that</u>

$$J_{PCT}\,\phi(x_1)\ldots\phi(x_n)\Psi = \phi(-x_1)\ldots\phi(-x_n)\Psi\,. \tag{26}$$

This theorem tells us essentially that the laws of Physics are invariant under the combined action of space-time inversions and particle-antiparticle interchanges. The next theorem concerns the action of the Lorentz transformations on the algebra of observables for the Rindler wedge $X^{(+)}$, defined as the algebra $\mathcal{A}^{(+)}$ generated by the smeared fields $\int\phi(x)\,f(x)\,dx$, with f running through the $\mathcal{S}$-class functions with support in $X^{(+)}$.

<u>Bisognano-Wichmann Theorem 13. Let iK be the infinitessimal generator of the one-parameter unitary group $\{V(s)/s\in\mathbb{R}\}$ of Lorentz boosts. Then</u>

$$e^{-\pi K}A\Psi = JA^*\Psi \quad \forall A \text{ in } \mathcal{A}^{(+)}, \tag{27}$$

<u>where J is the conjugation of $\mathcal{H}$ given by the formula</u>

$$J = J_{PCT}R \tag{28}$$

<u>and R is the unitary representation in $\mathcal{H}$ of the spatial rotation through π about $Ox^{(1)}$, i.e. of the transformation $x^{(2)} \to -x^{(2)}$, $x^{(3)} \to -x^{(3)}$.</u>

5. THERMALISATION OF QUANTUM FIELDS BY GRAVITATION

It is now a simple matter to infer from the results of the previous three Sections that the gravitational field corresponding to a uniform acceleration in flat space-time thermalises any ambient quantum field. The argument runs as follows. As noted in Section 2, time-translations ($\tau \to \tau + s$) for a uniformly accelerated observer in the Rindler wedge $X^{(+)}$ correspond to Lorentz boosts for an inertial observer. Hence it follows from the specification of K, in Section 3, that $\hbar K$ is the Hamiltonian governing time-translations on the τ-scale for the accelerated observer. Furthermore, by the Bisognano-Wichmann theorem, the restriction of the vacuum state Ψ to the observables $\mathcal{A}^{(+)}$ of $X^{(+)}$ satisfies the KMS condition (7) with respect to Lorentz boosts. Hence, as viewed by a uniformly accelerated observer in the Rindler wedge, this is a KMS state. Furthermore, on comparing the formulae (7) and (27), one sees that the temperature of this state, on a scale for which τ is the time, is $\hbar/2\pi k$. Therefore, by (14) and (15), the temperature corresponding to the proper time-scale of the accelerated observer is

$$T_\alpha = \hbar\alpha/2\pi kc,$$

in accordance with Unruh's result. By the Principle of Equivalence, this result implies that the gravitational field corresponding to a uniform acceleration α in flat space-time thermalises any ambient quantum field to the temperature T_α. Since this result is a consequence of the Bisognano-Wichmann theorem, its connection with the PCT theorem is evident.

Comments. (1) The essential significance of the result described here is that a state that is a vacuum for an inertial observer is thermal for a uniformly accelerated one. In other words, it corresponds to a relativity of temperature with respect of acceleration. It does *not* imply that the accelerated observer sees effects corresponding to what the inertial observer would regard as excitations (or particles), when the system is in its vacuum state: for the Hamiltonians, $\hbar K$ and H, of the two observers are quite different.
(2) From an operational point of view, this result signifies that a localised thermometer carried by a uniformly accelerated observer will register the finite temperature T_α. This follows from the thermal reservoir property of KMS states discussed in Section 2. In the present context, Bell et al [14] have pointed out that electrons subjected to uniform magnetic fields in linear accelerators can serve as thermometers, their depolarisation being a measure of the temperature they experience.
(3) As regards the action of gravitational fields in curved space-times such as those discussed in Section 3, the theory is more complex because time-translations there do not generally correspond to isometries. However, they do so on the event horizons, and by exploiting this fact, we have been able to extend the general axiomatic treatment of the Hawking-Unruh thermalisation effect to these manifolds [5].

REFERENCES

1. J. Glimm and A. Jaffe: 'Quantum Physics', Springer, New York, Heidelberg, Berlin, 1981.

2. S.W. Hawking: Commun. Math. Phys. 43, 199 (1975.

3. W.G. Unruh: Phys. Rev. D, 14, 870 (1976).

4. P.C.W. Davies: J. Phys. A 8, 609 (1975).

5. G.L. Sewell: Ann. Phys. 141, 201 (1982).

6. G.G. Emch: 'Algebraic Methods in Statistical Mechanics and Quantum Field Theory', Wiley-Interscience, London, New York, 1971.

7. G.L. Sewell: 'Quantum Theory of Collective Phenomena', to be published by Oxford University Press.

8. R. Haag, N.M. Hugenholtz and M. Winnink: Commun. Math. Phys. 5, 215 (1967).

9. G.L. Sewell: Phys. Rep. 57, 307 (1980).

10. A. Kossakowski, A. Frigerio, V. Gorini and M. Verri: Commun. Math. Phys. 57, 97 (1977).

11. C.W. Misner, K.S. Thorne and J.A. Wheeler: 'Gravitation', W.H. Freeman and Co., San Francisco, 1973.

12. R.F. Streater and A.S. Wightman: 'PCT, Spin and Statistics and All That', Benjamin, New York, Amsterdam, 1964.

13. J.J. Bisognano and E.H. Wichmann: J. Math. Phys. 16, 985 (1975).

14. J.S. Bell and J.M. Leinaas: Nucl. Phys. B212, 131 (1983).
J.S. Bell, R.J. Hughes and J.M. Leinaas: CERN-TH. 3948184 Preprint.

SELF-REPELLENT RANDOM WALKS AND POLYMER MEASURES IN TWO DIMENSIONS

A. Stoll, Abteilung für Mathematik, Ruhr-Universität, D-4630 Bochum 1

The statistical description of polymers requires a probability measure which takes into account the "excluded volume effect", i. e. the repulsive self-interaction of a polymer chain, which is caused by the fact that a polymer cannot loop back and cross itself. Thus Edwards [5] proposed the following polymer model: Equip the Wiener measure m on the path space $\mathcal{C}([0,1],\mathbb{R}^d)$ with the formal density

$$\frac{d\nu(g)}{dm}(\omega) := \frac{1}{Z(g)} \exp[-g\,J(\omega)] \ , \tag{1}$$

where the functional

$$J(\omega) := \int_0^1 ds \int_0^1 dt\, \delta(\omega(t)-\omega(s)) \tag{2}$$

is intended to measure the time which a Wiener path ω spends at its double points, the constant $g \in \mathbb{R}_+$ gives the strength of the self-repulsion, and $Z(g) := \int \exp(-g\,J)\,dm$ is the normalization constant. In dimensions $d = 2$ and $d = 3$, Varadhan [10] respectively Westwater [11] (see also Bovier, Felder, and Fröhlich [2]) could rigorously establish the polymer measure $\nu(g)$ as weak limit of polymer measures $\nu_n(g)$ where the δ-function is replaced by a continuous approximation f_n. Then the functionals J_n tend to infinity, but in dimension $d = 2$ the L^2-limit of $J_n - E[J_n]$ still exists and is m-integrable (see Varadhan [10]), whereas in dimension $d = 3$ the Westwater measure $\nu(g)$ is orthogonal to the Wiener measure m.

By nonstandard analysis, we can give a precise meaning to Edwards' heuristic approach. Choose a hyperfinite time line $T := \{ \eta\Delta t \mid \eta \in {}^*\mathbb{N}_0 \}$ with infinitesimal spacing Δt and a hyperfinite lattice $\Gamma := \{ (\eta_1\Delta x, \ldots, \eta_d\Delta x) \mid \eta_\ell \in {}^*\mathbb{Z} \ (\ell = 1,\ldots,d) \}$. The Brownian motion is represented by a hyperfinite simple random walk on Γ, as it was first done by Anderson [1]. Therefore we choose the internal probability space $(\Omega,\underline{\mathfrak{A}},\underline{P}) := \bigotimes_{t\in T} (E,P_u)$, where $E = \{+e_1,-e_1,\ldots,+e_d,-e_d\}$, $e_1,\ldots,e_d$ is the standard base of $\mathbb{R}^d$, E is equipped with its power set as σ-algebra, and P_u is the uniform distribution on E, i. e. $P_u(\{x\}) = \frac{1}{2d}$ $\forall x \in E$. Let $(\Omega,\mathfrak{A},P)$ be the Loeb space induced by $(\Omega,\underline{\mathfrak{A}},\underline{P})$, i. e. $\mathfrak{A} = L(\underline{\mathfrak{A}})$,

$P = L(\underset{\sim}{P})$. Each element $\omega \in \Omega$ can be considered as a function $\underset{\sim}{\omega}: T \to E$. By abuse of notation, we denote $\omega(t) := \sum'_{s:s<t} \Delta x \, \underset{\sim}{\omega}(s)$ $(t \in T)$, where $\sum'_s$ stands for $\sum_{s \in T}$ (hence $\omega(0) = 0$). Then the internal process $(\omega(t))_{t \in T}$ is a hyperfinite simple random walk on Γ . By Anderson [1], there is a standard Brownian motion $W: [0,\infty[\times \Omega \longrightarrow \mathbb{R}^d$ such that for P-a. a. $\omega \in \Omega$ we have

$$°\omega(t) = W(°t,\omega) \qquad (t \in ns(T)) \,, \tag{3}$$

supposed that we choose the appropriate scaling, i. e.

$$\Delta x = \sqrt{d \, \Delta t} \,. \tag{4}$$

Finally, we need a discrete version of the δ-function. If we choose

$$\delta(x) := \begin{cases} (\Delta x)^{-d} \,, & \text{if } x = 0 \\ 0 \,, & \text{otherwise} \end{cases} \qquad (x \in \Gamma) \,,$$

then the expressions (1), (2) are completely discrete, and the functional $J(\omega)$ actually counts the self-intersections of the discrete path $\omega: T \to \Gamma$ up to a constant $(\Delta t)^2/(\Delta x)^d$. That means that our nonstandard representation of Varadhan's polymer measure is nothing else than a hyperfinite "Domb - Joyce - model" (see [4]), i. e. the probability of a path decreases with the number of self-intersections:

$$P_G(\{\omega\}) = \frac{1}{Z(G)} \exp[-G \sum_{s<t\leq 1} \chi_{\{\omega(s) = \omega(t)\}}]$$

(χ_A denotes the indicator of a set A). If one chooses the appropriate scaling, i. e. $\Delta x = \sqrt{d \, \Delta t}$ and $G = 2 g d^{-d/2} (\Delta t)^{2-d/2}$, then in the case $d = 2$, we can show that the hyperfinite measure P_G has Varadhan's polymer measure $\nu(g)$ as standard part, i. e.

$$L(P_G) \circ \overline{W}^{-1} = \nu(g) \,, \tag{5}$$

where $L(P_G)$ is the Loeb measure on Ω induced by P_G and $\overline{W}: \Omega \longrightarrow \mathcal{C}([0,1],\mathbb{R}^d)$ is deduced from the Brownian motion W by $\overline{W}(\omega) := W(-,\omega)$. In the same manner as Anderson's nonstandard representation of Brownian motion implies Donsker's invariance principle, we obtain as a corollary that suitably scaled Domb - Joyce - models weakly converge to Varadhan's polymer measure. Recently, Brydges and Spencer [3] could show a similar limit theorem for the self-repellent random walk in dimension $d > 4$. As expected, the limit is then the Wiener measure.

For shortness, we have to omit some technical details. For a complete presentation and further references, we refer the reader to [9].

I. The intersection local time of independent Brownian motions in less than four dimensions

The crucial point in order to obtain (5) is to give a nonstandard representation of the local time

$$L(x,\omega) \;=\; \int_0^{1/2} ds \int_{1/2}^{1} dt\, \delta_x(W(t,\omega) - W(s,\omega))$$

of the difference of two independent Brownian motions. We can do this in dimension $d < 4$. In our nonstandard setting we are led to the internal process

$$\rho_\theta(x,\omega) := \sum_{s=\underline{a}}^{\overline{a}}{}' \sum_{t=\underline{b}}^{\overline{b}}{}' \frac{(\Delta t)^2}{(\Delta x)^d}\, \theta(s,t)\, \chi_{\{\omega(t) - \omega(s) = x\}} \qquad (x \in \Gamma,\ \omega \in \Omega)\ , \tag{6}$$

where $\underline{a}, \overline{a}, \underline{b}, \overline{b} \in T$ such that $0 \le \underline{a} \le \overline{a} \le \underline{b} \le \overline{b} \le w \in \mathbb{R}$ and $\theta \in \underline{\mathfrak{T}} := \{\theta\colon T \times T \to {}^*[0,1] \mid \theta \text{ internal}\}$. We use such a time weight θ by technical reasons and in order to demonstrate that our estimates reflect the fact that the internal simple random walk $(\omega(t))_{t \in T}$ can reach "even" lattice points $x \in \Gamma$ only at "even" time points $t \in T$. Put $T_e^2 := \{(s,t) \in T \times T \mid s + t \in 2\Delta t\, {}^*\mathbb{N}_0\}$, $\Gamma_e := \{(x_1,\dots,x_d) \in \Gamma \mid x_1 + \dots + x_d \in 2\Delta x\, {}^*\mathbb{Z}\}$, $T_o^2 := T^2 \setminus T_e^2$, $\Gamma_o := \Gamma \setminus \Gamma_e$, and define $\theta_e, \theta_o \in \underline{\mathfrak{T}}$ by $\theta = \theta_e + \theta_o$, $\theta_o \restriction T_e^2 \equiv 0$, $\theta_e \restriction T_o^2 \equiv 0$. Let $\underline{\lambda}$ be the natural uniform measure on T^2, i. e. $\underline{\lambda}\{(s,t)\} := (\Delta t)^2$. Then $\tilde{\underline{\lambda}} := L(\underline{\lambda}) \circ st^{-1}$ is the Lebesgue measure λ on $[0,\infty[^2$ and $L(\theta\underline{\lambda}) \circ st^{-1} = \vartheta\lambda$ with a measurable function $\vartheta\colon [0,\infty[^2 \to [0,1]$. Denote $\tilde{\theta} := \vartheta$.

<u>Definition 1.</u> $\theta\colon T^2 \to {}^*[0,1]$ is called <u>fair</u>, iff $\tilde{\theta}_e = \tilde{\theta}_o$ λ-a. s. ⌟

Remember that an internal function $F\colon \Gamma \to {}^*\mathbb{R}$ is called <u>S-continuous</u>, iff $x \approx y$ implies $F(x) \approx F(y)$ for all $x, y \in ns(\Gamma)$. Then we have the following theorem:

<u>Theorem 2.</u> Let $d < 4$. If $\theta \in \underline{\mathfrak{T}}$ is fair, then P-a. a. paths of the internal process $\rho_\theta\colon \Gamma \times \Omega \to {}^*\mathbb{R}_+$ defined by (6) are S-continuous.⌟

Let us first show up the consequences of theorem 2. Let $\mathcal{C}_0(\mathbb{R}^d,\mathbb{R}) := \{f\colon \mathbb{R}^d \to \mathbb{R} \mid f \text{ continuous, bounded, and } \sup\{|f(x)| \mid |x| > n\} \xrightarrow[n\to\infty]{} 0\}$ and $\mathfrak{F}_0(\Gamma,{}^*\mathbb{R}) := \{F \restriction \Gamma \mid F \in {}^*\mathcal{C}_0(\mathbb{R}^d,\mathbb{R})\}$. Let $\underline{st}\colon \mathfrak{F}_0(\Gamma,{}^*\mathbb{R}) \to \mathcal{C}_0(\mathbb{R}^d,\mathbb{R})$ be the partially defined function given by $\underline{st}(F) = f$, iff $\sup\{|F(x) - {}^*f(x)| \mid x \in \Gamma\} \approx 0$. Note that for any reasonable (i. e. S-continuity-preserving) interpolation $I: \mathfrak{F}_0(\Gamma,{}^*\mathbb{R}) \to {}^*\mathcal{C}_0(\mathbb{R}^d,\mathbb{R})$ we have that $\underline{st}\circ I$ is the usual standard part map ${}^*\mathcal{C}_0(\mathbb{R}^d,\mathbb{R}) \to \mathcal{C}_0(\mathbb{R}^d,\mathbb{R})$, coming from the sup-norm. Since for P-a. a. $\omega\in\Omega$ the internal path $T_w \to \Gamma$, $t \mapsto \omega(t)$ is nearstandard in $\mathcal{C}([0,w],\mathbb{R}^d)$, we have for such ω: $\omega(t)\in ns(\Gamma)$ $\forall t\in T_w := \{t\in T \mid t \le w\}$ and hence $\rho_\Theta(x,\omega)=0$ $\forall x\in\Gamma\setminus ns(\Gamma)$. With theorem 2, we infer that for P-a. a. $\omega\in\Omega$ the internal function $\Gamma \to {}^*\mathbb{R}_+$, $x\mapsto\rho_\Theta(x,\omega)$ belongs to $\underline{st}^{-1}(\mathcal{C}_0(\mathbb{R}^d,\mathbb{R}))$. Therefore, on the Loeb space $(\Omega,\mathfrak{A},P)$, a standard process $L: \mathbb{R}^d\times\Omega\to\mathbb{R}_+$ with paths in $\mathcal{C}_0(\mathbb{R}^d,\mathbb{R})$ is well defined by

$$L(-,\omega) = \underline{st}[\rho_\Theta(-,\omega)]\ ,\ \text{i. e.}\ [\,L({}^\circ x,\omega) = {}^\circ\rho_\Theta(x,\omega) \quad (x\in\Gamma)\,] \tag{7}$$

for P-a. a. $\omega\in\Omega$. Put $M := [{}^\circ\underline{a},{}^\circ\overline{a}]\times[{}^\circ\underline{b},{}^\circ\overline{b}]$, $\underset{\sim}{T} := \{(s,t)\in T^2 \mid \underline{a}\le s\le\overline{a}\ ,\ \underline{b}\le t\le\overline{b}\}$.

<u>Theorem 3.</u> L is the local time of the process $Z: M\times\Omega\to\mathbb{R}^d$, $((s,t),\omega)\mapsto W(t,\omega)-W(s,\omega)$ with respect to the time measure $(\tilde\theta\lambda)\restriction M$.

<u>Proof.</u> Fix $\omega\in\Omega$ such that (3) and (7) hold. Pick an arbitrary $f\in\mathcal{C}_K(\mathbb{R}^d,\mathbb{R}) := \{f: \mathbb{R}^d\to\mathbb{R} \mid f$ continuous with compact support$\}$. Then:

$$\int_{{}^\circ\underline{a}}^{{}^\circ\overline{a}} ds \int_{{}^\circ\underline{b}}^{{}^\circ\overline{b}} dt\, \tilde\theta(s,t)\, f(W(t,\omega)-W(s,\omega))$$

$= \int_{T^2\cap st^{-1}(M)} dL(\theta\underline{\lambda})(s,t)\, f(W({}^\circ t,\omega)-W({}^\circ s,\omega))$, since $L(\theta\underline{\lambda})\circ st^{-1} = \tilde\theta\lambda$;

$= \int_{\underset{\sim}{T}} dL(\theta\underline{\lambda})(s,t)\, f({}^\circ\omega(t)-{}^\circ\omega(s))$, by (3) and $L(\theta\underline{\lambda}) \ll L(\underline{\lambda})$;

$= {}^\circ\sum'^{\overline{a}}_{s=\underline{a}} \Delta t \sum'^{\overline{b}}_{t=\underline{b}} \Delta t\, \theta(s,t)\, {}^*f(\omega(t)-\omega(s))$, since the internal function $(s,t)\mapsto {}^*f(\omega(t)-\omega(s))$ is S-integrable with respect to $\theta\lambda\restriction\underset{\sim}{T}$;

$= {}^\circ\sum'^{\overline{a}}_{s=\underline{a}} \Delta t \sum'^{\overline{b}}_{t=\underline{b}} \Delta t\, \theta(s,t) \sum_{x\in\Gamma} \chi_{\{\omega(t)-\omega(s)=x\}}\, {}^*f(x)$, since $\omega(t)-\omega(s)\in\Gamma$ for all $(s,t)\in T^2$;

$= {}^\circ\sum_{x\in\Gamma} (\Delta x)^d \rho_\Theta(x,\omega)\, {}^*f(x)$, by the definition of ρ_Θ (see (6));

$= \int dL(\Delta x)^d(x)\, st[\rho_\Theta(x,\omega)\,{}^*f(x)]$, because the compactness of $\overline{\{f\neq 0\}}$ and the S-continuity of $\rho_\Theta(-,\omega)$ imply that the internal function $x\mapsto\rho_\Theta(x,\omega)\,{}^*f(x)$ is finitely bounded and S-integrable;

$= \int dL(\Delta x)^d(x)\, L({}^\circ x,\omega)\, f({}^\circ x)$, by (7) and the continuity of f ;

$= \int_{\mathbb{R}^d} d\lambda^d(x)\, L(x,\omega)\, f(x)$, since $L(\Delta x)^d\circ st^{-1} = \lambda^d$.

Since the function $f \in \mathcal{C}_K(\mathbb{R}^d,\mathbb{R})$ was arbitrary, it follows that $L(-,\omega)$ is the Lebesgue density of the occupation measure of the path $(M,\tilde{\partial}\lambda) \to \mathbb{R}^d$, $(s,t) \mapsto W(t,\omega) - W(s,\omega)$. This is true for all $\omega \in \Omega$ such that (3) and (7) hold, i. e. for P-a. a. $\omega \in \Omega$. □

By theorem 3, it is justified to write $L =: L_\vartheta$, where $\vartheta := \tilde{\theta}$. A corollary of theorem 3 is that for every Brownian motion $W'\colon [0,\infty[\times \Omega' \to \mathbb{R}^d$ the process $M \times \Omega' \to \mathbb{R}^d$, $((s,t),\omega) \mapsto W'(t,\omega) - W'(s,\omega)$ has a continuous local time in dimension $d < 4$. But theorem 3 contains more information, namely an approximation theorem that the local times of simple random walks converge in distribution to L_ϑ, as the lattice spacing tends to zero. In dimension $d = 1$, Perkins [7] obtained such a result by a nonstandard construction of Brownian local time. Let Q be the distribution of the random function $\Omega \to \mathcal{C}_0(M \times \mathbb{R}^d, \mathbb{R}^d \times \mathbb{R})$, $\omega \mapsto (\overline{Z}(\omega), \overline{L_\vartheta}(\omega))$, where $Z\colon M \times \Omega \to \mathbb{R}^d$, $((s,t),\omega) \mapsto W(t,\omega) - W(s,\omega)$. Obviously, every other Wiener process W' leads to the same distribution Q. For each $n \in \mathbb{N}$, let $(\Omega_n, \mathcal{A}_n, P_n)$ be a probability space, Δt_n a positive real, $\Delta x_n := \sqrt{d\,\Delta t_n}$, $T_n := \{ k\Delta t_n \mid k \in \mathbb{N}_0 \}$, $\Gamma_n := \{ (k_1\Delta x_n, \dots, k_d\Delta x_n) \mid k_1,\dots,k_d \in \mathbb{Z} \}$, $\underline{a}_n, \overline{a}_n, \underline{b}_n, \overline{b}_n \in T_n$ with $\underline{a}_n \le \overline{a}_n \le \underline{b}_n \le \overline{b}_n$, $\tilde{T}_n := \{ (s,t) \in T_n \times T_n \mid \underline{a}_n \le s \le \overline{a}_n, \underline{b}_n \le t \le \overline{b}_n \}$, $\theta_n\colon T_n \times T_n \to [0,1]$ an arbitrary function, and $(X_{n,t} \mid t \in T_n)$ a sequence of independent random variables on Ω_n with distribution $P_n(X_{n,t} = x) = 1/(2d)$ for all $x \in \{ +e_1, -e_1, \dots, +e_d, -e_d \}$, where $e_1,\dots,e_d$ is the standard base of $\mathbb{R}^d$. For each $n \in \mathbb{N}$, define the processes $W_n\colon T_n \times \Omega_n \to \mathbb{R}^d$, $\rho_n\colon \Gamma_n \times \Omega_n \to \mathbb{R}_+$ by

$$W_n(t,\omega) := \sum_{s \in T_n : s<t} X_{n,s}(\omega)\,\Delta x_n \text{, respectively}$$

$$\rho_n(x,\omega) := \sum_{(s,t)\in\tilde{T}_n} \theta_n(s,t)(\Delta t_n)^2(\Delta x_n)^{-d}\,\chi_{\{W_n(t,\omega) - W_n(s,\omega) = x\}},$$

and the random functions $\overline{Y_n}\colon \Omega_n \to \mathfrak{F}_0(\tilde{T}_n, \mathbb{R}^d)$, $\overline{\rho_n}\colon \Omega_n \to \mathfrak{F}_0(\Gamma_n, \mathbb{R})$ by $\overline{Y_n}(\omega)\colon \tilde{T}_n \to \mathbb{R}^d$, $(s,t) \mapsto W_n(t,\omega) - W_n(s,\omega)$, respectively $\overline{\rho_n}(\omega)\colon \Gamma_n \to \mathbb{R}$, $x \mapsto \rho_n(x,\omega)$. For each $n \in \mathbb{N}$, let Q_n be the probability measure on $\mathfrak{F}_0(\tilde{T}_n \times \Gamma_n, \mathbb{R}^d \times \mathbb{R})$, induced by the random function $(\overline{Y_n}, \overline{\rho_n})$.

<u>Theorem 4.</u> Suppose that

(i) $\lim_{n\to\infty} \Delta t_n = 0$;

(ii) $\lim_{n\to\infty} \underline{a}_n = {}^\circ\underline{a}$, $\lim_{n\to\infty} \overline{a}_n = {}^\circ\overline{a}$, $\lim_{n\to\infty} \underline{b}_n = {}^\circ\underline{b}$, $\lim_{n\to\infty} \overline{b}_n = {}^\circ\overline{b}$;

(iii) $(\theta_n)_{n\in\mathbb{N}}$ is a <u>fair approximation of ϑ</u>, i. e. $\theta_{n,e}\lambda_n \xrightarrow[n\to\infty]{\text{vaguely}} \frac{1}{2}\vartheta\lambda$ and $\theta_{n,o}\lambda_n \xrightarrow[n\to\infty]{\text{vaguely}} \frac{1}{2}\vartheta\lambda$ (or equivalently, $^*\theta_\eta$ is fair and $\vartheta = \widetilde{^*\theta_\eta}$ for all infinite η).

Then $(\overline{Y_n},\overline{\rho_n}) \xrightarrow[n\to\infty]{w} (\overline{Z},\overline{L_\vartheta})$, i. e. $Q_n \circ I_n^{-1} \xrightarrow[n\to\infty]{\text{weakly}} Q$ for every reasonable sequence $(I_n)_{n\in\mathbb{N}}$ of interpolations $I_n\colon \mathfrak{F}_0(\tilde{T}_n \times \Gamma_n, \mathbb{R}^d \times \mathbb{R}) \to \mathcal{C}_0(M \times \mathbb{R}^d, \mathbb{R}^d \times \mathbb{R})$. ("Reasonable" essentially means that $^*I_\eta$ preserves S-continuity for every infinite η .)

<u>Proof.</u> It is sufficient to show that for every infinite η , we have $L(^*Q_\eta)\circ \underline{st}^{-1} = Q$. But this follows immediately from (7) and theorem 3. □

Now let us return to the proof of theorem 2. According to the nonstandard version of Kolmogorov's continuity theorem, it would be sufficient to find positive reals α , β , C such that

$$\underline{E}[\,|\rho_\theta(x) - \rho_\theta(y)|^\alpha\,] \leq C\,|x-y|^{d+\beta} \tag{8}$$

for all $x, y \in ns(\Gamma)$. Unfortunately, such an estimate cannot be true, if e. g. $\theta = \theta_e$ (and $\tilde{\theta} \not\equiv 0$), because then $\rho_\theta \upharpoonright \Gamma_o \equiv 0$. But since θ is fair, we need the estimate (8) only in the case that $x - y \in \Gamma_e$, because then a similar argument as in the proof of theorem 3 shows that $\rho_\theta \upharpoonright \Gamma_e$ and $\rho_\theta \upharpoonright \Gamma_o$ have the same standard part L_ϑ with $\frac{1}{2}\vartheta = \widetilde{\theta_e} = \widetilde{\theta_o}$. Our proof of the estimate (8) is based on a discrete version of the Fourier inversion formula. To this end, let κ be the smallest even *integer such that $(\kappa/2)\Delta x > (1 \vee dw)/\Delta x$. Let $\Gamma_k := \{(n_1\Delta x,\dots,n_d\Delta x) \mid -k\kappa/2 < n_\ell \leq k\kappa/2\ ,\ n_\ell \in {}^*\mathbb{Z}\ \ (\ell = 1,\dots,d)\}$ $(k \in \mathbb{N})$ and put $\gamma := 2\pi/[\kappa(\Delta x)^2] < {}^\circ\gamma = \pi/(1 \vee dw)$. Note that $\omega(t) - \omega(s) \in \Gamma_1$ for all $s, t \in T$ with $s, t \leq w$.

<u>Lemma 5.</u> Let $\sigma\colon \Gamma_1 \to {}^*\mathbb{C}$ be internal. Define $\hat{\sigma}\colon \Gamma_1 \to {}^*\mathbb{C}$ by

$$\hat{\sigma}(y) := \sum_{x\in\Gamma_1} (\Delta x)^d \exp[\gamma i\, x\cdot y] \qquad (y \in \Gamma_1)\ ,$$

where $x\cdot y$ denotes the scalar product of the vectors $x, y \in \Gamma_1$. Then

$$\sigma(z) = \left(\frac{\gamma}{2\pi}\right)^d \sum_{y\in\Gamma_1} (\Delta x)^d\, \hat{\sigma}(y) \exp[-\gamma i\, y\cdot z] \qquad \forall z \in \Gamma_1\ .$$

<u>Proof.</u> $\left(\frac{\gamma}{2\pi}\right)^d \sum_{y\in\Gamma_1} (\Delta x)^d\, \hat{\sigma}(y) \exp[-\gamma i\, y\cdot z] =$

$$\left(\frac{\gamma(\Delta x)^2}{2\pi}\right)^d \sum_{x\in\Gamma_1} \sigma(x) \Big(\sum_{y\in\Gamma_1} \exp[\gamma i\, y\cdot(x-z)] \Big)\ .$$

So it is sufficient to show that

$$\tau(x,z) := \sum_{y\in\Gamma_1} \exp[\gamma i\, y\cdot(x-z)] = \begin{cases} \left(\frac{2\pi}{\gamma(\Delta x)^2}\right)^d = \kappa^d & \text{, if } x = z \\ 0 & \text{, otherwise} \end{cases}$$

$(x , z \in \Gamma_1)$. Since

$$\tau(x,z) = \sum_{n_1=1-\kappa/2}^{\kappa/2} \cdots \sum_{n_d=1-\kappa/2}^{\kappa/2} \exp[\sum_{\ell=1}^{d} \gamma i n_\ell \Delta x\,(x_\ell - z_\ell)]$$

$$= \prod_{\ell=1}^{d} \sum_{n_\ell=1-\kappa/2}^{\kappa/2} \exp[\gamma i n_\ell \Delta x\,(x_\ell - z_\ell)] ,$$

we only have to consider the case $d = 1$. Fix arbitrary $x , z \in \Gamma_1$. Then $x - z = \zeta\Delta x$ with $\zeta \in {}^*\mathbb{Z}$ such that $-(\kappa-1) \leq \zeta \leq (\kappa-1)$. Hence $\kappa|\zeta$, iff $\zeta = 0$, i. e. $x = z$. This implies

$$\tau(x,z) = \sum_{\eta=1-\kappa/2}^{\kappa/2} \exp[\gamma i \eta\zeta(\Delta x)^2] = \sum_{\eta=1-\kappa/2}^{\kappa/2} \exp[2\pi i \tfrac{\eta\zeta}{\kappa}]$$

$$= \begin{cases} \kappa & \text{, if } \kappa|\zeta \text{, i. e. } x = z \\ 0 & \text{, otherwise} \end{cases} . \square$$

By lemma 5, we have

$$\rho_\theta(z,\omega) = \left(\frac{\gamma}{2\pi}\right)^d \sum_{y\in\Gamma_1} (\Delta x)^d \hat{\rho}_\theta(y,\omega) \exp(-\gamma i\, y\cdot z) \qquad (z \in \Gamma_1 , \omega \in \Omega)$$

with

$$\hat{\rho}_\theta(y,\omega) = \sum_{x\in\Gamma_1} (\Delta x)^d \rho_\theta(x,\omega) \exp(\gamma i\, x\cdot y)$$

$$= \sum_{s=\underline{a}}^{\overline{a}}{}' \Delta t \sum_{t=\underline{b}}^{\overline{b}}{}' \Delta t\, \theta(s,t) \exp(\gamma i[\omega(t) - \omega(s)]\cdot y) \text{ , by (6).}$$

Therefore, we get for every $k \in \mathbb{N}$ and $x , y \in \Gamma_1$:

$$\underline{E}([\rho_\theta(x) - \rho_\theta(y)]^k) = \left(\frac{\gamma}{2\pi}\right)^{dk} \sum_{u_1\in\Gamma_1} \cdots \sum_{u_k\in\Gamma_1} (\Delta x)^{dk} \Big(\prod_{j=1}^{k} [\exp(-\gamma i u_j x) -$$

$$\exp(-\gamma i u_j y)]\Big)\Big(\sum_{s_1=\underline{a}}^{\overline{a}}{}' \cdots \sum_{s_k=\underline{a}}^{\overline{a}}{}' \sum_{t_1=\underline{b}}^{\overline{b}}{}' \cdots \sum_{t_k=\underline{b}}^{\overline{b}}{}' (\Delta t)^{2k} [\prod_{j=1}^{k} \theta(s_j,t_j)] \underline{E}[\exp(\gamma i \sum_{j=1}^{k}$$

$$[\omega(t_j) - \omega(s_j)]\cdot u_j)]\Big) .$$

Using the fact that $(\omega(t))_{t\in T}$ has independent, identically distributed increments and employing Hölder's inequality, we after some substitutions arrive at

$$|\underline{E}([\rho_\Theta(x) - \rho_\Theta(y)]^k)| \leq (\tfrac{\gamma}{2\pi})^{dk}(k!)^2 \{ \sum_{v_1\in\Gamma_1} \cdots \sum_{v_k\in\Gamma_k} (\Delta x)^{dk} (\prod_{j=1}^{k} |\exp[-\gamma i (v_j-v_{j-1})\cdot x] - \exp[-\gamma i (v_j-v_{j-1})\cdot y]|) \prod_{j=1}^{k} [\sum_{t=0}^{\bar{a}-a}{}^{\tau} \Delta t \, |\underline{E}(\exp[\gamma i\, \omega(t)\cdot v_j])|]^2 \}^{1/2}$$

$$\{ \sum_{v_1\in\Gamma_1} \cdots \sum_{v_k\in\Gamma_k} (\Delta x)^{dk} (\prod_{j=1}^{k} |\exp[-\gamma i (v_j-v_{j-1})\cdot x] - \exp[-\gamma i (v_j-v_{j-1})\cdot y]|)$$

$$\prod_{j=1}^{k} [\sum_{t=0}^{\bar{b}-b}{}^{\tau} \Delta t \, |\underline{E}(\exp[\gamma i\, \omega(t)\cdot v_j])|]^2 \}^{1/2} \quad , \text{ where } v_0 := 0 .$$

If $x - y \in \Gamma_e$, we can choose a suitable transformation $v \mapsto \tilde{v}$ such that we can use the estimate

$$|\exp[-\gamma i (v_2-v_1)\cdot x] - \exp[-\gamma i (v_2-v_1)\cdot y]| \leq 2\gamma^\alpha (|\tilde{v}_1| + |\tilde{v}_2|)^\alpha |x-y|^\alpha$$

($\alpha \in]0,1[$) in order to obtain the following lemma after some very technical manipulations:

Lemma 6. If $k \in \mathbb{N}$, $\alpha \in [0,1[$, and $s \in {}^*\mathbb{R}_+$, let

$$S_d(k,\alpha,s) := \sum_{v_1\in\tilde{\Gamma}} \cdots \sum_{v_k\in\tilde{\Gamma}} (\Delta x)^{dk} \prod_{j=1}^{k} [(|v_j| + |v_{j-1}|)^\alpha \frac{1}{|v_j|^4} (1-\exp[-\tfrac{\gamma^2}{6}|v_j|^2\cdot s])^2 ,$$

where $v_0 := 0$, $0^0 := 1$, and $\tilde{\Gamma} := \{x/\sqrt{2} \mid x \in \Gamma_1\}$, if $d = 2$, respectively $\tilde{\Gamma} := \{(x_1/\sqrt{2},\dots,x_{d-1}/\sqrt{2},x_d) \mid (x_1,\dots,x_d) \in \Gamma_1\}$, if d is odd. Then for every $k \in \mathbb{N}$, there exists a constant $c_1 = c_1(d,k,w) \in \mathbb{R}_+$ with the following properties:

(i) Whenever $\alpha \in]0,1[$ and $x , y \in \Gamma_1$ with $x - y \in \Gamma_e$, then

$$|\underline{E}([\rho_\Theta(x) - \rho_\Theta(y)]^k)| \leq c_1 |x-y|^{\alpha k} \sqrt{S_d(k,\alpha,\bar{a}-\underline{a})\, S_d(k,\alpha,\bar{b}-\underline{b})} \; ;$$

(ii) for all $x \in \Gamma_1$, we have

$$|\underline{E}([\rho_\Theta(x)]^k)| \leq c_1 \sqrt{S_d(k,0,\bar{a}-\underline{a})\, S_d(k,0,\bar{b}-\underline{b})} \; .\lrcorner$$

By a straightforward calculation, the quantities $S_d(k,\alpha,s)$ can be estimated as follows:

Lemma 7. (i) There exists a constant $c_2 = c_2(d) \in \mathbb{R}_+$ such that

$$S_d(k,0,s) \leq (c_2 s^{2-d/2})^k \quad \text{for all } k \in \mathbb{N},\ s \in {}^*\mathbb{R},\ 0 \leq s \leq w .$$

(ii) Fix $k \in \mathbb{N}$. Let $s \in {}^*\mathbb{R}_+$ be finite. If $\alpha \in]0,1[\wedge d \in \{1,2\}$ or $\alpha \in]0,1/2[\wedge d = 3$, then $S_d(k,\alpha,s)$ is finite.⌋

If we now choose e. g. $\alpha = 1/3$ and $k = 12$, then lemma 6(i) and lemma 7(ii) give the desired estimate (8) for all $x, y \in ns(\Gamma)$ with $x - y \in \Gamma_e$. This completes the proof of theorem 2. □

Since by theorem 3 the process Z has a local time L_ϑ we can give sense to

$$\int_{{}^\circ\underline{a}}^{{}^\circ\overline{a}} ds \int_{{}^\circ\underline{b}}^{{}^\circ\overline{b}} dt\, \vartheta(s,t)\, \varphi(W(t,\omega) - W(s,\omega)) := \int_{\mathbb{R}^d} L_\vartheta(x,\omega)\, \mu_\varphi(dx) \tag{9}$$

even if φ is only a "generalized function", i. e. $\mu_\varphi := \varphi\lambda^d$ is not absolutely continuous. In this case, the heuristic left side of (9) nevertheless has a precise meaning in the nonstandard universe (as well as the heuristic formula $\mu_\varphi = \varphi\lambda^d$). Let $\Phi \in \mathcal{P}_1 := \{\Phi: \Gamma \to {}^*\mathbb{R}_+ \mid \Phi$ internal, $\sum_{x \in \Gamma} (\Delta x)^d \Phi(x) \leq 1\}$. Define the internal measure μ_Φ on Γ by $\mu_\Phi(\{x\}) := \Phi(x)(\Delta x)^d$ $(x \in \Gamma)$. Then $\widetilde{\mu_\Phi} := L(\mu_\Phi) \circ st^{-1}$ is a measure on $\mathbb{R}^d$ with $\widetilde{\mu_\Phi}(\mathbb{R}^d) \leq 1$. Denote $\tilde{\Phi} := \varphi$ with $\mu_\varphi = \varphi\lambda^d = \widetilde{\mu_\Phi}$. Then we have:

Proposition 8. According to (9), let $X: \Omega \to \mathbb{R}$ be the random variable

$$\omega \longmapsto \int_{{}^\circ\underline{a}}^{{}^\circ\overline{a}} ds \int_{{}^\circ\underline{b}}^{{}^\circ\overline{b}} dt\, \vartheta(s,t)\, \varphi(W(t,\omega) - W(s,\omega)) .$$

Define the internal random variable $\underline{X}: \Omega \to {}^*\mathbb{R}$ by

$$\omega \longmapsto \sum_{s=\underline{a}}^{\overline{a}}{}' \Delta t \sum_{t=\underline{b}}^{\overline{b}}{}' \Delta t\, \theta(s,t)\, \Phi(\omega(t) - \omega(s)) .$$

Then $\underline{X}$ is a SL^p-lifting of X for all $p \in]0,\infty[$. In particular, we have for P-a. a. $\omega \in \Omega$:

$${}^\circ\sum_{s=\underline{a}}^{\overline{a}}{}' \Delta t \sum_{t=\underline{b}}^{\overline{b}}{}' \Delta t\, \theta(s,t)\Phi(\omega(t) - \omega(s)) = \int_{{}^\circ\underline{a}}^{{}^\circ\overline{a}} ds \int_{{}^\circ\underline{b}}^{{}^\circ\overline{b}} dt\, \vartheta(s,t)\varphi(W(t,\omega) - W(s,\omega)) .$$

Proof. By the definition of ρ_θ (see (6)), we have

$$\underline{X}(\omega) = \sum_{x \in \Gamma} (\Delta x)^d\, \Phi(x)\, \rho_\theta(x,\omega) \quad \text{for all } \omega \in \Omega . \tag{10}$$

$$\underline{E}[\underline{X}^k] = \sum_{x_1\in\Gamma} \cdots \sum_{x_k\in\Gamma} (\Delta x)^{dk}\, \phi(x_1)\ldots\phi(x_k)\, \underline{E}[\rho_\theta(x_1)\ldots\rho_\theta(x_k)]$$

$$\leq \sum_{x_1\in\Gamma} \cdots \sum_{x_k\in\Gamma} (\Delta x)^{dk}\, \phi(x_1)\ldots\phi(x_k) \prod_{j=1}^{k} (\underline{E}[(\rho_\theta(x_j))^k])^{1/k}$$

$$\leq [\sum_{x\in\Gamma} (\Delta x)^d\, \phi(x)]^k\, k\, c_1 c_2 [(\overline{a}-\underline{a})(\overline{b}-\underline{b})]^{(2-d/2)/2}, \text{ by lemma 6(ii) and lemma 7(i);}$$

$$\leq k\, c_1 c_2\, w^{2-d/2} .$$

Therefore $\underline{E}[\underline{X}^k]$ is finite for all $k \in \mathbb{N}$. It remains to show that $X = \text{st} \circ \underline{X}$ P-a. s. For P-a. a. $\omega \in \Omega$ we have:

$$°\underline{X}(\omega) = °\sum_{x\in\Gamma} (\Delta x)^d\, \phi(x)\, \rho_\theta(x,\omega), \text{ by (10);}$$

$$= \int_\Gamma °\rho_\theta(x,\omega)\, L(\mu_\phi)(dx), \text{ since } \rho_\theta(-,\omega) \text{ is S-integrable with respect to } \mu_\phi;$$

$$= \int_{ns(\Gamma)} L_\vartheta(°x,\omega)\, L(\mu_\phi)(dx), \text{ by (7);}$$

$$= \int_{\mathbb{R}^d} L_\vartheta(x,\omega)\, \mu_\varphi(dx), \text{ by } L(\mu_\phi) \circ \text{st}^{-1} = \mu_\varphi;$$

$$= X(\omega), \text{ by (9).} \ \square$$

II. Nonstandard construction of polymer measures in two dimensions

Now we turn to the construction of polymer measures as standard parts of hyperfinite self-repellent random walks. For the rest of this article, we restrict ourselves to the case $d = 2$. Internally, we want to give each walk $\omega \in \Omega$ a probability proportional to $\exp[-G\, \tau^v_{\theta,\phi}(\omega)]$, where the self-repulsion is given by the internal random variable

$$\tau^v_{\theta,\phi}: \Omega \longrightarrow {}^*\mathbb{R}_+, \quad \omega \longmapsto \sum_{s,t\in T_v} (\Delta t)^2\, \theta(s,t)\, \phi(\omega(t)-\omega(s)),$$

with fixed $\theta \in \underline{\mathcal{T}}$, $\phi \in \underline{\mathcal{P}}_1$, $v \in ns({}^*\mathbb{R}_+)$, and $T_v := \{t \in T \mid t \leq v\}$. Because of the obvious connection

$$\tau^v_{\theta,\phi}(\omega) = \sum_{x\in\Gamma} (\Delta x)^2\, \phi(x)\, \tau^v_\theta(x,\omega) \qquad (\omega \in \Omega)$$

with

$$\tau_\Theta^v(x,\omega) := \sum_{s,t\in T_v} \Theta(s,t)\left(\frac{\Delta t}{\Delta x}\right)^2 \chi_{\{\omega(t)-\omega(s)=x\}} \qquad (x\in\Gamma\ ,\ \omega\in\Omega),$$

we now have to study the full internal intersection local time $\tau_\Theta : \Gamma\times\Omega \to {}^*\mathbb{R}_+$, $(x,\omega)\mapsto \tau_\Theta(x,\omega) := \tau_\Theta^v(x,\omega)$ instead of ρ_Θ . We cannot expect that the internal process τ_Θ has a standard part as it was the case for ρ_Θ . An easy calculation shows that the expectation of $\tau_1(0)$ is infinite because of the contributions with $s\approx t$. However, we can show:

<u>Proposition 9.</u> $\underline{\mathrm{Var}}[\tau_\Theta^v(x)] \le 2^{16}v^2$ for all $x\in\Gamma$.┘

The proof of proposition 9 consists of very technical, but elementary estimates. One can work with the following estimates for the random walk probabilities, which are a straightforward consequence of Stirling's formula.

<u>Lemma 10.</u> Fix $t = n\Delta t \in T\setminus\{0\}$ and $x = (k\Delta x, \ell\Delta x)\in\Gamma$. Then

(i) $\underline{P}(\omega:\omega(t)=x) \le \frac{1}{t}\exp\left(-\frac{|x|^2}{2t}\right)(\Delta x)^2$;

(ii) if $|x|^2 \le t$ and $k+\ell+n$ is even, we have

$$\frac{1}{\pi t}\exp\left(-\frac{|x|^2}{2t} - 2\frac{\Delta t}{t}\right)(\Delta x)^2 \le \underline{P}(\omega:\omega(t)=x) \le \frac{1}{\pi t}\exp\left(-\frac{|x|^2}{2t} + 2\frac{\Delta t}{t}\right)(\Delta x)^2 \text{ ┘}$$

Obviously proposition 9 implies

$$\underline{\mathrm{Var}}(\tau_{\Theta,\Phi}^v) \le 2^{16}v^2 \quad \text{for all } \Phi\in\underline{\mathcal{P}}_1 . \tag{11}$$

This estimate allows us to apply Nelson's trick (see [6]) in order to show the finiteness of

$$\underline{E}[\exp(-G[\tau_{\Theta,\Phi}^v - \underline{E}\,\tau_{\Theta,\Phi}^v])] \qquad (G\in \mathrm{ns}({}^*\mathbb{R}_+)) .$$

To this end, we split the domain T_v^2 by Westwater's manner (cf. [11]). Put

$$\kappa := \kappa^v := \min\{\eta\in{}^*\mathbb{N}_0 \mid 2^\eta\Delta t > v\} . \tag{12}$$

Without restriction, we may assume $v\ge\Delta t$, i. e. $\kappa\in{}^*\mathbb{N}$. For every $\xi\in\{0,1,\dots,\kappa\}$, we define the following domains: If $\eta\in\{0,1,\dots,2^\xi-1\}$, put $\Delta_\xi(\eta) := \{(s,t)\in T^2 \mid$ there are $s_0, t_0\in T$ such that $s_0, t_0 < 2^{\kappa-\xi}\Delta t$ and $s = \eta 2^{\kappa-\xi}\Delta t + s_0$, $t = \eta 2^{\kappa-\xi}\Delta t + t_0\}$, $\Lambda_\xi(\eta,1) := \{(s,t)\in T^2 \mid$ there are $s_0, t_0\in T$ such that $s_0, t_0 < 2^{\kappa-\xi-1}\Delta t$ and $s = (\eta+\frac{1}{2})2^{\kappa-\xi}\Delta t + s_0$, $t = \eta 2^{\kappa-\xi}\Delta t + t_0\}$, $\Lambda_\xi(\eta,2) := \{(s,t)\in T^2 \mid$ there are $s_0, t_0\in T$ such that $s_0, t_0 < 2^{\kappa-\xi-1}\Delta t$ and $s = \eta 2^{\kappa-\xi}\Delta t + s_0$, $t =$

$(n+\frac{1}{2})2^{\kappa-\xi}\Delta t + t_0\}$, $\Lambda_\xi(n) = \Lambda_\xi(n,1) \cup \Lambda_\xi(n,2)$; let $\Delta_\xi := \bigcup\{\Delta_\xi(n) \mid n = 0,1,\ldots,2^\xi-1\}$, $\Lambda_\xi := \bigcup\{\Lambda_\xi(n) \mid n = 0,1,\ldots,2^\xi-1\}$. Note that for all $\xi \in \{0,1,\ldots,\kappa-1\}$, we have $\Delta_\xi(n) = \Lambda_\xi(n) \cup \Delta_{\xi+1}(2n) \cup \Delta_{\xi+1}(2n+1)$ $(n = 0,1,\ldots,2^\xi-1)$ and $\Delta_{\xi+1} = \Delta_\xi \setminus \Lambda_\xi$. Moreover, $\Delta_0 = \{(s,t) \in T^2 \mid s,t < 2^\kappa \Delta t\}$, $\Delta_\kappa = \{(s,t) \in T^2 \mid s = t < 2^\kappa\Delta t\}$, and $\Lambda_\kappa = \emptyset$. If Δ is an arbitrary internal subset of T^2 , define the internal function $\theta(\Delta)\colon T^2 \to {}^*[0,1]$ by $\theta(\Delta) := \theta\cdot\chi_\Delta$. In particular, put for all $\xi \in \{0,1,\ldots,\kappa\}$:

$$\sigma^v_{\theta,\phi,\xi} := \tau^v_{\theta(\Delta_\xi),\phi} \;;\quad \rho^v_{\theta,\phi,\xi} := \tau^v_{\theta(\Lambda_\xi),\phi} \;;\quad \tau^v_{\theta,\phi,\xi} := \tau^v_{\theta(\Delta),\phi}$$

with $\Delta = \bigcup\{\Lambda_\zeta \mid \zeta = 0,1,\ldots,\xi-1\}$. Note that

$$\tau^v_{\theta,\phi} = \sigma^v_{\theta,\phi,\xi} + \tau^v_{\theta,\phi,\xi} = \sigma^v_{\theta,\phi,\xi} + \sum_{\zeta=0}^{\xi-1} \rho^v_{\theta,\phi,\zeta} \tag{13}$$

for all $\xi \in \{0,1,\ldots,\kappa\}$. Our application of Nelson's trick is based on the following two estimates:

<u>Lemma 11.</u> $\underline{\mathrm{Var}}(\sigma^v_{\theta,\phi,\xi}) \le 2^{18} v^2 2^{-\xi}$ for all $\xi \in \{0,1,\ldots,\kappa\}$.

<u>Lemma 12.</u> For each real $w \in\,]0,\infty[$ there exists a positive real constant $c_3 = c_3(w)$ such that $\underline{E}\,\tau^v_{\theta,\phi,\xi} \le c_3 v \xi$ for all $\xi \in \{0,1,\ldots,\kappa\}$, $\theta \in \underline{\mathcal{T}}$, $\phi \in \underline{\mathcal{P}}_1$, and $v \in {}^*\mathbb{R}$ with $\Delta t \le v \le w$.┘

<u>Proof of lemma 11.</u> We have

$$\underline{\mathrm{Var}}(\sigma^v_{\theta,\phi,\xi}) = \sum_{n=0}^{2^\xi-1} \underline{\mathrm{Var}}(\tau^v_{\theta(\Delta_\xi(n)),\phi}) \tag{14}$$

by *independency. On the other hand, we have for each $n \in \{0,1,\ldots,2^\xi-1\}$:

$$\underline{\mathrm{Var}}(\tau^v_{\theta(\Delta_\xi(n))}) = \underline{\mathrm{Var}}(\tau^{\bar v}_{\bar\theta,\phi}) , \tag{15}$$

with $\bar v = 2^{\kappa-\xi}\Delta t$ and $\bar\theta\colon T\times T \to {}^*[0,1]$, $(s,t) \mapsto [\theta(T^2_v)](s + n2^{\kappa-\xi}\Delta t, t + n2^{\kappa-\xi}\Delta t)$, since the internal process $(\omega(t))_{t\in T}$ has *identically distributed increments. (11) implies

$$\underline{\mathrm{Var}}(\tau^{\bar v}_{\bar\theta,\phi}) \le 2^{16}\bar v^2 = 2^{16}(2^{\kappa-\xi}\Delta t)^2 . \tag{16}$$

From (14), (15), (16), we obtain the claim:

$$\underline{\mathrm{Var}}(\sigma^v_{\theta,\phi,\xi}) \le 2^{16}(2^\kappa\Delta t)^2 2^{-\xi} \le 2^{18} v^2 2^{-\xi} ,$$

by the choice of κ (see (12)). □

Proof of lemma 12. We shall show that one can choose

$$c_3 := c_3(w) := 1 + 2\, c_1(2,1,2w)\, c_2(2) \,, \tag{17}$$

where $c_1 = c_1(2,1,2w)$, $c_2 = c_2(2)$ are the constants occurring in lemma 6(ii) respectively lemma 7(i). To this end, fix admissible $\theta , \Phi , w , v , \xi$. Then we have:

$$\underline{E}\,\tau^{v}_{\theta,\Phi,\xi} = \sum_{\zeta=0}^{\xi-1} \sum_{\eta=0}^{2^\zeta-1} \sum_{j=1}^{2} \underline{E}\,\tau^{v}_{\theta(\Lambda_\zeta(\eta,j)),\Phi} \,. \tag{18}$$

Now fix $\zeta \in \{0,1,\dots,\xi-1\}$ and $\eta \in \{0,1,\dots,2^\zeta-1\}$. Put $\underline{a} := \eta 2^{\kappa-\zeta}\Delta t$, $\overline{a} := (\eta+\frac{1}{2})2^{\kappa-\zeta}\Delta t =: \underline{b}$, $\overline{b} := (\eta+1)2^{\kappa-\zeta}\Delta t \leq 2^\kappa \Delta t \leq 2v$. Then:

$$\underline{E}\,\tau^{v}_{\theta(\Lambda_\zeta(\eta,1)),\Phi} \leq \underline{E}\,\tau^{2v}_{\theta(\Delta),\Phi} \,, \text{ with } \Delta := \{(s,t) \in T^2 \mid \underline{a} \leq s \leq \overline{a} , \ \underline{b} \leq t \leq \overline{b}\} \,;$$

$$= \underline{E}\Big[\sum_{x\in\Gamma} (\Delta x)^2\, \Phi(x)\, \rho_\theta(x)\Big] \,, \text{ where the internal random variable } \rho_\theta(x) := \rho_\theta^{2w}(\underline{a},\overline{a},\underline{b},\overline{b};x) \text{ is defined as in (6);}$$

$$\leq \max_{x\in\Gamma} \underline{E}[\rho_\theta(x)] \,, \text{ since } \sum_{x\in\Gamma}(\Delta x)^2\, \Phi(x) \leq 1 \,;$$

$$\leq c_1 c_2 2^{\kappa-\zeta-1}\Delta t \,, \text{ by lemma 6(ii) and lemma 7(i).}$$

With an analogous estimate for $\Lambda_\zeta(\eta,2)$, we obtain from (18):

$$\underline{E}\,\tau^{v}_{\theta,\Phi,\xi} \leq c_1 c_2 2^\kappa \Delta t\, \xi \leq c_3 v \xi \,, \text{ by (12) and (17).} \ \square$$

Proposition 13. There exist positive real constants c_4 , c_5 such that

$$\underline{E}[\exp(-G[\tau^{v}_{\theta,\Phi} - \underline{E}\,\tau^{v}_{\theta,\Phi}])] \leq c_5$$

for all $\theta \in \mathfrak{T}$, $\Phi \in \mathfrak{P}_1$, and $v , G \in {}^*\mathbb{R}_+$ with $v\cdot G \leq c_4$ and $v \leq 1$.

Proof. Put $c_4 := (\log 2)/(2c_3)$ with the constant $c_3 = c_3(1)$ of lemma 12, and pick admissible θ , Φ , v , G . Without restriction, we may assume that $G > 0$ and $v \geq \Delta t$. Then for all $\xi \in \{1,2,\dots,\kappa\}$, we have:

$$\underline{P}(\tau^{v}_{\theta,\Phi} - \underline{E}\,\tau^{v}_{\theta,\Phi} \leq -2c_3 v\xi) \leq \underline{P}(\sigma^{v}_{\theta,\Phi,\xi} - \underline{E}\,\sigma^{v}_{\theta,\Phi,\xi} \leq -2c_3 v\xi + \underline{E}\,\tau^{v}_{\theta,\Phi,\xi})$$

$$\leq \underline{P}(\sigma^{v}_{\theta,\Phi,\xi} - \underline{E}\,\sigma^{v}_{\theta,\Phi,\xi} \leq -c_3 v\xi) \,, \text{ by lemma 12;}$$

$$\leq (c_3 v)^{-2}\, \xi^{-2}\, \underline{\mathrm{Var}}(\sigma^{v}_{\theta,\Phi,\xi}) \,, \text{ by Chebyshev's inequality;}$$

$$\leq 2^{18}(c_3)^{-2}\, 2^{-\xi}\, \xi^{-2} \,, \text{ by lemma 11.} \tag{19}$$

On the other hand, by integral transformation we get:

$$\underline{E}[\exp(-G[\tau^{v}_{\theta,\Phi} - \underline{E}\tau^{v}_{\theta,\Phi}])] \le 1 + \int_0^\infty \underline{P}(\tau^{v}_{\theta,\Phi} - \underline{E}\tau^{v}_{\theta,\Phi} \le -t)\, e^{Gt}\, {}^*dt$$

$$\le 1 + \int_0^{2c_3 v} e^{Gt}\, {}^*dt + \sum_{\xi=1}^{\kappa-1} \underline{P}(\tau^{v}_{\theta,\Phi} - \underline{E}\tau^{v}_{\theta,\Phi} \le -2c_3 v\xi)\, 2c_3 v\, e^{2c_3 Gv(\xi+1)} ,$$

since $\tau^{v}_{\theta,\Phi} - \underline{E}\tau^{v}_{\theta,\Phi} \ge -\underline{E}\tau^{v}_{\theta,\Phi,\kappa} \ge -c_3 v\kappa$, by lemma 12;

$$\le 1 + \frac{1}{G}(e^{2c_3 vG} - 1) + 2^{18} c_3^{-2} (2c_3 v)\, e^{2c_3 vG} \sum_{\xi=1}^{\kappa-1} 2^{-\xi} \xi^{-2} e^{2c_3 Gv\xi} , \text{ by (19)};$$

$$\le 1 + \frac{1}{G}(2c_3 vG\, e^{2c_3 c_4}) + 2^{19} \frac{v}{c_3} e^{2c_3 c_4} \sum_{\xi=1}^{\kappa-1} \xi^{-2} \exp[-\xi(\log 2 - 2c_3 c_4)] ,$$
by $vG \le c_4$;

$$= 1 + 4c_3 v + 2^{20} \frac{v}{c_3} \sum_{\xi=1}^{\kappa-1} \xi^{-2} , \text{ by } 2c_3 c_4 = \log 2 ;$$

$$\le 1 + 4c_3 + 2^{21}/c_3 =: c_5 . \square$$

Corollary 14. $\underline{E}[\exp(-G[\tau^{v}_{\theta,\Phi} - \underline{E}\tau^{v}_{\theta,\Phi}])]$ is finite for all finite $v, G \in {}^*\mathbb{R}_+$ and all $\theta \in \underline{T}$, $\Phi \in \underline{\mathfrak{P}}_1$.

Proof. Pick finite $v, G \in {}^*\mathbb{R}_+$. By proposition 13, it is sufficient to consider the case that v is not infinitesimal, in particular $\kappa \in {}^*\mathbb{N} \setminus \mathbb{N}$ (cf. (12)). Choose $n \in \mathbb{N}$ such that $2^n > 2v(1 \vee 2G/c_4)$. Hölder's inequality gives

$$\underline{E}[\exp(-G[\tau^{v}_{\theta,\Phi} - \underline{E}\tau^{v}_{\theta,\Phi}])] \le$$

$$(\underline{E}[\exp(-2G[\tau^{v}_{\theta,\Phi,n} - \underline{E}\tau^{v}_{\theta,\Phi,n}])])^{1/2} (\underline{E}[\exp(-2G[\sigma^{v}_{\theta,\Phi,n} - \underline{E}\sigma^{v}_{\theta,\Phi,n}])])^{1/2} .$$

Moreover

$$(\underline{E}[\exp(-2G[\tau^{v}_{\theta,\Phi,n} - \underline{E}\tau^{v}_{\theta,\Phi,n}])])^{1/2} \le \exp(G\,\underline{E}\tau^{v}_{\theta,\Phi,n})$$

$\le \exp(G c_3 v n)$, by lemma 12, where $c_3 = c_3(w)$ with $w = {}^\circ(v+1)$.

Thus it remains to show the finiteness of

$$\underline{E}[\exp(-2G[\sigma^{v}_{\theta,\Phi,n} - \underline{E}\sigma^{v}_{\theta,\Phi,n}])] =$$

$$\prod_{\eta=0}^{2^n-1} \underline{E}[\exp(-2G[\tau^{v}_{\theta(\Delta_n(\eta)),\Phi} - \underline{E}\tau^{v}_{\theta(\Delta_n(\eta)),\Phi}])] . \quad (20)$$

Fix $\eta \in \{0, 1, \dots, 2^n-1\}$. Put $\bar{v} := 2^{\kappa-n}\Delta t$. Note that

$$\bar{v} \leq 2^{-(n-1)} v \leq 1 \quad \text{and} \quad 2\bar{v}G \leq 2^{-(n-2)} vG \leq c_4 , \tag{21}$$

by the choice of n. Define the random variable $\bar{\theta}: T^2 \to {}^*[0,1]$ by

$$\bar{\theta}(s,t) := [\theta(T_v^2)](s + \eta 2^{\kappa-n}\Delta t , t + \eta 2^{\kappa-n}\Delta t) \qquad (s , t \in T) .$$

Then we have

$$\underline{E}[\exp(-2G[\tau^v_{\theta(\Delta_n(\eta)),\Phi} - \underline{E}\,\tau^v_{\theta(\Delta_n(\eta)),\Phi}]))]$$

$= \underline{E}[\exp(-2G[\tau^{\bar{v}}_{\bar{\theta},\Phi} - \underline{E}\,\tau^{\bar{v}}_{\bar{\theta},\Phi}]))]$, since the internal process $(\omega(t))_{t\in T}$ is stationary;

$\leq c_6$, by proposition 13 because of (21).

Since $\eta \in \{0, 1, \ldots, 2^n-1\}$ was arbitrary, (20) implies

$$\underline{E}[\exp(-2G[\sigma^v_{\theta,\Phi,n} - \underline{E}\,\sigma^v_{\theta,\Phi,n}]))] \leq (c_6)^{(2^n)} < \infty . \square$$

Next we want to introduce the standard entity corresponding to $\tau^v_{\theta,\Phi} - \underline{E}\,\tau^v_{\theta,\Phi}$. For the rest of this article, we fix a real $w > 0$. Let m be the Wiener measure on the path space $\mathcal{C}([0,w],\mathbb{R}^2)$. Then $m = P \circ \overline{W}^{-1}$ with $\overline{W}: \Omega \to \mathcal{C}([0,w],\mathbb{R}^2)$, $\omega \mapsto [t \mapsto W(t,\omega)]$. Let $\mathcal{T} := \{\vartheta: [0,\infty[^2 \to [0,1] \mid \vartheta$ measurable$\}$ (identify λ-a. s. identical functions) and $\mathcal{P}_1 := \{\varphi \mid \mu_\varphi$ ("$= \varphi\lambda^d$") arbitrary measure on $\mathbb{R}^2$ with $\mu_\varphi(\mathbb{R}^2) \leq 1\}$.

<u>Theorem 15.</u> 1. There exists exactly one operator $\mathcal{R}: \mathcal{T} \times \mathcal{P}_1 \to L^2(\mathcal{C}([0,w],\mathbb{R}^2),m)$ which has the following properties:

(i) If $\vartheta \in \mathcal{T}$, $\varphi: \mathbb{R}^2 \to \mathbb{R}_+$ is continuous, bounded, and $\int \varphi(x)\,\lambda^2(dx) \leq 1$, then for m-a. a. $\omega \in \mathcal{C}([0,w],\mathbb{R}^2)$ we have

$$[\mathcal{R}(\vartheta,\varphi)](\omega) = \int_0^w ds \int_0^w dt\,\vartheta(s,t)\,\varphi(\omega(t)-\omega(s)) - E_m\Big[\int_0^w ds \int_0^w dt\,\vartheta(s,t)\,\varphi(\omega(t)-\omega(s))\Big] .$$

(ii) $\mathcal{R}$ is linear in the following sense: If $\varphi, \psi \in \mathcal{P}_1$ and $\alpha, \beta \in \mathbb{R}$ such that $\alpha\varphi + \beta\psi \in \mathcal{P}_1$, then $\mathcal{R}(\vartheta, \alpha\varphi + \beta\psi) = \alpha\mathcal{R}(\vartheta,\varphi) + \beta\mathcal{R}(\vartheta,\psi)$ for all $\vartheta \in \mathcal{T}$. An analogous property holds for the first argument ϑ.

(iii) $\mathcal{R}$ is continuous in the following sense: If $(\vartheta_n)_{n\in\mathbb{N}_0}$, $(\varphi_n)_{n\in\mathbb{N}_0}$ are sequences in $\mathcal{T}$ respectively $\mathcal{P}_1$ such that $\vartheta_n\lambda \xrightarrow[n\to\infty]{\text{vaguely}} \vartheta_0\lambda$ and $\mu_{\varphi_n} \xrightarrow[n\to\infty]{\text{vaguely}} \mu_{\varphi_0}$, then $\mathcal{R}(\vartheta_n,\varphi_n) \xrightarrow[n\to\infty]{L^2(m)} \mathcal{R}(\vartheta_0,\varphi_0)$.

2. Fix arbitrary $\Phi \in \underline{\mathcal{P}}_1$, $\Theta \in \underline{\mathcal{T}}$, and $v \in {}^*\mathbb{R}_+$ with ${}^\circ v = w$. If Θ is fair, then $\tau^v_{\Theta,\Phi} - \underline{E}\,\tau^v_{\Theta,\Phi}$ is a $SL^2(\underline{P})$ - lifting of $\mathcal{R}(\tilde{\Theta},\tilde{\Phi}) \circ \overline{W}$.

Proof. A density argument shows that the operator $\mathcal{R}$ is uniquely determined by (i) and (iii). For proving the existence of $\mathcal{R}$, we fix $v \in {}^*\mathbb{R}_+$ such that ${}^\circ v = w$. We want to define $\mathcal{R}(\vartheta,\varphi)$ by using "Westwater - domains" as before, depending on v respectively κ . To this end, put $a_{\ell,k} := {}^\circ(k\,2^{\kappa-\ell}\,\Delta t) \wedge v$. Given $(\vartheta,\varphi) \in \mathcal{T} \times \mathcal{P}_1$, define the random variables $U^{\ell,k,j}_{\vartheta,\varphi}$, $U^{(\ell)}_{\vartheta,\varphi}$, $U_{\vartheta,\varphi,\ell}$: $\Omega \longrightarrow \mathbb{R}_+$ by

$$U^{\ell,k,1}_{\vartheta,\varphi} := \int_{a_{\ell,k}}^{a_{\ell,k+1/2}} ds \int_{a_{\ell,k+1/2}}^{a_{\ell,k+1}} dt\,\vartheta(s,t)\,\varphi(W(t,-) - W(s,-)) \,, \qquad (k = 0 , 1 , \dots , 2^\ell-1)$$

$$U^{\ell,k,2}_{\vartheta,\varphi} := \int_{a_{\ell,k+1/2}}^{a_{\ell,k+1}} ds \int_{a_{\ell,k}}^{a_{\ell,k+1/2}} dt\,\vartheta(s,t)\,\varphi(W(t,-) - W(s,-)) \,;$$

$$U^{(\ell)}_{\vartheta,\varphi} := \sum_{k=0}^{2^\ell-1} (U^{\ell,k,1}_{\vartheta,\varphi} + U^{\ell,k,2}_{\vartheta,\varphi}) \,, \qquad U_{\vartheta,\varphi,\ell} := \sum_{k=0}^{\ell-1} U^{(\ell)}_{\vartheta,\varphi} \qquad (\ell \in \mathbb{N}_0) \,,$$

where the integrals are understood as in (9), by means of theorem 3. Note that the random variables $U_{\vartheta,\varphi,\ell}$ $(\ell \in \mathbb{N}_0)$ are p - integrable ($p \in [1,\infty[$) by proposition 8. A consequence of theorem 3 is that we have

$$U_{\vartheta,\varphi,\ell} = V_{\vartheta,\varphi,\ell} \circ \overline{W} \,, \tag{22}$$

where $V_{\vartheta,\varphi,\ell}$ is a random variable on $\mathcal{C}([0,w],\mathbb{R}^2)$. Put $\underset{\sim}{U}_{\vartheta,\varphi,\ell} := U_{\vartheta,\varphi,\ell} - E\,U_{\vartheta,\varphi,\ell}$ and $\underset{\sim}{V}_{\vartheta,\varphi,\ell} := V_{\vartheta,\varphi,\ell} - E_m V_{\vartheta,\varphi,\ell}$ $(\ell \in \mathbb{N}_0)$. We claim that

$$(\underset{\sim}{V}_{\vartheta,\varphi,\ell})_{\ell \in \mathbb{N}_0} \text{ is a Cauchy sequence in } L^2(m) \,. \tag{23}$$

To this end, choose $\Phi \in \underline{\mathcal{P}}_1$, $\Theta \in \underline{\mathcal{T}}$ such that $\tilde{\Phi} = \varphi$, $\tilde{\Theta} = \vartheta$, and Θ is fair. By proposition 8,

$$\tau^v_{\Theta,\Phi,\ell} \text{ is a } SL^p\text{ - lifting of } U_{\vartheta,\varphi,\ell} \text{ for all } p \in [1,\infty[\,, \tag{24}$$

particularly

$$\underset{\sim}{U}_{\vartheta,\varphi,\ell} = st \circ [\tau^v_{\Theta,\Phi,\ell} - \underline{E}\,\tau^v_{\Theta,\Phi,\ell}] \quad P\text{ - a. s.} \quad (\ell \in \mathbb{N}_0) \,. \tag{25}$$

By (11), a random variable $\underset{\sim}{U}_{\Theta,\Phi} : \Omega \longrightarrow \mathbb{R}$ is well defined by

$$\underset{\sim}{U}_{\Theta,\Phi}(\omega) := {}^\circ[\tau^v_{\Theta,\Phi}(\omega) - \underline{E}\,\tau^v_{\Theta,\Phi}] \quad \text{for } P\text{ - a. a. } \omega \in \Omega \,. \tag{26}$$

Furthermore, for all $\ell \in \mathbb{N}_0$ we have

$$E[(\underset{\sim}{U}_{\Theta,\Phi} - \underset{\sim}{U}_{\Theta,\Phi,\ell})^2] = E[(st \circ [(\tau^v_{\Theta,\Phi} - \tau^v_{\Theta,\Phi,\ell}) - \underline{E}(\tau^v_{\Theta,\Phi} - \tau^v_{\Theta,\Phi,\ell})])^2] \,,$$

by (25), (26);

$$= E[(st \circ [\sigma^V_{\Theta,\Phi,\ell} - \underline{E}\, \sigma^V_{\Theta,\Phi,\ell}])^2]$$

$$\leq \; {}^\circ\underline{Var}(\sigma^V_{\Theta,\Phi,\ell}) \; \leq \; 2^{18} w^2 2^{-\ell} \text{ , by lemma 11 and } {}^\circ v = w .$$

Therefore $\underset{\sim}{U}_{\vartheta,\varphi,\ell} \xrightarrow[\ell\to\infty]{L^2(P)} \underset{\sim}{U}_{\Theta,\Phi}$. In particular, $(\underset{\sim}{U}_{\vartheta,\varphi,\ell})_{\ell \in \mathbb{N}_0}$ is a Cauchy sequence in $L^2(P)$. This proves our claim (23) because of (22) and $m = P \circ \overline{W}^{-1}$. Let $\mathcal{R}(\vartheta,\varphi)$ be the L^2 - limit of the sequence $(\underset{\sim}{V}_{\vartheta,\varphi,\ell})_{\ell \in \mathbb{N}_0}$. Then we have

$$\mathcal{R}(\vartheta,\varphi) \circ \overline{W} = \underset{\sim}{U}_{\Theta,\Phi} \quad P\text{-a. s.} \tag{27}$$

by the uniqueness of the L^2 - limit. Moreover,

$$\begin{aligned} Var(\mathcal{R}(\vartheta,\varphi) \circ \overline{W}) &= \lim_{\ell\to\infty} Var(U_{\vartheta,\varphi,\ell}) \\ &= \lim_{\ell\to\infty} {}^\circ\underline{Var}(\tau^V_{\Theta,\Phi,\ell}) \text{ , by (24);} \\ &= {}^\circ\underline{Var}(\tau^V_{\Theta,\Phi}) \text{ , by lemma 11.} \end{aligned} \tag{28}$$

By (26), (27), (28), $\tau^V_{\Theta,\Phi} - \underline{E}\,\tau^V_{\Theta,\Phi}$ is a SL^2 - lifting of $\mathcal{R}(\vartheta,\varphi) \circ \overline{W}$. The crucial point is that this is true, whenever $\tilde{\Phi} = \varphi$, $\tilde{\Theta} = \vartheta$, and Θ is fair. This fact implies that the operator $\mathcal{R}$ has the required continuity property (iii), as a simple application of the permanence principle shows. If $\varphi: \mathbb{R}^2 \to \mathbb{R}_+$ is continuous, bounded, and $\int \varphi(x)\, \lambda^2(dx) \leq 1$, then from our construction of $\mathcal{R}(\vartheta,\varphi)$ one immediately obtains the formula of theorem 15.1(i) by dominated convergence. The linearity of the operator $\mathcal{R}$ is also obvious, since $\tau^V_{\Theta,\Phi}$ has a corresponding property. □

From corollary 14 and theorem 15 we obtain:

__Proposition 16.__ $E_m(\exp[-g\,\mathcal{R}(\vartheta,\varphi)])$ is finite for all $\vartheta \in \mathfrak{T}$, $\varphi \in \mathcal{P}_1$, $g \in \mathbb{R}_+$. Moreover, if $p \in [1,\infty[$, $\Theta \in \underline{\mathfrak{T}}$, $\Phi \in \underline{\mathcal{P}}_1$, $v \in {}^*\mathbb{R}_+$, $G \in ns({}^*\mathbb{R}_+)$ with ${}^\circ v = w$, ${}^\circ G = g$, and Θ is fair, then $\exp[-G(\tau^V_{\Theta,\Phi} - \underline{E}\,\tau^V_{\Theta,\Phi})]$ is a SL^p - lifting of $\exp[-g\,\mathcal{R}(\tilde{\Theta},\tilde{\Phi}) \circ \overline{W}]$ ⌟

In the following, we fix $g \in \mathbb{R}_+$ and put $\mathcal{M}_w := \{ \nu \mid \nu$ probability measure on $\mathcal{C}([0,w],\mathbb{R}^2) \}$. Moreover, we fix $v , G \in {}^*\mathbb{R}_+$ such that ${}^\circ v = w$ and ${}^\circ G = g$. Define the operator $\mathcal{V}_g: \mathfrak{T} \times \mathcal{P}_1 \to \mathcal{M}_w$ by $(\vartheta,\varphi) \mapsto \nu_{\vartheta,\varphi,g}$ with

$$\frac{d\nu_{\vartheta,\varphi,g}}{dm} := \exp[-g\,\mathcal{R}(\vartheta,\varphi)] \,/\, E_m(\exp[-g\,\mathcal{R}(\vartheta,\varphi)]) \; . \tag{29}$$

The associated measures $P_{\Theta,\Phi,G}$ on $(\Omega,\underline{\mathfrak{a}})$ are given by

$$P_{\Theta,\Phi,G}(\{\omega\}) := \exp[-G\,\tau^{V}_{\Theta,\Phi}(\omega)] / \underline{E}(\exp[-G\,\tau^{V}_{\Theta,\Phi}]) \qquad (\omega \in \Omega) \tag{30}$$

for every $(\Theta,\Phi) \in \mathcal{T} \times \mathcal{P}_1$. All our results on polymer measures can be deduced from the following fundamental relationship between the internal probability measures $P_{\Theta,\Phi,G}$ and the standard probability measures $\nu_{\vartheta,\varphi,g}$.

Theorem 17. For all fair $\Theta \in \underline{\mathcal{T}}$ and all $\Phi \in \underline{\mathcal{P}}_1$, we have

$$L(P_{\Theta,\Phi,G}) \circ \overline{W}^{-1} = \nu_{\tilde{\Theta},\tilde{\Phi},g}.$$

Proof. Fix a fair $\Theta \in \underline{\mathcal{T}}$ and $\Phi \in \underline{\mathcal{P}}_1$. Then from proposition 16 it follows immediately that $\exp[-G\,\tau^{V}_{\Theta,\Phi}] / \underline{E}(\exp[-G\,\tau^{V}_{\Theta,\Phi}])$ is a SL^1-lifting of $\exp[-g\,\mathcal{R}(\tilde{\Theta},\tilde{\Phi}) \circ \overline{W}] / E_m(\exp[-g\,\mathcal{R}(\tilde{\Theta},\tilde{\Phi})])$. Therefore

$$\frac{dL(P_{\Theta,\Phi,G})}{dP} = \exp[-g\,\mathcal{R}(\tilde{\Theta},\tilde{\Phi}) \circ \overline{W}] / E_m(\exp[-g\,\mathcal{R}(\tilde{\Theta},\tilde{\Phi})]), \text{ by (30).} \tag{31}$$

Because of $m = P \circ \overline{W}^{-1}$, this implies

$$\frac{d\,L(P_{\Theta,\Phi,G}) \circ \overline{W}^{-1}}{d\,m} = \exp[-g\,\mathcal{R}(\tilde{\Theta},\tilde{\Phi})] / E_m(\exp[-g\,\mathcal{R}(\tilde{\Theta},\tilde{\Phi})]),$$

i. e. $L(P_{\Theta,\Phi,G}) \circ \overline{W}^{-1} = \nu_{\tilde{\Theta},\tilde{\Phi},g}$ by (29). □

Define the internal function $\underline{W}: \Omega \longrightarrow \mathcal{F}_0(T_V, {}^*\mathbb{R}^2)$ by $\underline{W}(\omega): T_V \longrightarrow {}^*\mathbb{R}^2$, $t \longmapsto \omega(t)$ $(\omega \in \Omega)$. Obviously (3) means $\underline{st} \circ \underline{W} = \overline{W}$ P-a. s. Since $L(P_{\Theta,\Phi,G}) \ll P$ by (31), the statement of theorem 17 is equivalent to

$$L(P_{\Theta,\Phi,G} \circ \underline{W}^{-1}) \circ \underline{st}^{-1} = \nu_{\tilde{\Theta},\tilde{\Phi},g}. \tag{32}$$

Therefore a simple application of the permanence principle shows:

Theorem 18. The operators $\mathcal{V}_g: \mathcal{T} \times \mathcal{P}_1 \longrightarrow \mathcal{M}_W$ $(g \in \mathbb{R}_+)$ are continuous in the following sense: If $(\vartheta_n,\varphi_n)_{n\in\mathbb{N}_0}$, $(g_n)_{n\in\mathbb{N}_0}$ are sequences in $\mathcal{T} \times \mathcal{P}_1$ respectively $\mathbb{R}_+$ such that $\vartheta_n\lambda \xrightarrow[n\to\infty]{\text{vaguely}} \vartheta_0\lambda$, $\mu_{\varphi_n} \xrightarrow[n\to\infty]{\text{vaguely}} \mu_{\varphi_0}$, and $g_n \xrightarrow[n\to\infty]{} g_0$, then $\mathcal{V}_{g_n}(\vartheta_n,\varphi_n) \xrightarrow[n\to\infty]{W} \mathcal{V}_{g_0}(\vartheta_0,\varphi_0)$ ⌟

If $\varphi: \mathbb{R}^2 \longrightarrow \mathbb{R}_+$ is continuous, bounded, and $\int \varphi(x)\,\lambda^2(dx) \leq 1$, then because of theorem 15.1(i) and (29), we can explicitly write down the measures $\nu_{\vartheta,\varphi,g}$ by

$$d\nu_{\vartheta,\varphi,g}(\omega) = \frac{1}{Z(\vartheta,\varphi,g)} \exp[-g \int_0^w ds \int_0^w dt\, \vartheta(s,t)\, \varphi(\omega(t)-\omega(s))]\, dm(\omega) \quad (33)$$

with

$$Z(\vartheta,\varphi,g) := E_m(\exp[-g \int_0^w ds \int_0^w dt\, \vartheta(s,t)\, \varphi(\omega(t)-\omega(s))]) \ .$$

In general, i. e. if μ_φ is not absolutely continuous, the right side of (33) is only a heuristic expression. But by theorem 18, it makes sense to consider (29) as the exact definition of the heuristic expression (33). In the case that φ is the δ - function at zero, we recover Varadhan's construction of the two-dimensional polymer measure (see [10]).

An essential advantage of the nonstandard approach is that one obtains not only the existence of standard polymer measures, but also a convergence theorem, how the constructed polymer measures can be approximated by self-repellent random walks. This is the main standard result of our work. In the free case, Anderson's nonstandard construction of the Wiener measure implicitly contains Donsker's invariance principle (see [1]). The formulation of our main theorem requires some additional notation. For every $n \in \mathbb{N}$, let Δt_n be a positive real, $\Delta x_n := \sqrt{d\,\Delta t_n}$, $\Gamma_n := \{(k\Delta x_n, \ell\Delta x_n) \mid k, \ell \in \mathbb{Z}\}$, $T_n := \{k\Delta t_n \mid k \in \mathbb{N}_0\}$, $T_{n,w} := \{t \in T_n \mid t \le w\}$, $\theta_n : T_n \times T_n \to [0,1]$ an arbitrary function, $\Phi_n : \Gamma_n \to \mathbb{R}_+$ a function such that $\sum_{x\in\Gamma_n} (\Delta x_n)^2\, \Phi_n(x) \le 1$, and $g_n \in \mathbb{R}_+$. For every $n \in \mathbb{N}$, define a probability measure P_n on $\Omega_n := \{\omega : T_{n,w} \to \Gamma_n \mid \omega(0) = 0,\ |\omega(t) - \omega(t-\Delta t)| = \Delta x_n \text{ for all } t \in T_{n,w} \setminus \{0\}\}$ by

$$P_n(\{\omega\}) := \frac{1}{Z_n} \exp[-g_n \sum_{s,t\in T_{n,w}} (\Delta t_n)^2\, \theta_n(s,t)\, \Phi_n(\omega(t)-\omega(s))]$$

$(\omega \in \Omega_n)$, where Z_n is the normalization constant, i. e.

$$Z_n := \sum_{\omega\in\Omega_n} \exp[-g_n \sum_{s,t\in T_{n,w}} (\Delta t_n)^2\, \theta_n(s,t)\, \Phi_n(\omega(t)-\omega(s))] \ .$$

<u>Main theorem 19.</u> If $\Delta t_n \xrightarrow[n\to\infty]{} 0$, $g_n \xrightarrow[n\to\infty]{} g$, $\mu_{\Phi_n} \xrightarrow[n\to\infty]{\text{vaguely}} \mu_\varphi$, and $(\theta_n)_{n\in\mathbb{N}}$ is a fair approximation of ϑ (as defined in theorem 4), with $g \in \mathbb{R}_+$, $\vartheta \in \mathfrak{T}$, $\varphi \in \mathcal{P}_1$, then we have $P_n \circ I_n^{-1} \xrightarrow[n\to\infty]{w} \nu_{\vartheta,\varphi,g}$, where $I_n : \mathcal{F}_0(T_{n,w}, \mathbb{R}^2) \to \mathcal{C}([0,w], \mathbb{R}^2)$ is e. g. linear interpolation.

<u>Proof.</u> It is sufficient to show that for every infinite η, we have $L({}^*P_\eta) \circ \underline{st}^{-1} = \nu_{\vartheta,\varphi,g}$. But this follows immediately from (32). □

In the case that $\varphi = \delta_0$, i. e. $\mu_\varphi = \varepsilon_0$ is the Dirac measure at zero, and $\vartheta \equiv 1$, the measure $\nu_g := \nu_{\vartheta,\varphi,g}$ $(g \in \mathbb{R}_+)$ is of particular interest as it represents Edwards' model for polymers (see [5]). If we choose

$$\Phi_n : \Gamma_n \longrightarrow \mathbb{R}_+ , \ x \longmapsto \begin{cases} (\Delta x_n)^{-2} , & \text{if } x = 0 \\ 0 , & \text{otherwise} \end{cases} ,$$

our main theorem gives the following approximation theorem:

<u>Corollary 20.</u> For every $n \in \mathbb{N}$, define the measure P_n' on Ω_n by

$$P_n'(\{\omega\}) := \frac{1}{Z_n'} \exp[-g_n \sum_{s,t \in T_{n,w} : s<t} \chi_{\{\omega(t)=\omega(s)\}}] ,$$

where

$$Z_n' := \sum_{\omega \in \Omega_n} \exp[-g_n \sum_{s,t \in T_{n,w} : s<t} \chi_{\{\omega(t)=\omega(s)\}}]$$

is the normalization constant $(n \in \mathbb{N})$. Note that the measure P_n' represents the so-called "Domb-Joyce-model" (see [4]). If the limit $\lim_{n\to\infty} g_n/\Delta t_n = 2g/d$ exists in $\mathbb{R}_+$ and $\lim_{n\to\infty} \Delta t_n = 0$, then $P_n' \circ I_n^{-1} \xrightarrow[n\to\infty]{w} \nu_g$. ┘

Thus we have proved that properly scaled Domb - Joyce - models weakly converge to Edwards' model. But our main theorem shows a much stronger result than corollary 20, concerning the domain of attraction for convergence to Edwards' model. Instead of the specific representation (34) of the δ - function, which is used in the definition of the Domb - Joyce - model, one can take an arbitrary approximation $(\Phi_n)_{n \in \mathbb{N}}$ of the δ - function and simultaneously an arbitrary fair approximation $(\Theta_n)_{n \in \mathbb{N}}$ of the full square $[0,w]^2$, e. g. one may omit a diagonal part. Thus the concrete kind of the "potential" Φ_n is irrelevant. Notice that our main theorem 19 not only treats the case of a δ - function in the limit. In the nonstandard language: No matter which potential Φ and which "region of interaction" Θ one has, it is always possible to take and determine the standard part in the sense of theorem 17, supposed that the coupling g is in the proper domain.

References

[1] Anderson, R. M.: A non-standard representation for Brownian motion and Itô integration. Isr. J. Math. 25, 15 - 46 (1976).

[2] Bovier, A., Felder, G., Fröhlich, J.: On the critical properties of the Edwards and the self-avoiding walk model of polymer chains. Nucl. Phys. B 230, 119 - 147 (1984).

[3] Brydges, D., Spencer, T.: Self-avoiding walk in 5 or more dimensions. Commun. Math. Phys. 97, 125 - 148 (1985).

[4] Domb, C., Joyce, G. S.: Cluster expansion for a polymer chain. J. Phys. C 5, 956 - 976 (1975).

[5] Edwards, S. F.: The statistical mechanics of polymers with excluded volume. Proc. Phys. Soc. 85, 613 - 624 (1965).

[6] Nelson, E.: A quartic interaction in two dimensions. In: Mathematical Theory of Elementary Particles (eds. Goodman, R., Segal, I.), pp. 69 - 74. Cambridge, Mass. (1966).

[7] Perkins, E.: Weak invariance principles for local time. Z. Wahrscheinlichkeitstheorie verw. Geb. 60, 437 - 451 (1982).

[8] Rosen, J.: A local time approach to the self-intersections of Brownian paths in space. Commun. Math. Phys. 88, 327 - 338 (1983).

[9] Stoll, A.: Self-repellent random walks and polymer measures in two dimensions. Ph. D. thesis, Ruhr-Universität Bochum (1985).

[10] Symanzik, K.: Euclidean quantum field theory. Appendix by Varadhan, S. R. S. In: Local Quantum Theory (ed. Jost, R.), pp. 152 - 226. New York: Academic Press (1969).

[11] Westwater, J.: On Edwards' model for long polymer chains. Commun. Math. Phys. 72, 131 - 174 (1980). On Edwards' model for polymer chains III, Borel Summability. Commun. Math. Phys. 84, 459 - 470 (1982).

ON THE UNIQUNESS OF THE MARKOVIAN SELF-ADJOINT EXTENSION

Masayoshi-TAKEDA
Department of Mathematics
Faculty of Science
Osaka University
Toyonaka 560,Osaka,Japan

§1. Intrduction

Albeverio, Høegh-Krohn, Streit [1] reformulate the Schrödinger dynamics in terms of energy forms. In their approach, the dynamical information sits in the ground state measure ν. Under the condition of closability, Hamiltonian H_ν is given as the self-adjoint operator repmsenting the following energy form $(\mathcal{E}_\nu, \mathcal{F}_\nu)$ on $L^2(\nu)$;

$$(1.1)\qquad \mathcal{E}_\nu(u,v) = \frac{1}{2}\int_{R^d}(\text{grad } u, \text{grad } v)_{R^d}\, d\nu = (\sqrt{-H_\nu}u, \sqrt{-H_\nu}v)_\nu$$

$\mathcal{F}_\nu$: the closure of $C_0^\infty(R^d)$ with respect to $\mathcal{E}_{\nu,1}(=\mathcal{E}_\nu + (\ ,\ )_\nu)$.

Here, the ground state measure ν is Radon measure with support R^d. When ν is absolutely continuous with respect to Lebesgue measure, say $\nu=\rho dx$, H_ρ is of form

$$(1.2)\qquad H_\rho = \frac{1}{2\rho}\sum_{i=1}^{d}\frac{d}{dx_i}(\rho\frac{d}{dx_i}).$$

An important property of H_ρ is its unitary equivalence to some Schrödinger operator. In fact, by setting $\psi = \sqrt{\rho}$ we have

$$(1.3)\qquad \psi H_\rho \psi^{-1} = \frac{1}{2}\Delta - \frac{\Delta\psi}{\psi}.$$

By (1.3), we can conceive a general Schrödinger operator whose potential may be singular. And this formulation can be naturally extended to the infinite dimensional case. Let Q be countably normed space such that there exist a separable Hilbert space H and continuous injection $Q \hookrightarrow H \hookrightarrow Q'$ (the dual space of Q). Let $\{e_i\}_{i=1}^\infty \subset Q$ be a complete orthonormal basis and define FC_0^∞ by

$$FC_0^\infty = \left\{ f;\ \begin{array}{l} f \text{ is a function on } Q' \text{ which is written as } f(\xi)=\tilde{f}(\langle e_{i_1},\xi\rangle,\cdots, \\ \langle e_{i_n},\xi\rangle) \text{ for some } n \text{ and a real } C_0^\infty(R^n)\text{-function } \tilde{f} \end{array} \right\}.$$

Then, for quasi-invariant probability measure ν on Q', the energy form on $L^2(Q',\nu)$ is given by

$$(1.4)\qquad \mathcal{E}_\nu(u,v) = \frac{1}{2}\int_{Q'}(Du,Dv)_H\, d\nu, \quad u,v \in FC_0^\infty .$$

Here, $Du = \sum_{i=1}^{\infty} D_i u \otimes e_i \in L^2(\nu) \otimes H$ and $D_i u$ is a derivative in the direction of e_i. In the quantum field theory, we take $\mathcal{S}(R^d)$ (real Schwartz space) and $L^2(R^d)$ as Q' and H.

The closability condition of $(\mathcal{E}_\nu, FC_0^\infty)$ was investigated by Albeverio-Høegh-Krohn, Kusuoka ([2],[7]). For example, if identity function 1 belongs to the domain of $\pi(q)$ for any $q \in Q$, $(\mathcal{E}_\nu, FC_0^\infty)$ is closable, where $\pi(q)$ is the generator of one-parameter unitary group $(V(tq)f)(\xi) = \sqrt{\frac{d\nu(\xi+tq)}{d\nu(\xi)}} f(\xi+tq)$. We denote by $\mathcal{P}_1$ the class of quasi-invariant probabality measure on Q' satisfying the above condition. In this case, Hamiltonian is written as

$$(1.5) \qquad (H_\nu f)(\xi) = \frac{1}{2}\sum_{i=1}^{\infty} (D_i^2 f + \beta(e_i) D_i f)(\xi), \quad \text{for } f \in FC_0^\infty ,$$

where $\beta(e_i)(\xi) = 2i(\pi(e_i)1)(\xi)$. From now, we assume that ν belong to $\mathcal{P}_1$.

Next, we must ask the following uniqueness question.

(I) All closed extensions of $(\mathcal{E}_\nu, FC_0^\infty)$ coincide ?

This is equivalent to the question whether H_ν already essentially self-adjoint operator on FC_0^∞ or not. But in this note, we treat the Markovian uniqueness question which is weaker than the above, that is,

(II) For which ν do all Dirichlet extensions $(\mathcal{E}_\nu, FC_0)$ coincide ? Namely, when do all self-adjoint extensions of $S_\nu = H_\nu \upharpoonright FC_0^\infty$ generating Markov semi-groups coincide ?

In §3, we give two classes of quasi-invariant probability measures such that S_ν has the Markovian uniqueness. In particular, Theorem 3 gives an extension of the result of [8].

§2. The characterization of maximum Dirichlet space

We denote by $\mathcal{A}_M(S_\nu)$ the totality of Markovian self-adjoint extensions of S_ν. Let us introduce a semi-order $\prec$ in $\mathcal{A}_M(S_\nu)$ by

$$(2.1) \qquad A_1 \prec A_2 \iff \mathcal{D}(\sqrt{-A_1}) \subset \mathcal{D}(\sqrt{-A_2})$$
$$(\sqrt{-A_1}u, \sqrt{-A_1}u)_\nu \geqq (\sqrt{-A_2}u, \sqrt{-A_2}u)_\nu \quad \text{for } u \in \mathcal{D}(\sqrt{-A_1}).$$

In order to show the Markovian uniqueness, we usually need two steps. The first step is to show that there exists a maximum element, say A_R, in $\mathcal{A}_M(S_\nu)$ and to characterize the space $\mathcal{D}(\sqrt{-A_R})$. The second step is to identify the operator A_R with the Friedrics extension H_ν. Recently for a fairly general class of quasi-invariant measures, Albeverio-

Kusuoka [4] settled the first step.

Theorem 1 (Albeverio-Kusuoka) If ν belongs to $\mathcal{P}_1$ and strictly positive, there exists the maximum element of $\mathcal{A}_{\mathcal{H}}(S_\nu)$, say A_R, and the space $\mathcal{D}(\sqrt{-A_R})$ is represented as

$$(2.2) \qquad \mathcal{D}(\sqrt{-A_R}) = \left\{ u \in L^2(\nu);\ \begin{array}{l} \text{u is stochastic H Gateaux differentiable} \\ \text{with respect to } \nu\text{, ray absolutely conti-} \\ \text{nuous and the stochastic Gateaux deriva-} \\ \text{tive } \tilde{D}u \text{ of u satisfies that } \|\tilde{D}u(\nu)\|_H \in L^2(\nu) \end{array} \right\}$$

To prove Theorem 1, they prepare the next lemma.

Lemma 1 (Albeverio-Kusuoka) If ν belongs to $\mathcal{P}_1$, then it holds that for any $A \in \mathcal{A}_{\mathcal{H}}(S_\nu)$,

$$(2.3) \qquad \mathcal{D}(\sqrt{-A}) \subset \mathcal{H}_\nu = \left\{ u \in L^2(\nu);\ \begin{array}{l} \text{there exists some } g \in L^2(\nu)\otimes H \text{ such that} \\ (u, D_\nu^*\phi)_\nu = (g,\phi)_{L^2(\nu)\otimes H} \text{ for any } \in FC_0^\infty \otimes Q \end{array} \right\}$$

Here, D_ν^* is the dual operator D from FC_0^∞ to $L^2(\nu)\otimes H$, and $FC_0^\infty \otimes Q = \{ f;$ f is written as $f = \sum_{i=1}^{n} f_i \otimes e_i$ for some n and $f_i \in FC_0^\infty\}$.

Under the condition that ν is strictly positive, they obtained

$$(2.4) \qquad \mathcal{H}_\nu = \mathcal{D}(\sqrt{-A_R}).$$

§3. Markovian uniqueness

Let (H,B,μ) be the abstruct Wiener space.

Theorem 2 If ν is absolutely continuous with respect to μ, say $\nu = \psi^2\mu$, and ψ satisfies

i) $\psi > 0$, μ-a.e. ii) $\psi \in D_4^1$,

then $S_{\psi^2} (= S_{\psi^2\mu})$ has a unique Markovian self-adjoint extension. Here, D_4^1 is Sobolev space of order r and degree p on the Wiener space (see [6]).

Proof Let $\{a_\ell(t)\}_{\ell=1}^\infty$ be a sequence of $C_0^\infty(R^1)$-functions satisfying that

i) $0 \leq a_\ell(t) \leq 1$ ii) $a_\ell(t) = \begin{cases} 1 & \text{on } \frac{1}{2^\ell} \leq t \leq 2^\ell \\ 0 & \text{on } t \leq \frac{1}{2^{\ell+1}},\ t \geq 2^{\ell+1} \end{cases}$

iii) $|a_\ell'(t)| \leq \begin{cases} c\ 2^{\ell+1} & \text{on } t \leq \frac{1}{2^\ell} \\ c\ \frac{1}{2^\ell} & \text{on } 2^\ell \leq t \leq 2^{\ell+1} \end{cases}$,where c is some constant.

Put $\phi_\ell(x) = a_\ell \circ \psi(x)$ and set

$$\widetilde{\mathcal{H}}_\psi 2 = \{u \in L^2(\psi^2\mu);\ \phi_\ell \cdot u \in \bigcup_{1<p<2} D^1_p \text{ for any } \ell \text{ and } \int (Du,Du)_H \psi^2 d\mu < \infty\}.$$

Then, by [8;Lemma 2.1], the closure of FC_0^∞ with respect to $\mathcal{E}_{\psi}2,_1$ includes the space $\widetilde{\mathcal{H}}_\psi 2$. On the other hand, for $u \in \mathcal{H}_\psi 2$, there exists some $g \in L^2(\psi^2\mu)\otimes H$ such that $(u, D^*_{\psi^2\mu}\phi)_{\psi^2\mu} = (g,\phi)_{L^2(\psi^2\mu)\otimes H}$. Since the left hand side equals $(u\psi^2, D^*_\mu\phi)_\mu - (2u\psi D\psi,\phi)_{L^2(\mu)\otimes H}$, we obtain

$$(3.1) \qquad (u\psi^2, D^*_\mu \phi)_\mu = (2u\psi D\psi + \psi^2 g, \phi)_{L^2(\mu)\otimes H}.$$

Since

$$\int (u\psi^2)^{4/3} d\mu \leq \left(\int ((u\psi)^{4/3})^{3/2} d\mu\right)^{2/3} \cdot \left(\int (\psi^{4/3})^3 d\mu\right)^{1/3} = \left(\int u^2\psi^2 d\mu\right)^{2/3} \cdot \left(\int \psi^4 d\mu\right)^{1/3} < \infty$$

and $\|u\psi D\psi\|_H$ and $\|\psi^2 g\|_H$ also belong to $L^{4/3}(d\mu)$, we have $u\psi^2 \in \bigcup_{1<p<2} D^1_p$. From this, we see that $\phi_\ell u \in \bigcup_{1<p<2} D^1_p$ and $Du = g$. therefore, by Lemma 1, we have

$$(3.2) \qquad \mathcal{D}(\sqrt{-A}) \subset \mathcal{H}_{\psi^2\mu} \subset \widetilde{\mathcal{H}}_{\psi^2\mu} \subset \overline{FC_0^\infty}^{\mathcal{E}_{\psi}2,1}, \text{ for any } A \in \mathcal{A}_M(S_{\psi}2).$$

Remark 1 This theorem, of course, holds in finite dimensional by replacing μ and D^1_4 with the Lebesgue measure and H^1_4 respectively. And when $\psi \in H^1_2$ and $\psi \geq c(K) > 0$ for any compact set K, we take a sequence $\{\phi_\ell\}$ independently of the function ψ such that

i) $\phi_\ell(x) = \begin{cases} 1 & \text{on } |x| \leq \ell \\ 0 & \text{on } |x| \geq \ell+1 \end{cases}$ ii) $\|\text{grad } \phi_\ell\| \leq M < \infty$ for any ℓ,

and set

$$\widetilde{\mathcal{H}}'_\psi 2 = \left\{u \in L^2(\psi^2\mu);\ \phi_\ell u \in H^1_1 \text{ for any } \ell \text{ and } \int (\text{grad } u, \text{grad } u)\psi^2 dx < \infty\right\}$$

Then, in the same way, we can shaw

$$\mathcal{D}(\sqrt{-A}) \subset \mathcal{H}_{\psi^2 dx} \subset \widetilde{\mathcal{H}}'_{\psi^2 dx} \subset \overline{C_0^\infty(R^d)}^{\mathcal{E}_{\psi}2,1}, \text{ for any } A \in \mathcal{A}_M(S_{\psi^2 dx}),$$

and conclude the Markovian uniqueness.

Next we denote by $\mathcal{P}^\infty$ the class of quasi-invariant measure satisfying

i) strictly positive and smooth density with respect to finite dimensional subspace.

ii) for sequence $\{K_n\}$ of finite dimensional subspace of Q such that $K_n \uparrow Q$, it holds

(3.3) $$\lim_{n\to\infty} \sum_{e_i \in K_n} \|\beta(e_i) - E_{K_n}\beta(e_i)\|^2_{L^2(\nu)} = 0 .$$

Here, E_K is the projection to $L^2(\Sigma_K,\nu)$ and Σ_K is σ-field generated by $\{\langle e_i,\xi\rangle; e_i \in K\}$.

For the definition of i), see [2]. Then, we have

Theorem 3 if $\nu \in \mathcal{P}^\infty$, then S_ν has unique Markovian self-adjoint extension.

For a finite dimensional subspace of Q, we set

(3.4) $$\mathcal{F}^1_K = \left\{ u \in L^2(\Sigma_K,\nu);\ \begin{matrix} \text{there exists some } g \in L^2(\Sigma_K,\nu)\otimes K \text{ such that} \\ (u, D^*_\nu\phi)_\nu = (g,\phi)_{L^2(\nu)\otimes H} \text{ for any } \phi \in FC^\infty_0(K)\otimes K \end{matrix} \right\}$$

where $FC^\infty_0(K)$ is totality of Σ_K-measurable functions in FC^∞_0. Denote by $D^1_K u$ the above g and define the closed bilinear form $(\mathcal{E}^1_K, \mathcal{F}^1_K)$ as

(3.5) $$\mathcal{E}^1_K(u,v) = \frac{1}{2}(D^1_K u, D^1_K v)_{L^2(\nu)\otimes H} \quad \text{for } u,v \in \mathcal{F}^1_K .$$

Identifying the measurable space (Q',Σ_K,ν) with $(R^d,\mathcal{B}(R^d),\rho_K(x)dx)$ $(d=\dim K)$, $\rho_K(x)$ is strictly positive and C^∞-function by the assumption i). Hence, Hamiltonian H_{ρ_K} corresponding to $\rho_K(x)dx$ is essentially self-adjoint on $C^\infty_0(R^d)$. Therefore, we have

Lemma 2 $\overline{FC^\infty_0(K)}^{\mathcal{E}^1_{K,1}} = \mathcal{F}^1_K$.

By the essential self-adjointness of H_{ρ_K}, we have

Lemma 3 $\mathcal{F}^1_K = \mathcal{D}(\sqrt{-A_R}) \cap L^2(\Sigma_K)$ and $D^1_K = \widetilde{D}$ on $\mathcal{F}^1_K$.

Lemma 4 If $u \in \mathcal{D}(\sqrt{-A_R}) \cap L^\infty$, then $E_K u \in \mathcal{D}(\sqrt{-A_R})$ and

(3.7) $$D_i(E_K u) = E_K D_i u + E_K(u(\beta(e_i) - E_K\beta(e_i))) .$$

Proof Let D^*_i be the dual operator of $D_i : FC^\infty_0 \longrightarrow L^2(\nu)$. Since for any ϕ $FC^\infty_0(K)$,

$$\begin{aligned}(E_K u, D_i\phi)_\nu &= (E_K u, -D_i\phi - \beta(e_i)\phi)_\nu \\ &= (u, -D_i\phi - E_K\beta(e_i)\phi)_\nu\end{aligned}$$

$$= (u, D_i\phi + (\beta(e_i) - E_K\beta(e_i)), \phi)_\nu$$
$$= (D_i u + u(\beta(e_i) - E_K\beta(e_i)), \phi)_\nu$$
$$= (E_K D_i u + E_K(u(\beta(e_i) - E_K\beta(e_i))), \phi)_\nu$$

and $E_K D_i u + E_K(u(\beta(e_i) - E_K\beta(e_i))) \in L^2(\nu)$, we see that $E_K u \in \mathcal{F}_K^1$ and $D_K^1 E_K u = E_K D_i u + E_K(u(\beta(e_i) - E_K\beta(e_i)))$. By Lemma 3, we obtain the lemma.

(Proof of Theorem 3)

For $u \in \mathfrak{D}(\sqrt{-A_R}) \cap L^\infty$, we have

$$\|\tilde{D}u - \tilde{D}E_K u\|_H^2 = \sum_{e_i \in K} (D_i u - D_i E_K u)^2 + \sum_{e_i \notin K} (D_i u)^2$$
$$= \sum_{e_i \in K} (D_i u - E_K D_i u - E_K(u(\beta(e_i) - E_K\beta(e_i))))^2 + \sum_{e_i \notin K} (D_i u)^2$$
$$2 \sum_{e_i \in K} (D_i u - E_K D_i u)^2 + 2\|u\|_\infty \sum_{e_i \in K} E_K((\beta(e_i) - E_K\beta(e_i))^2) + \sum_{e_i \notin K} (D_i u)^2.$$

Hence, $\int_{Q'} \|Du - DE_K u\|_H^2 d\nu \longrightarrow 0$ $(K \uparrow Q)$ by assumption (3.3). Therefore, by Lemma 2 and Lemma 3, we see that $\overline{FC_0^\infty}^{\mathcal{E}_\nu,1} = \mathfrak{D}(\sqrt{-A_R})$.

In [8], we studied the case that ψ in theorem 2 satisfies that i) $\psi > 0$ ii) $\psi \in D^\infty$. In this case, $\beta(e_i)$ becomes $\beta(e_i) = \langle e_i, \xi \rangle + \frac{D_i\psi}{\psi}$ and

$$\sum_{e_i \in K_n} \|\beta(e_i) - E_{K_n}\frac{D_i\psi}{\psi}\|_{L^2(\nu)}^2 = \sum_{e_i \in K} \|\frac{D_i\psi}{\psi} - E_{K_n}\frac{D_i\psi}{\psi}\|_{L^2(\nu)}^2$$
$$\leq \|\frac{D\psi}{\psi} - E_{K_n}\frac{D\psi}{\psi}\|_{L^2(\nu)\otimes H}^2$$
$$= \|D\psi - E_{K_n} D\psi\|_{L^2(\mu)\otimes H}^2 \xrightarrow[n \to \infty]{} 0.$$

Therefore, $\nu = \psi^2\mu$ belongs to the class $\mathcal{S}^\infty$.

Acknowledgement.

I thank Prof. S. Kusuoka for teaching me the results of §2. I am also grateful to Prof. S. Albeverio for his interest and helpful discussions. I gratefully acknowledge the hospitality at ZiF, Universitat Bielefeld and Forschungszentrum BiBoS.

References

[1] S. Albeverio, R. Høegh-Krohn and L. Streit, Energy forms, Hamiltonians and distorted Brownian paths,J. Math. Fhys., 18, 907-917, 1977.

[2] S. Albeverio and R. Høegh-Krohn, Dirichlet forms and diffusion processes on rigged Hilbert space, Z. Wahr. 40, 1-55, 1977.

[3] s. Albeverio and R. H egh-Krohn, Diffusion fields, quantum fields, fields with values in Lie groups, in Stochastic Analysis and Applications, ed. M. Pinsky, M. Dekker, New York, 1-98, 1980.

[4] S. Albeverio and S. Kusuoka, in preparation.

[5] M. Fukushima, Dirichlet forms and Markov processes, North Holland, Kodansha, 1980.

[6] N. Ikeda and S. Watanabe, An introduction to Malliavin's calculus, Proc. Taniguchi Symposium on Stochastic Analysis, Katata, 1982.

[7] S. Kusuoka, Dirichlet forms and diffusion processes on Banach spaces, J. Fac. Sci. Univ. Tokyo, 79-95, 1982.

[8] M. Takeda, On the uniqueness of self-adjoint extension of diffusion operators on infinite dimensional spaces, to appear in Osaka J. Math..

REPRESENTATIONS OF THE GROUP OF EQUIVARIANT LOOPS IN SU(N)

by

D. TESTARD

Département de Physique
Faculté des Sciences d'Avignon
AVIGNON - FRANCE

Centre de Physique Théorique
MARSEILLE - FRANCE

and

Research Center Bielefeld-Bochum-Stochastics
University of Bielefeld
BIELEFELD - F.R.G.

ABSTRACT

We construct and study representations of the multiplicative group of equivariant loops with values in SU(N), the equivariance being with respect to an order-two automorphism.

This talk is a report of a joint work with S. ALBEVERIO and R. HØEGH-KROHN.

In the last years, new results about non commutative distributions theory "à la GELFAND, GRAIEV, VERSHIK" were obtained. In particular, the so-called energy representation of groups of mappings with values in a semi-simple compact Lie Group was the object of most of the work on the subject.

Let us recall the main ingredients in the theory. We start with a compact, semi-simple Lie Group G and a riemanian manifold X, with a volume measure denoted dx. Let G^X the multiplicative group of C^∞-mappings, with compact support from X to G. The energy representation is the one constructed via the GELFAND, NAIMARK, SEGAL procedure with the positive type function on G^X :

$$\psi \in G^X \to \exp - \frac{|d\psi\ \psi^{-1}|^2}{2} \tag{1}$$

where $|\ |$ is the norm associated to the prehilbert structure on the set $\Omega^1(X,G)$ of one-forms on X with values in the Lie algebra g of G, defined by : $(\forall\omega, \omega' \in \Omega^1\ (X,G))$

$$< \omega, \omega' > = \int Tr\ (\omega_x^* \ \omega'_x)\ dx.$$

In this expression the $*$ refers to the canonical structure of euclidian spaces on the tangent spaces of X and on the Lie-algebra g equipped with the opposite of the Killing-form as a scalar product.

As mentionned before, these representations are non commutative distributions : in the non abelian-case, consideration of continuous linear forms or equivalently continuous characters has to be replaced by the one of equivalence classes of irreducible representations. Precisely the energy representation is an irreducible representation if $\dim X > 1$ (with a supplementary condition in case $\dim X = 2$) and its class is a non commutative distribution of order 1. Moreover different riemanian structures on X give rise to inequivalent energy representations (for more details, see : |1|, |2|, |3|, |6|, |8|).

The case of $\dim X = 1$ (and the general case of $\dim X = 2$) is also very interesting. It is transparent already in (1) that the Brownian motion of G can be used in order to describe the energy representation at least in the case of $X = \mathbb{R}^{*,+}$ (the positive real numbers). The representation space can be described as the L^2-space of the Brownian motion, the representation being the left (or the right) translation by smooth pathes with values in G. The same kind of interpretation can be obtained for $X = \mathbb{R}$ (see below).

The case $X = S$ (the unit circle

$$S = \mathbb{R}/_{2\pi\mathbb{Z}}\ )$$

is obtained using a conditionning of the brownian measure of the case $X = \mathbb{R}$. The main idea, of this work is the simple remark that one can even make stronger conditionning, without changing the essence of the argument. As we will see, this allows

In the next section, we proceed to the construction and we state the main results. The last section is devoted to the presentation of rather detailed indications on proofs.

AKNOWLEDGMENTS

The author want to thank the Scientific Direction of Bielefeld-Bochum-Stochastics for his invitation in participating to the research group and the Deutsche Forschungsgemeinschaft, the Centre National de la Recherche Scientifique and the Volkwagen-Schtiftungwerk for financial support. The author thanks S. ALBEVERIO and R. HØEGH-KROHN for usefull comments and suggestions about the content of this paper.

I. DEFINITIONS, NOTATIONS AND STATEMENT OF RESULTS

Let G be a compact, semi-simple Lie Group of the form $SU(N)$ for $N > 1$ and τ an order-two automorphism of G.

Definition I.1.

A loop in $SU(N)$ is a continuous mapping from S into $SU(N)$. An equivariant loop in $SU(N)$ is a loop f satisfying :

$$\tau(f(\theta)) = f(\theta + \pi). \tag{2}$$

If $T \subset G$ is a subgroup, invariant by τ, we can speak about equivariant loops in T (resp : $G/_T$, $_T\backslash G$, $_T\backslash G/_T$) as continuous mappings f with values in T (resp : $G/_T$, $_T\backslash G$, $_T\backslash G/_T$) satisfying (2) which makes sense with an evident meaning for τ (acting on T or on the quotient).

We will denote by $C_e(G)$, $C_e(T)$, $C_e(G/_T)$, $C_e(_T\backslash G)$, $C_e(_T\backslash G/_T)$ the set of equivariant pathes in G , T , $G/_T$, $_T\backslash G$, $_T\backslash G/_T$ respectively.

Let us remark, as a consequence of the classification of finite order automorphism of $SU(N)$ |5|, that there always exists a maximal torus T which is invariant by τ .

Let us proceed to the construction of the representation we are concerned with. Let $C(\mathbb{R},G)$ be the set of continuous functions on $\mathbb{R}$ with values in G. Let $\eta(t,h)$ $(h \in G)$ be the process such that

$$\eta(0,h) = h \, ,$$
$$d\eta \, \eta^{-1} = \xi \, ,$$

where ξ is the white-noise process in the Lie algebra g of G. The corresponding measure is actually supported by $C(\mathbb{R},G)$. It is well known that, the process

$$t \to \eta(t - s,h)$$

converges in probability for $s \to \infty$ to a process $\eta(t)$ which is actually independant of h. The corresponding measure μ is supported by $C(\mathbb{R},G)$: it is called the standard brownian measure on G.

Conditionning the measure μ by the constraint :

$$\tau(\eta(0)) = \eta(\pi) \tag{3}$$

one gets a measure μ_o which is called "the conditionned brownian measure" and which is supported by those pathes satisfying (3). Now for any equivariant loop η in G, its restriction to $[0,\pi]$ is in the support of μ_o and we can transport the measure μ_o in such a way that one can consider that μ_o is supported by the set $C_e(G)$ of equivariant loops. This will be done in the sequel.

By desintegration theory of measures, one easily sees that the mapping $\eta \to \eta\psi$ (pointwise multiplication of equivariant loops) where ψ belongs to the set G_e^S of C^∞-equivariant loops in G leaves μ_o quasi-invariant (actually less regularity of ψ is needed) and let $\frac{d\mu_o(\eta\psi)}{d\mu_o(\eta)}$ denotes the corresponding Radon-Nykodim cocycle. Then the following formula $(\forall \eta \in C_e(G))$

$$(U^R(\psi)F)\,(\eta) = \left(\frac{d\mu_o(\eta\psi)}{d\mu_o(\eta)}\right)^{1/2} F(\eta\psi) \tag{4}$$

defines a unitary operator $U^R(\psi)$ for $\psi \in G_e^S$ acting on $F \in H \equiv L^2(C_e(G), \mu_o)$.

In the same way, one sees that μ_o is quasi-invariant by $\eta \to \psi^{-1}\eta$ for ψ a C^∞-equivariant map. This allows to define another unitary operator :

$$(U^L(\psi)F)\ (\eta) = \left(\frac{d\mu_o(\psi^{-1}\ \eta)}{d\mu_o(\eta)}\right)^{1/2} F(\psi^{-1}\ \eta) \tag{5}$$

Clearly, U^R and U^L are mutually commuting representations of the group G_e^S with respect to pointwise multiplication. The mapping $\eta \to \eta^{-1}$ (which leaves μ_o invariant) allows to define an intertwiner for the pair U^R, U^L of representations and U^L, U^R are actually equivalent.

An easy computation shows that the function :

$$\Pi\ (\eta) = 1 \qquad\qquad (\forall \eta \in C_e(G))$$

is cyclic in H. The positive type function associated to each representation U^R or U^L and to the cyclic vector Π is :

$$\psi \in G_e^X \to \exp - \frac{|d\psi\ \psi^{-1}|^2}{2}$$

due to the definition of the measure μ_o .

As already said in the last section, we want to study the algebraic properties of U^R and U^L. The first result describe the situation when one diagonalizes the set of operators $U^R(T_e^S)$ where T is a maximal torus in G which is invariant by τ .

Theorem I.1.

Let G,τ as before and T a maximal torus with $\tau(T) = T$. Then

$$\text{i)} \quad H = \int^{\oplus} H^\alpha \, d\mu_T(\alpha)$$

where

- in the decomposition $U^R(T_e^S)$ is the set of diagonalizable operators (see for instance |10|) ;

- μ_T is the equivariant brownian measure on T ;

- for almost all $\alpha \in C_e(T)$:

$$H^{\alpha} = L^2(C_e(^G/_T), \mu_1)$$

where μ_1 is the canonical image of μ_0 by the quotient mapping $\eta \to \dot{\eta}$ from $C_e(G)$ in $C_e(^G/_T)$ ("*the equivariant brownian measure in* $^G/_T$).

$$\text{ii)} \quad U^L(\psi) = \int^{\oplus} U^{\alpha}(\psi)\, d\,\mu_T(\alpha)$$

and where for almost all α (with respect to μ_T) :

$$(U^{\alpha}(\psi)F)(\xi) = \left(\frac{d\mu_1(\psi^{-1}\xi)}{d\mu_1(\xi)}\right)^{1/2} F(\psi^{-1}\xi)\, \exp\,(-i < \alpha^{-1}\, d\alpha,\ \phi^{-1}\, d\psi\ \psi^{-1}\,\phi >) \qquad (6)$$

($\forall\ \xi \in C_e(^G/_T)$, $\forall\ F \in L^2\ (C_e(^G/_T), \mu_1)$, $\forall\ \psi \in G^S_e$ and ϕ any path in ξ).

Let us remark that *Theorem I.1.* is actually true for any semi-simple Lie Group G. In contrast, the next theorem is only proven for $G = SU(N)$ for some $N > 1$.

Theorem I.2.

Assume $G = SU(N)$ for $N > 1$. τ and T as before :

i) $U^R(T^S_e)''$ is maximal abelian in $U^R(G^S_e)''$,

ii) $U^L(T^S_e)''$ is maximal abelian in $U^L(G^S_e)''$,

iii) U^L and U^R are factor representations generating Von NEUMANN algebras which are the commutant of each other ,

iv) U^{α} (as defined in (6)) is irreducible, for almost all α (with respect to μ_T).

Before going to the proofs of these results, let us indicate that they are unsatisfactory in two different contexts. It would be nice to get results similar to those of *Theorem I.3.* for other classical groups than SU(N). As we will see, our method is unefficient in these cases.

A very interesting extension of these results would be to consider also projective representations. Actually equivariant loops-groups are the building-blocks of the construction of the so-called Kac-Moody Groups |9|. The only difference between loops-groups and Kac-Moody Groups is that one has to make a central extension with a well-defined 2-cocycle. In order to construct new representations (or to obtain a new interpretation of known representations) of the so-called Kac-Moody algebras, one has to consider projective representations of loops-groups. In the last years, measure theoretic interpretations of important fact of the theory of Kac-Moody were obtained |4| |5| ; our approach may also be considered as a tentative in the same direction as in |5| where the connection between the Basic Representation of Kac-Moody Groups and the energy representation was already stressed (|5| proposition 3.3.).

II. INDICATIONS ON PROOFS

In this section G is the considered group, g its Lie-algebra, τ is an order-two automorphism. T is a maximal torus invariant by τ and t the corresponding CARTAN subalgebra in g . The notations $C_e(G)$, G_e^S, μ_0 , μ_1 , μ_T are the same as before.

The main tool in the proof of *Theorem I.1.* is the decomposition of $C_e(G)$ which appears in the next lemma and which is the exact space counterpart of the algebraic decomposition described in the statement of *Theorem I.1.* .

Lemma II.1.

G is considered as a principal fiber bundle with the right action of T and equipped with the connection with xt as vertical subspace and its orthogonal as horizontal subspace at the point x. Then :

i) For μ_0-almost all equivariant path η, there is a decomposition (the so-called *horizontal decomposition of* η) :

$$\eta = \phi\, \alpha,$$

such that α is equivariant, continuous T-valued and such that ϕ has horizontal logarithmic increments i.e. $\forall\, \delta$, C^{∞}-function from S into t :

$$\int_S < \delta\, ,\, \phi^{-1}\, d\, \phi > \; = 0 \quad . \tag{7}$$

ii) The mapping $\eta \to (\dot{\eta}, \alpha)$, where $\dot{\eta}$ is the image of η by the canonical projection from $C_e(G)$, onto $C_e({}^G/_T)$, is bimeasurable and transforms μ_0 in $\mu_1 \otimes \mu_T$.

iii) If $\eta = \phi\alpha$ in the horizontal decomposition of η, then the horizontal decomposition of $\psi^{-1}\, \eta$ is :

$$\psi^{-1}\, \eta = (\psi^{-1}\, \phi\, \beta)\; (\beta^{-1}\, \alpha)$$

where β is equivariant in T and satisfies :

$$\beta^{-1}\, d\, \beta = P(\phi^{-1}\, d\psi\; \psi^{-1}\, \phi)$$

where P is the orthogonal projection on t in g.

Before giving indications on the proof, let us remark that the right-action $\eta \to \eta\alpha$ is very simple in the description $(\dot{\eta},\alpha)$ as in ii). Well known properties of Gaussian measures imply that if f is a function on $C_e(G)$ such that, if :

$$f(\eta) = f(\eta\, \alpha)$$

for $\forall\, \alpha \in T_e^S$, then it is a constant on each coset of $C_e({}^G/_T)$. Using conjugacies one sees that the right-action of G_e^S into $C_e(G)$ is ergodic. The same is also true for the left action of G_e into $C_e(G)$ and $C_e({}^G/_T)$.

Proof of *Lemma II.1.*

A formal version of (7) is :

$$P((\eta\,\alpha^{-1})^{-1}\, d(\eta\,\alpha^{-1})) = 0\ .$$

So, we have to solve the differential equation :

$$\alpha^{-1}\, d\alpha = P(\eta^{-1}\, d\eta) \tag{8}$$

with $\alpha(0) = e$ (the identify in G). This, in general does not give an equivariant solution. In order to get an equivariant solution, one can use the following trivial trick. Since $\eta^{-1}\, d\eta$ is equivariant in g with respect to the canonical action of τ in g, the same happens for $P(\eta^{-1}\, d\eta)$. Then solving for α_o, the equations :

$$\alpha_o(0) = e \qquad\qquad \alpha_o{}^{-1}\, d\,\alpha_o = \frac{1}{2}\ (P(\eta^{-1}\, d\eta))$$

one easily sees that :

$$\alpha(\theta) = \alpha_o(\theta)\ \tau\ (\alpha_o(\theta + \pi))$$

due to the abelian character of T, is an equivariant solution of (8). The properties in i) and ii) are easy consequences of (8).

In order to understand the appearance of the β in iii), let us first remark that if ϕ has horizontal logarithmic increments then $\psi^{-1}\,\phi$ has certainly not the same property. So, one has to correct by a path with vertical logarithmic increments the term $\psi^{-1}\,\phi$ in order to get a path with horizontal logarithmic increments. This is the effect of β in iii). The equation for β comes from the very simple remark which follows. Since $\psi^{-1}\,\phi\,\beta$ has horizontal logarithmic increments, one has in distributions sence :

$$P\{(\beta^{-1}\,\phi^{-1}\,\psi)\ d\ (\psi^{-1}\,\phi\,\beta)\} = 0$$

which is equivalent, using horizontality of $\phi^{-1}\, d\,\phi$ to :

$$\beta^{-1}\, d\,\beta = P(\phi^{-1}\, d\psi\ \psi^{-1}\,\phi)\ . \tag{9}$$

For the equivariance, the same trick as before applies.

Now, the *Theorem I.1.* can be proved along the following argument.

As noticed before, the right action of $\gamma \in T^S_e$ onto $\eta \in C_e(G)$ is just the multiplication of the vertical part α in the horizontal decomposition $\eta = \phi\,\alpha$ of η :

$$\eta\gamma = \phi\alpha\gamma \ .$$

So, the diagonalization of $U^R(T^S_e)$ can be performed by Fourier-transform and the spectral measure associated to this representation of the (abelian) group T^S_e is just the classical Gaussian distribution. The elements in the desintegration of this representation are just characters acting multiplicatively :

$$U^R(\beta) = \int^{\oplus} \exp\,(i < \alpha^{-1}\, d\alpha,\ \beta^{-1}\, d\,\beta >)\ d\mu_T(\alpha) \qquad (10)$$

with respect to the α-coordinate and changing nothing with respect to the $\dot{\eta}$ component in the coordinates of *Lemma II.1.* ii).

Let us explain the appearence of the factor $\exp - i < \alpha^{-1}\, d\alpha,\ \phi^{-1}\, d\psi\ \psi^{-1}\, \phi >$ in the component α of $U^L(\psi)$. Left action by $\psi \in G^S_e$ on η not only imposes a change in the class $\dot{\eta}$ which becomes $\psi^{-1}\, \dot{\eta}$ but also a translation of the vertical component by the element β^{-1} satisfying (9).

In the component α of the desintegration (10), this is performed by evaluating the character $\exp i < \alpha^{-1}\, d\alpha,\ .\ >$ on $-\ \beta^{-1}\, d\beta = \beta\, d(\beta^{-1})$. This gives :

$$\exp - i < \alpha^{-1}\, d\alpha,\ \beta^{-1}\, d\beta > = \exp - i < \alpha^{-1}\, d\alpha,\ P(\phi^{-1}\, d\psi\ \psi^{-1}\ \phi > =$$
$$\exp - i < \alpha^{-1}\, d\alpha,\ \phi^{-1}\, d\psi\ \psi^{-1}\ \phi >$$

by orthogonality of the projection P and since $\alpha^{-1}\, d\alpha \in t$.

The proof of *Theorem I.2.* uses a further decomposition of the representation space. It is performed by diagonalizing a bigger abelian Von NEUMANN algebra generated by the left and the right-representations of T^S_e . As before, this decomposition has a space counterpart in $C_e(G)$, which is now described.

Lemma II.2.

G is considered as a principal fiber bundle with the left-right action of $T \times T$: $(\forall\, x \in G)$

$$(\alpha, \beta) \, . \, x = \alpha^{-1} \, x \, \beta$$

and equipped with the connection with $tx + xt$ as vertical subspace and the orthogonal space in xg as horizontal subspace. Then :

i) for μ_o-almost all equivariant path η, there is a decomposition (the so-called *bihorizontal decomposition of* η) :

$$\eta = \alpha^{-1} \, \phi\beta$$

where α and β are continuous, T-valued and equivariant and where ϕ has horizontal logarithmic increments i.e. $\forall \, \delta$, t-valued C^∞-function on S :

$$\int_S < \delta, \phi^{-1} \, d\phi > \; = \int_S < \delta, \, d\phi \, \phi^{-1} > \; = 0 \quad ; \tag{11}$$

ii) $\eta \to (\alpha, \ddot{\eta}, \beta)$ is a bimeasurable isomorphism from $C_e(G)$ onto $C_e(T) \times C_e({}_T\backslash^G/_T) \times C_e(T)$ where $\ddot{\eta}$ denotes the canonical image of η in $C_e({}_T\backslash^G/_T)$.

This isomorphism transports μ_o in

$$\int_{\rho \in C_e({}_T\backslash^G/_T)} \nu^\rho \, d\ddot{\mu} \, (\rho)$$

where $\ddot{\mu}$ is the canonical image of μ_o by the projection of $C_e(G)$ onto $C_e({}_T\backslash^G/_T)$ and, for $\rho \in C_e({}_T\backslash^G/_T)$, ν^ρ is the measure on $C_e(T \times T)$ such that the pairs $(\alpha^{-1} \, d\alpha, \, \beta^{-1} \, d\beta)$ have a Gaussian distribution with covariance matrix :

$$\begin{bmatrix} \mathbb{I} & - P \, \mathrm{Ad} \, \phi \\ - P \, \mathrm{Ad} \, \phi & \mathbb{I} \end{bmatrix}$$

for some (arbitrary) $\phi \in \rho$ acting on L^2-equivariant functions on the circle S with values in $t \times t$.

Remark that in the coordinates $(\alpha, \ddot{\eta}, \beta)$ introduced in *Lemma II.2.* ii), the left and right translations by an element γ of T_e^S are very easy to describe. These are the mappings :

$$(\alpha, \ddot{\eta}, \beta) \to (\alpha\gamma, \ddot{\eta}, \beta) ,$$
$$(\alpha, \ddot{\eta}, \beta) \to (\alpha, \ddot{\eta}, \beta\gamma) .$$

Let us give an indication of the proof of this lemma. Given η in $C_e(G)$, one has to find α and β in such a way that $\phi^{-1} d\phi$ and $d\phi\, \phi^{-1}$ are horizontal where $\phi = \alpha\eta\, \beta^{-1}$. Formally (11) is equivalent to :

$$P(\phi^{-1} d\phi) = P(d\phi\, \phi^{-1}) = 0 .$$

This gives :

$$\begin{aligned} \alpha^{-1} d\alpha - \eta\, \beta^{-1} d\beta\, \eta^{-1} &= - P(d\eta\, \eta^{-1}) \\ \beta^{-1} d\beta - \eta^{-1} \alpha^{-1} d\alpha\, \eta &= + P(\eta^{-1} d\eta) \end{aligned} \qquad (12)$$

(12) can be solved for $\alpha^{-1} d\alpha$ and $\beta^{-1} d\beta$ for almost all η with respect to μ_o because μ_o-almost everywhere :

$$(\eta\, t\, \eta^{-1}) \cap t = \{0\}$$

t being a proper subspace of g for $G = SU(N)$, $N > 1$, due to brownian character of pathes.

The same trick as in the proof of *Theorem I.1.* allows to construct α, β satisfying (12) and equivariant.

Since :

$$\ddot{\eta} = \alpha^{-1} \phi\beta$$

one gets

$$\eta^{-1} d\eta = \beta^{-1} d\beta + \beta^{-1} \phi^{-1} d\phi\beta - \beta^{-1} \phi^{-1} \alpha^{-1} d\alpha\, \phi\, \beta$$
$$|\eta^{-1} d\eta|^2 = |\alpha^{-1} d\alpha|^2 + |\beta^{-1} d\beta|^2 - 2 < \phi^{-1} \alpha^{-1} d\alpha\, \phi, \beta^{-1} d\beta > + |\phi^{-1} d\phi|^2$$

the other mixed-terms being zero due to the horizontality of the logarithmic increments of ϕ.

The assertion corresponding to measures are now clear : the term $|\phi^{-1} d\phi|^2$ is responsible of the appearance of $d\ddot{\mu}$ and the other terms are just the ones which give the stated Gaussian measure.

Due to the remark after the statement of *Lemma II.2.*, one can give a simple description of the restriction to T_e^S of U^R and U^L. This is the content of the following.

Proposition II.1.

For $\forall\, \alpha, \beta \in T_e^S$

$$U^L(\alpha) = \int_{\rho \in C_e(_T\backslash^G/_T)} U^{L,\rho}(\alpha)\; d\ddot{\mu}(\rho)$$

$$U^R(\beta) = \int_{\rho \in C_e(_T\backslash^G/_T)} U^{R,\rho}(\beta)\; d\ddot{\mu}(\rho) \tag{13}$$

and for a given $\rho \in C_e(_T\backslash^G/_T)$, the mapping :

$$(\alpha,\beta) \to U^\rho(\alpha,\beta) = U^{L,\rho}(\alpha)\; U^{R,\rho}(\beta) \tag{14}$$

is the cyclic representation of $(T \times T)_e^S$ acting on $L^2(C_e(T \times T), \nu^\rho)$ where ν^ρ is as in *Lemma II.2.* and with Π as cyclic vector. In other words U^ρ is defined by the positive type function :

$$(\Pi\, ,\, U^\rho(\alpha,\beta)\Pi\,) = \exp\{-\frac{1}{2}(|\alpha^{-1}\, d\alpha|^2 + |\beta^{-1}\, d\beta|^2 - 2 < \alpha^{-1}\, d\alpha,\, \phi^{-1}\, \beta^{-1}\, d\beta\phi >)\} \tag{15}$$

where ϕ is an (arbitrary) element in ρ,

We can now prove *Theorem I.2.* . The main point in the proof is to realize that the decomposition exhibited in *Proposition II.1.* is disjoint. This means that for each partition of $C_e(_T\backslash^G/_T)$ in two subsets say A and A^c, the representations obtained by integrating U^ρ on A and A^c respectively are mutually disjoint.

The first step is the following lemma.

Lemma II.3.

i) The set of functions

$$\eta \to \exp\,(-\,i < \alpha^{-1}\, d\alpha,\ \eta^{-1}\, \beta^{-1}\, d\beta\, \eta >)$$

where α, β run in T_e^S separates points in $C_e({}_T\backslash{}^G/{}_T)$.

ii) The set of functions

$$\eta \to \exp\,(-\,i < \alpha^{-1}\, d\alpha,\ \eta^{-1}\, \psi^{-1}\, \beta^{-1}\, d\beta\, \psi\, \eta >)$$

when α, β run in T_e^S and ψ in G_e^S generates the Borel structure of $C_e({}^G/{}_T)$.

The part ii) is an easy consequence of i) using conjugacy of toruses in G. The result i) is the origin of the specification of $G = SU(N)$ $N > 1$. Actually, in this case an algebraic lemma |3| says that for generic pairs of Cartan subalgebra t_1, t_2 then t_1, t_2 and $[t_1, t_2]$ generate *linearly* the Lie-algebra su(N) of SU(N) (a simple counting of dimensions makes this result plausible). This algebraic lemma is the infinetesimal version of i) (see |3|). Clearly for other classical groups, the dimension argument does not work and *Lemma II.3.* cannot be true.

Now, taking A a Borel set in $C_e({}_T\backslash{}^G/{}_T)$ one sees that for $\rho_1 \in A$ and $\rho_1 \in A^C$, the spectral measures of $\mathcal{U}^{\rho_1}$ and of $\mathcal{U}^{\rho_2}$ are Gaussian measures supported by disjoint sets only depending on A. This means the disjointness of the direct integrals (13).

The proof of *Theorem I.2.* is now easy. The commutant of $A = \{\mathcal{U}^L(T_e^S),\ \mathcal{U}^R(T_e^S)\}''$ only contains decomposable operators of the direct integral (13) or, in other words, A is maximal abelian in the bounded operators on H.

Considering the operator, for $\gamma \in T_e^S$, $\xi \in C_e({}^G/{}_T)$:

$$(W(\gamma)F)\ (\xi) = \left(\frac{d\mu_1\ (\gamma^{-1}\ \xi)}{d\ \mu_1\ (\xi)}\right)^{1/2} F(\gamma^{-1}\ \xi)$$

acting on $F \in L^2(C_e({}^G/{}_T), \mu_1)$ and

$$W = \int^{\oplus} W(\gamma)\ d\mu_T\ (\alpha)$$

(the integral of the constant field of operation $\alpha \to W(\gamma)$), one easily sees that W commutes with $U^L(T_e^S)$ and $U^R(T_e^S)$. It follows from the simplicity of spectrum of A, that :

$$W(\gamma) \in U^\alpha(T_e^S)''$$

for μ_T-almost all α and consequently, $\forall\psi \in G_e^S$:

$$U^\alpha(\psi)\ W\ (\gamma)\ U^\alpha\ (\psi)^{-1} \in U^\alpha\ (G_e^S)''$$

this with *Lemma II.3. ii)* implies that $U^\alpha(G_e^S)''$ contains the multiplication by all measurable bounded functions on $C_e({}^G/_T)$. By the ergodicity of the left action of G_e^S on $C_e({}^G/_T)$, one gets that $U^L(T_e^S)''$ is maximal abelian in $U^L(G_e^S)''$.

The same happens for $U^R(T_e^S)''$ in $U^R(G_e^S)''$ by conjugacy by the mapping inducing $\eta \to \eta^{-1}$. Irreducibility of U^α , for μ_T-almost all α comes now from |10| (*Theorem 8-32*).

For the factoriality of U^R , taking Z in the center, one has that Z is decomposable in the decomposition of *Theorem I.1.* . By irreducibility of U^α for almost all α , Z is diagonalizable. Using a conjugacy and ergodicity as in the case of W before, one obtains that Z is a scalar.

The proof of *Theorem I.2.* is completed.

REFERENCES

|1| S. ALBEVERIO, R. HØEGH-KROHN, D. TESTARD.
J. Funct. Anal. 41, 378, (1981).

|2| S. ALBEVERIO, R. HØEGH-KROHN, D. TESTARD, A. VERSHIK.
J. Funct. Anal. 51, 115, (1983).

|3| S. ALBEVERIO, R. HØEGH-KROHN, D. TESTARD.
J. Funct. Anal. 57, 49, (1984).

|4| T. DUNCAN.
Brownian motion and Affine Lie Algebras. Preprint University of Kansas, Lawrence, Kansas, (1984).

|5| I. FRENKEL, V. KAC.
Invent. Math. 62, 23, (1980).

|6| Ya. GELFAND, M. GRAEV, A. VERSHIK.
Uspehi. Mat. Nauk, 28, 5, (1973).
(Translation : Russian Math. Surveys, 28, 83, (1973)).

|7| V. KAC.
Funkt. Analys i ego prilozh, 3, 252, (1969).
(Translation : Funct. Anal. Appl. 3, 252, (1969)).

|8| J. MARION.
Anal. Pol. Math., 43, 79, (1983).

|9| G. SEGAL.
Commun. Math. Phys., 80, 301, (1981).

|10| M. TAKESAKI.
"Theory of Operator Algebras, Vol. I".
Springer-Verlag, New-York, (1979).

Proof of an algebraic central limit theorem by moment generating functions

Wilhelm von Waldenfels

Institut für Angewandte Mathematik
Universität Heidelberg
Im Neuenheimer Feld 294
D-6900 Heidelberg 1

In a previous paper [2] for any integer $s > 0$ a non-commutative central limit theorem was established. For $s = 1$ the theorem was the analogue to the weak law of large numbers, for $s = 2$ it was the analogue to the usual central limit theorem. For $s > 2$ there does not exist a classical analogue because the vanishing of all the first and second order moments implies that for probability measures the theorem is trivial. The proof in [2] was based on combinatorial considerations. We establish a new proof by the use of moment generating functions. The proof uses similar ideas as [3].

Consider a set of non-commuting indeterminates and the free complex algebra $\mathfrak{F} = \mathbb{C}\langle x_1, \dots, x_n \rangle$ generated by them. $\mathfrak{F}$ consists of the complex linear combinations of 1 and of monomials of the form $x_{i_1} \dots x_{i_k}$. Remark that the order of factors is important, e.g., $x_i x_j \neq x_j x_i$ if $i \neq j$.

Assume a s-dimensional matrix $Q = ((Q_{i_1 \dots i_s}))$. The Gaussian functional γ_Q on $\mathfrak{F}$ is a linear function $\mathfrak{F} \to \mathbb{C}$ given by

$$\gamma_Q(1) = 1$$

$$\gamma_Q(x_{i_1} \cdots x_{i_k}) = 0 \quad \text{if } k \text{ is not a multiple of } s$$

$$\gamma_Q(x_{i_1} \cdots x_{i_s}) = Q_{i_1 \dots i_s}$$

$$\gamma_Q(x_{i_1} \dots x_{i_{ps}}) = \sum_{\{S_1, \dots, S_p\}} Q_{i_{S_1}} \cdots Q_{i_{S_p}}$$

where $\{S_1, \dots S_p\}$ runs through all partitions of $\{1, \dots, ps\}$ into p sets with exactly s elements and where

$$Q_{i_S} = Q_{i_{j_1} \dots i_{j_s}}$$

if $S = \{j_1, \dots, j_s\}$, $j_1 < \dots < j_s$.

Let $\mathcal{A}$ be an algebra over $\mathbb{C}$ with unit 1. Let $(a_i)_{i \in I}$ be a family of elements of $\mathcal{A}$ indexed by the same set as the indeterminates of $\mathcal{F}$ and let $\pi : \mathcal{A} \to \mathbb{C}$ be a linear functional. Consider the tensor algebra $\mathcal{A}^{\otimes N}$ and the linear functional $\pi^{\otimes N}$ on $\mathcal{A}^{\otimes N}$ given by

$$\pi^{\otimes N}(f_1 \otimes \dots \otimes f_N) = \pi(f_1) \cdots \pi(f_N).$$

Define the imbeddings

$$u_1 : \mathcal{A} \to \mathcal{A}^{\otimes N}, \quad g \mapsto g \otimes 1 \otimes \dots \otimes 1; \quad \dots$$

$$u_N : \mathcal{A} \to \mathcal{A}^{\otimes N}, \quad g \mapsto 1 \otimes \dots \otimes 1 \otimes g.$$

<u>Central limit theorem of degree s:</u> Assume $\pi(1) = 1$ and $\pi(a_{i_1} \cdots a_{i_\ell}) = 0$ for all $i_1 \in I, \dots, i_\ell \in I$ and $1 \leq \ell \leq s-1$. Let $f \in \mathcal{F}$. Then for $N \to \infty$

$$\pi^{\otimes N}\left(f(x_i \longmapsto (u_1(a_i) + \dots + u_N(a_i))N^{-1/s}; i \in I)\right)$$

converges to

$$\gamma_Q(f) \quad \text{with Q given by} \quad Q_{i_1 \dots i_s} = \pi(a_{i_1} \cdots a_{i_s}).$$

<u>Remark:</u> Without loss of generality we may assume that $\mathcal{A} = \mathcal{F}$ and $a_i = x_i$. For, if this is not the case, consider the canonical homomorphism $\eta : \mathcal{F} \to \mathcal{A}$, $x_i \mapsto a_i$ and the linear functional $\pi \circ \eta$ on $\mathcal{F}$.

The free algebra $\mathcal{F}$ is a bialgebra. The coproduct is the homomorphism

$$\Delta : \mathcal{F} \to \mathcal{F} \otimes \mathcal{F}$$

$$x_i \longmapsto x_i \otimes 1 + 1 \otimes x_i$$

and the counit is the homomorphism

$$\delta : \mathfrak{f} \longrightarrow \mathbb{C}$$
$$x_i \longmapsto 0$$

for $i \in I$. The N-th iterate Δ_N of Δ is given by

$$\Delta_N : \mathfrak{f} \longrightarrow \mathfrak{f}^{\otimes N}$$

$$x_i \longmapsto x_i \otimes 1 \otimes \cdots \otimes 1 + \cdots + 1 \otimes 1 \otimes \cdots \otimes 1 \otimes x_i$$
$$= \sum_{j=1}^{N} u_j(x_i).$$

Define the homomorphism $\alpha(t) : \mathfrak{f} \to \mathfrak{f}$ by $\alpha(t) x_i = t x_i$, $t \in \mathbb{C}$.

Assuming as mentioned above $\mathfrak{a} = \mathfrak{f}$ and $x_i = a_i$ the central limit theorem may be formulated

(1) $$\pi^{\otimes N} \circ \Delta_N \circ \alpha(N^{-1/s}) \longrightarrow \gamma_Q .$$

Let W be a monomial $W = x_{i_1} \cdots x_{i_k}$, then

(2) $$\Delta_p W = \sum_{(S_1, \dots, S_p)} x_{i_{S_1}} \otimes \cdots \otimes x_{i_{S_p}}$$

where $(S_1, \dots, S_p)$ runs through all <u>sequences</u> of subsets $S_1, \dots, S_p \subset \{1, \dots, k\}$ such that $S_\ell \cap S_m = \emptyset$ for $\ell \neq m$ and $\bigcup S_\ell = \{1, \dots, k\}$.

Put

$$x_{i_\emptyset} = 1$$

and

$$x_{i_S} = x_{i_{j_1}} \cdots x_{i_{j_r}}$$

if

$$S = \{j_1, \dots, j_r\}, \quad j_1 < \cdots < j_r.$$

The convolution $\mu * \nu$ of two linear functionals on $\mathcal{F}$ is given by

$$\mu * \nu = (\mu \otimes \nu) \circ \Delta .$$

Define the linear functional δ_Q on $\mathcal{F}$ by

$$\delta_Q (X_{i_1} \cdots X_{i_s}) = Q_{i_1 \cdots i_s}$$

$\delta_Q(W) = 0$ if $W = 1$ or if W is a monomial of degree $\neq s$.

Then

$$\gamma_Q = \exp_* \delta_Q = \delta + \frac{1}{1!} \delta_Q + \frac{1}{2!} \delta_Q * \delta_Q + \cdots \tag{3}$$

For by (2)

$$\frac{1}{p!} \delta_Q^{*p}(W) = \frac{1}{p!} (\delta_Q^{\otimes p} \circ \Delta_p)(W)$$

$$= \frac{1}{p!} \sum_{(S_1, \ldots, S_p)} \delta_Q (X_{i_{S_1}}) \cdots \delta_Q (X_{i_{S_p}})$$

This expression vanishes if the degree of W is not equal to ps. If the degree of W is equal to ps then only those $(S_1, \ldots, S_p)$ give a contribution where all S_i have s elements. There are p! different sequences $(S_1, \ldots, S_p)$ determining the same partition $\{S_1, \ldots, S_p\}$. So finally if the degree of W is equal to ps

$$\frac{1}{p!} \delta_Q^{*p}(W) = \sum_{\{S_1, \ldots, S_p\}} Q_{i_{S_1}} \cdots Q_{i_{S_p}}$$

In classical probability theory the function

$$(t_1, \ldots, t_n) \longmapsto \mathbb{E}\, e^{i(t_1 X_1 + \cdots + t_n X_n)}$$

determines all the moments of the stochastic variables $X_1, \ldots, X_n$ provided the moments exist. As pointed out in [2] the theory presen ed here is a theory of moments. One is tempted to try

$$f : (t_1, \dots, t_n) \longmapsto \pi\left(e^{i(t_1 x_1 + \dots + t_n x_n)}\right)$$

as generating function for all the moments, i.e. the expressions

$$\pi\left(x_{i_1} \cdots x_{i_k}\right)$$

This, however, is not convenient. Firstly, because no convergence assumptions have been made and secondly, because, even when everything converges, we can hope to get out of f only expressions with are symmetric in the x_i, e.g., $x_i x_j + x_j x_i$, but not $x_i x_j$ alone if $i \neq j$.

A generating function f will be a formal power series in $\mathfrak{f}[[t]]$, i.e.

$$f = f_0 + f_1 t + f_2 t^2 + \cdots \quad , \; f_i \in \mathfrak{f}$$

(we assume t to commute with the elements of $\mathfrak{f}$) of the form

$$f = e^{t y_1} \cdots e^{t y_k}$$

where $y_1, \dots, y_k$ are <u>linear</u> combinations of the x_i. It is clear that the functional π is determined by all the power series of the form

$$\pi\left(e^{t y_1} \cdots e^{t y_k}\right)$$

No convergence problems arise because of the introduction of the indeterminate t.

Let

$$f = e^{t y_1} \cdots e^{t y_k}$$

then

(4) $$\Delta f = f \otimes f$$

Hence by (3)

(5) $$\gamma_Q(f) = \exp \delta_Q(f).$$

Going back to (1) we obtain

$$\left(\pi^{\otimes N} \circ \Delta_N \circ \alpha(N^{-1/s})\right)(f) = \pi^{\otimes N}\left((\alpha(N^{-1/s})f)^{\otimes N}\right)$$
$$= \left(\pi\left(\alpha(N^{-1/s})f\right)\right)^N$$

We have

$$f = 1 + t f_1 + t^2 f_2 + \cdots$$

where f_n is a homogeneous polynomial of degree n. So

$$\alpha(N^{-1/s})f = 1 + N^{-1/s} t f_1 + N^{-2/s} t^2 f_2 + \cdots .$$

Taking into account the assumption of the theorem

$$\pi(\alpha(N^{-1/s}f)) = 1 + \frac{t^s}{N}\pi(f_s) + \frac{t^{s+1}}{N^{1+1/s}}\pi(f_{s+1}) + \cdots$$

Hence

(6) $$\left(\pi^{\otimes N} \circ \Delta_N \circ \alpha(N^{-1/s})\right)(f) \longrightarrow \exp t^s \pi(f_s)$$

As

$$\delta_Q(f) = t^s \pi(f_s)$$

equations (5) and (6) prove (1).

Literature

[1] N. Bourbaki, Eléments des Mathématiques, Algèbre, Chap. III, Hermann, Paris 1970.

[2] N. Giri and W. v. Waldenfels, An algebraic version of the central limit theorem. Z. Wahrscheinlichkeitstheorie verw. Gebiete 42, 129-134 (1978).

[3] M. Schürmann, Positiv definite Funktionen auf der freien Lie-Gruppe und stabile Grenzverteilungen. Diplomarbeit Heidelberg 1979.

AVERAGING AND FLUCTUATIONS OF CERTAIN STOCHASTIC EQUATIONS

Hisao Watanabe
Department of Applied Science
Faculty of Engineering
Kyushu University 36
Fukuoka 812, Japan

§ 1. Introduction

Let us consider ordinary differential equations with parameter ε

$$\frac{dx^{\varepsilon}}{dt} = \varepsilon F(t,x^{\varepsilon}), \qquad \varepsilon > 0. \tag{1.1}$$

In the study of asymptotic behavior of trajectories of (1.1), the averaging method was introduced by Krylov and Bogolubov.

Let $F(t,x)$ be a real valued function of $(x,t)\in[0,\infty)\times R^d$, uniformly bounded, and satisfied a Lipschitz condition in x, with constant independent of t and x. If the limit

$$\lim_{T\to\infty} \frac{1}{T}\int_0^T F(x,t)dt = \overline{F}(x)$$

exist uniformly $x\in R^d$, then the trajectory of (1.1) is in some neighborhood of the trajectory of

$$\frac{dx^0(t)}{dt} = \varepsilon\overline{F}(x^0(t)) \tag{1.2}$$

on $0 \leq t \leq L/\varepsilon$, where L is an arbitrary constant. Also if $x^{\varepsilon}(0) - x^0(0) = 0(\varepsilon)$, then it holds

$$x^{\varepsilon}(t) - x^0(t) = 0(\varepsilon), \qquad 0 \leqq t \leqq L/\varepsilon.$$

The assertion that the trajectory $x^{\varepsilon}(t)$ is close to $x^0(t)$ is called the averaging principle. In this paper, we discuss the averaging principle in some stochastic equations.

§ 2. Stochastic ordinary differential equations

If the equation (1.1) is disturbed by noises, we describe $F(t,x)$ as a random function. By making the change of variable, we can write (1.1) as follows:

$$\frac{dx^{\varepsilon}(t)}{dt} = F(\frac{t}{\varepsilon}, x^{\varepsilon}(t), \omega) \tag{2.1}$$

where $\omega \in \Omega$ and $(\Omega, \mathcal{B}, P)$ is a probability space. The first results about averaging principle for (2.1) is obtained by Khas'minskii. He assumed that F satisfies uniform Lipschitz conditions; namely, for any $x_1, x_2 \in R^d$,

$$|F(t,x_1,\omega) - F(t,x_2,\omega)| \leq K|x_1 - x_2|, \qquad \text{for any} \quad t > 0,$$

where K is independent of t, x, ω. Also, he assumed that $F(s,x,\infty)$ is integrable in finite interval with respect to s, with probability one. If

$$\sup_{t_0>0} E\left|\frac{1}{T}\int_{t_0}^{t_0+T} F(t,x,\omega)dt - \bar{F}(x)\right| \to 0, \qquad (T \to \infty)$$

for any $x \in R^d$, then he showed that $x^{\varepsilon}(t)$ converges in the mean to $x^0(t)$, uniformly for $0 \leq t \leq L$ as $\varepsilon \to 0$, where $x^0(t)$ is the solution of the equation

$$(2.2) \qquad \frac{dx^0(t)}{dt} = \bar{F}(x^0(t)).$$

Furthermore, under the more restrictive conditions he showed that $(x^{\varepsilon}(t) - x^0(t))/\sqrt{\varepsilon}$ converges weakly to a Gaussian process $y^0(t)$. Roughly speaking, it implies that $x^{\varepsilon}(t)$ can be approximated by $x^0(t) + \sqrt{\varepsilon}\, y^0(t)$. After Khas'minskii [6], there appeared several papers, namely, Brodskiy and Lakacher [2], Geman [4], and Kushner [8].

Consider the following example,

$$\frac{dx_{\lambda}(t)}{dt} = -(ae^{-\xi_{\lambda}(t)} + be^{\xi_{\lambda}(t)})x_{\lambda}(t) + be^{\xi_{\lambda}(t)},$$

where $\xi_{\lambda}(t)$ is a stationary Gaussian processes with means $E(\xi_{\lambda}(t)) = 0$ and covariance $E(\xi_{\lambda}(t)\xi_{\lambda}(0)) = e^{-\lambda|t|}$, $a > 0$ and $b > 0$. Now if we put $a(t,\omega) = -(ae^{-\xi_1(t)} + be^{\xi_1(t)})$, $b(t,\omega) = be^{\xi_1(t)}$ and $F(t,x,\omega) = a(t,\omega)x + b(t,\omega)$, by making use of the relation $\xi_{\lambda}(t) = \xi_1(\lambda t)$, we have the equation:

$$\frac{dx_{\lambda}(t)}{dt} = F(\lambda t, x_{\lambda}(t), \omega).$$

Put $\lambda = 1/\varepsilon$ and $x_{\lambda}(t) = x^{\varepsilon}(t)$, then we have

$$\frac{dx^{\varepsilon}(t)}{dt} = F(\frac{t}{\varepsilon}, x^{\varepsilon}(t), \infty).$$

Here, $F(t,x,\omega)$ does not satisfy the condition of Khas'minskii which require the Lipschitz constant independent of t, x, ω. Brodskiy and Lakacher, Geman and Kushner assume mixing conditions which is not satisfied by $\xi_{\lambda}(t)$. In H. Watanabe [7], the author proposed the new result which covers the above example.

For convenience, we state the main result. We assume:

(A.I) (2.1) has a unique bounded solution (bounds may depend on $\omega \in \Omega$) on each interval $[0,T]$ almost surely, and (2.2) has a unique bounded solution on $[0,\infty)$.

(A.II) $F(t,x,\omega)$ is jointly measurable in all argument $t \geq 0$, $x \in R^d$, $\omega \in \Omega$, and $C^2(R^d)$ as a function of x almost surely.

(A.III) The process $\{F(t,x,\omega)\}$ is stationary in t, for each fixed $x \in R^d$.

(A.IV) Let $\mathcal{M}_s^t(M) = \sigma\{F(u,x,\omega) \mid s \le u \le t,\ |x| \le M\}$ and

$$\beta(t,M) = \sup_{s\ge 0} \sup_{A\in\mathcal{M}_0^s(M),\, B\in\mathcal{M}_{s+t}^\infty(M)} |P(A\cap B) - P(A)P(B)|.$$

Then for some $\delta > 0$,

$$\int_0^\infty \beta(t,M)^{\delta/(4+2\delta)} dt < \infty$$

for each $M > 0$.

(A.V) For each $M < \infty$, there exists a constant C independent of t.

a) $E(\sup_{|x|\le M} |D^\beta F(t,x,\omega)|^{(4+2\delta)}) \le C, \qquad 0 \le |\beta| \le 1,$

b) $E(\sup_{|x|\le M} |D^\beta F(t,x,\omega)|^2) \le C, \qquad |\beta| = 2,$

where $\beta = (\beta_1, \beta_2, \ldots, \beta_d)$, β_i is a non-negative integer, and $|\beta| = \beta_1 + \beta_2 + \ldots \beta_d$.

Theorem 2.1. We assume that (A.I)-(A.V). Let R^ε be the measure on $C([0,\infty);R^d)$ induced by the stochastic process $\{x^\varepsilon(t)\}$ and R be the measure on $C([0,\infty);R^d)$ induced by the stochastic process $y^0(t)$ which is the unique solution of the following stochastic differential equation:

$$y_i^0(t) = w_i(t) + \int_0^t \sum_{j=1}^d \frac{\partial \bar{F}_i}{\partial x_j}(x^0(\tau))y_j^0(\tau)d\tau, \qquad (i = 1,2,\ldots,d).$$

where $\bar{F}_i(x) = E(F_i(t,x,\omega))$ and $\{w(t)\}$ is a Gaussian process with independent increments. Then R^ε converges weakly to R on $C([0,\infty);R^d)$.

§ 3. Averaging principle for parabolic equation

In this section, we are concerned with the following parabolic equation with rapidly oscillating random coefficients which is the natural extension to the partial differential equation of (2.1).

$$(3.1)\quad \begin{cases} \dfrac{\partial u^\varepsilon(t,x)}{\partial t} = \dfrac{1}{2}\sum_{i,j=1}^d a_{ij}(\frac{t}{\varepsilon},x,\omega)\dfrac{\partial^2 u(t,x)}{\partial x_i \partial x_j} \\ \qquad\qquad + \sum_{i=1}^d b_i(\frac{t}{\varepsilon},x,\omega)\dfrac{\partial u^\varepsilon(t,x)}{\partial x_i} \\ u^\varepsilon(0,x) = f(x). \end{cases}$$

The corresponding averaging equation to (3.1) is the following.

$$(3.2)\quad \begin{cases} \dfrac{\partial u^0(t,x)}{\partial t} = \dfrac{1}{2}\sum_{i,j=1}^d \bar{a}_{ij}(x)\dfrac{\partial^2 u^0(t,x)}{\partial x_i \partial x_j} + \sum_{i=1}^d \bar{b}_i(x)\dfrac{\partial u^0(t,x)}{\partial x_i} \\ u^0(0,x) = f(x), \end{cases}$$

where $\overline{a}_{ij}(x) = E(a_{ij}(t,x,\omega))$ and $\overline{b}_i(x) = E(b_i(t,x,\omega))$.

We claim that $u^{\varepsilon}(t,x)$ is close to $u^0(t,x)$ as ε tends to zero. Averaging principle in parabolic equation is considered by many authors. Here, we cite Khas'minskii [5], Bensoussan, Lions and Papanicolaou [1] and Zhikov, Kozlov and Oleinik [8]. Our methods is close to Bensoussan, Lions and Papanicolaou [1] which is concerned with parabolic equation with rapidly oscillating periodic coefficients. Khas'minskii [5] and Zhikov, Kozlov and Oleinik [8] base on analytic methods.

To describe the statements, we introduce the following assumptions. Let $(\Omega, \mathcal{B}, P)$ be a probability space.

(B.I) $a_{ij}(t,x,\omega)$, $b_i(t,x,\omega)$ $(i,j=1,\ldots,d)$ are measurable function as a function of $(t,x,\omega) \in [0,\infty) \times R^d \times \Omega$.

(B.II) There exist constants λ, μ (independent of t, x, ω) such that $\lambda \geq \mu > 0$ and it holds if $t \in [0,\infty)$, $x \in R^d$ and $\xi \in R^d$

$$\mu|\xi|^2 \leq \sum_{i,j=1}^{d} a_{ij}(t,x,\omega)\xi_i\xi_j \leq \lambda|\xi|^2$$

with probability one.

(B.III) $D^{\beta}a_{ij}(t,x,\omega)$ and $D^{\beta}b_i(t,x,\omega)$ $(|\beta| \leq 6)$ are bounded and continuous with respect to (t,x) uniformly $x \in R^d$.

(B.VI) $\{a_{ij}(t,x,\omega)\}$, $\{b_i(t,x,\omega)\}$ $(i,j=1,\ldots,d)$ are stationary correlated stochastic processes, for each fixed $x \in R^d$.

(B.V) Let $\mathcal{M}_s^t = \sigma\{a_{ij}(t,x,\omega), b_i(t,x,\omega)$ $(i,j=1,\ldots,d | s \leq u \leq t,\ x \in R^d\}$ and

$$\beta(t) = \sup_{s \geq 0} \sup_{A \in \mathcal{M}_0^s, B \in \mathcal{M}_{s+t}^{\infty}} |P(A \cap B) - P(A)P(B)|.$$

Then

$$\int_0^{\infty} \beta^{1/4}(t)dt < \infty.$$

(B.VI) For $\varepsilon \geq 0$, $u^{\varepsilon}(0,x) = f(x)$ is in $C^6(R^d) = \{f; D^{\alpha}f(x)$ exists and $\sup_{x \in R^d}|D^{\alpha}f(x)| < \infty$ for $|\alpha| \leq 6\}$.

<u>Theorem 3.1.</u> Under the assumptions (B.I)-(B.VI), we have

$$\lim_{\varepsilon \to 0} \sup_{x \in R^d} E(|u^{\varepsilon}(t,x) - u^0(t,x)|^2) = 0 \quad \text{for each} \quad t \geq 0.$$

Furthermore, we introduce the following condition.

(B.VII) Equation (3.1) can be written in the following way.

$$D_t u^{\varepsilon}(t,x) = \frac{1}{2}\sum_{ij} D_{x_j}(a_{ij}(\tfrac{t}{\varepsilon},x,\omega)D_{x_i})u^{\varepsilon}(t,x).$$

Then, we have

<u>Theorem 3.2.</u> Under the assumptions (B.I)-(B.VII) for each fixed $0 \leq s \leq t$

and $x,y \in R^d$, we have

$$\lim_{\varepsilon \to 0} E((p^{\varepsilon}(t,x;s,y) - p^0(t,x;s,y))^2) = 0,$$

where $p^{\varepsilon}(t,x;s,y)$ and $p^0(t,x;s,y)$ are fundamental solutions of (3.1) and (3.2), respectively.

§ 4. Fluctuation for parabolic equation with averaging

In this section, we state the central limit theorem for parabolic equation which is analogous to Theorem 2.1 in § 2. Let $u^{\varepsilon}(t,x)$ and $u^0(t,x)$ be the solution of parabolic Cauchy problems (3.1) and (3.2), respectively. We consider the fluctuation process

$$y^{\varepsilon}(t,x) := \frac{u^{\varepsilon}(t,x) - u^0(t,x)}{\sqrt{\varepsilon}}.$$

By definition, $y^{\varepsilon}(t,x)$ satisfies the equation:

$$\begin{cases} \dfrac{\partial y^{\varepsilon}(t,x)}{\partial t} = A^{\varepsilon,\omega}_{t,x} y^{\varepsilon}(t,x) + \dfrac{1}{\sqrt{\varepsilon}} F^{\varepsilon}(t,x,\omega) & \text{for } 0 < t < \infty \\ y^{\varepsilon}(0,x) = 0 & \text{for } x \in R^d, \end{cases}$$

where

$$A^{\varepsilon,\omega}_{t,x} = \frac{1}{2}\sum_{i,j=1}^{d} a_{ij}(\tfrac{t}{\varepsilon},x,\omega)\frac{\partial^2}{\partial x_i \partial x_j} + \sum_{i=1}^{d} b_i(\tfrac{t}{\varepsilon},x,\omega)\frac{\partial}{\partial x_i},$$

$$\bar{A}_x = \frac{1}{2}\sum_{i,j=1}^{d} \bar{a}_{ij}(x)\frac{\partial^2}{\partial x_i \partial x_j} + \sum_{i=1}^{d} \bar{b}_i(x)\frac{\partial}{\partial x_i},$$

and

$$F^{\varepsilon}(t,x,\omega) = (A^{\varepsilon,\omega}_{t,x} - \bar{A}_x)u^0(t,x).$$

Therefore, $y^{\varepsilon}(t,x)$ can be expressed in the following way (cf. Eidel'man [3])

$$y^{\varepsilon}(t,x) = \frac{1}{\sqrt{\varepsilon}}\int_0^t ds \int_{R^d} p^{\varepsilon,\omega}(t,x;s,y)F^{\varepsilon}(s,y,\omega)dy.$$

Now we split $y^{\varepsilon}(t,x)$ into two parts

$$y^{\varepsilon}(t,x) = z^{\varepsilon}(t,x) + w^{\varepsilon}(t,x),$$

where

$$z^{\varepsilon}(t,x) = \frac{1}{\sqrt{\varepsilon}}\int_0^t ds \int_{R^d} p^0(t,x;s,y)F^{\varepsilon}(s,y,\omega)dy$$

and

$$w^{\varepsilon}(t,x) = \frac{1}{\sqrt{\varepsilon}}\int_{R^d} (p^{\varepsilon,\omega}(t,x;s,y) - p^0(t,x;s,y))F^{\varepsilon}(s,y,\omega)dy.$$

Let $\phi \in \mathcal{S}(R^d)$ = {the space of rapidly decreasing functions}. We define

$$z_\phi^\varepsilon(t) = \int_{R^d} z^\varepsilon(t,x)\phi(x)dx \qquad \text{for} \quad \phi \in L^1(R^d).$$

Lemma 4.1. For $|\alpha| \leq 2$

$$\sup_{\substack{x\in R^d \\ \varepsilon>0 \\ 0\leq s\leq t}} E(|D^\alpha z^\varepsilon(s,x)|^2) < \infty .$$

By definition, we can easily show that $z_\phi^\varepsilon(t)$ satisfies the following equation

$$\frac{dz_\phi^\varepsilon(t)}{dt} = z_{\bar{A}*}(t) + \frac{1}{\sqrt{\varepsilon}} F_\phi(t),$$

where

$$\bar{A}^* = \frac{1}{2}\sum_{i,j=1}^d \frac{\partial^2}{\partial x_i \partial x_j}(\bar{a}_{ij}(x)\phi(x)) - \sum_{i=1}^d \frac{\partial}{\partial x_i}(\bar{b}_i(x)\phi(x))$$

and

$$F_\phi(t) = \int_{R^d} F(t,x,\omega)\phi(x)dx.$$

We can obtain the following lemma.

Lemma 4.2. There exists constant C such that

$$E(|z_\phi^\varepsilon(s_1) - z_\phi^\varepsilon(s_2)|^4) \leq C|s_1 - s_2|^2$$

for any s_1 and s_2 with $0 \leq s_1 \leq s_2 \leq t$.

Therefore, by noting that $P(z_\phi^\varepsilon(0) = 0) = 1$, we can see that a family of probability measures Q^ε which is induced by stochastic processes $\{z_\phi^\varepsilon(t)\}$ $\phi \in \mathcal{S}$ on $C([0,\infty), \mathcal{S}'(R^d))$ is tight. Furthermore, we can show the following proposition.

Proposition 4.3. Any limiting measure of Q^ε is the solution of the following martingale problem; for any $\phi \in \mathcal{S}$,

(4.1) $$f(z_\phi(s)) - \int_0^s \frac{1}{2} f''(z_\phi(\sigma))B(\phi,\phi)_\sigma d\sigma - \int_0^s f'(z_\phi(\sigma))z_{\bar{A}^*\phi}(\sigma)d\sigma$$

is martingale,

where

$$\frac{1}{2}B(\phi,\phi)_\sigma = \sum_{k,l,m,n} B^{a,a}_{k,l,m,n}(\phi,\phi)_\sigma + \sum_{k,l,m} B^{a,b}_{k,l,m}(\phi,\phi)_\sigma$$

$$+ \sum_{k,m,n} B^{b,a}_{k,m,n}(\phi,\phi)_\sigma + \sum_{k,m} B^{b,b}_{k,m}(\phi,\phi)_\sigma$$

$$B^{a,a}_{k,l,m,n}(\phi,\phi)_\sigma = \int_{R^d}\int_{R^d} \rho^{a,a}_{k,l,m,n}(x,y)\frac{\partial^2 u^0}{\partial x_k \partial x_l}(\sigma,x)\phi(x) \times \frac{\partial^2 u^0}{\partial y_m \partial y_n}(\sigma,y)\phi(y)dxdy,$$

$$\rho^{a,a}_{k,l,m,n}(x,y) = \int_0^\infty E[(a_{kl}(t,x,\omega) - \bar{a}_{kl}(x)) \times (a_{m,n}(0,y,\omega) - \bar{a}_{m,n}(y))]\,dt,$$

and other quantities are defined similarly.

The next stage is to show that $w^\varepsilon_\phi(t)$ converges in probability to zero as ε tends to zero.

By definition, we can see that $w^\varepsilon(t,x)$ satisfies the following equation.

$$\frac{\partial w^\varepsilon(t,x)}{\partial t} = A^{\varepsilon,\omega}_{t,x}\, w^\varepsilon(t,x) + G^\varepsilon(t,x),$$

where

$$G^\varepsilon(t,x) = (A^{\varepsilon,\omega}_{t,x} - \bar{A}_x) z^\varepsilon(t,x).$$

Therefore $w^\varepsilon_\phi(t)$ can be written in the following manner.

$$w^\varepsilon_\phi(t) = \int_0^t ds \int_{R^d} \phi(x)dx \int_{R^d} (p^{\varepsilon,\omega}(t,x;s,y) - p^0(t,x;s,y)) \times G^\varepsilon(s,y)dy$$

$$+ \int_0^t ds \int_{R^d} \phi(x)dx \int_{R^d} p^0(t,x;s,y) G^\varepsilon(s,y)dy$$

$$= K^\varepsilon_1 + K^\varepsilon_2.$$

Since the coefficients of differential operator $A^{\varepsilon,\omega}_{t,x}$ and $\bar{A}_x$ are bounded, by using of Theorem 3.2 and Lemma 4.1, we can see that

$$E(|K^\varepsilon_1|) \to 0 \qquad (\varepsilon \to 0).$$

For K_2, by making use of the mixing properties and the useful estimate of Eidel'man (pp. 101-102 in [3]), we can show that

$$E(|K^\varepsilon_2|^2) \to 0 \qquad (\varepsilon \to 0).$$

If $\bar{A} = \bar{A}^*$, the martingale problem (4.1) has a unique solution and it is given by the measure determined by the following process:

$$z_\phi(t) = \int_0^t e^{\bar{A}(t-s)} dL_\phi(s), \tag{4.2}$$

where $L_\phi(t)$ is a Gaussian process uniquely determined by the relation

$$E(e^{iL_\phi(x)}) = \exp(-\frac{1}{2}\int_0^t B(\phi,\phi)_\sigma d\sigma).$$

In summarizing the above disucssion, we obtain the following theorem.

<u>Theorem 4.4.</u> If the conditions (B.I)-(B.VI) are satisfied, then $y^\varepsilon_\phi(t)$

converges weakly to the stochastic process $z_\phi(t)$ of (4.2) on $C([0,\infty);\mathcal{S}')$.

References

[1] A. Bensoussan, J.L. Lions and G. Papanicolaou, Asymptotic analysis for periodic structures, 1978, 516-533, North-Holland, Amsterdam.

[2] Ya.S. Brodskiy and B.Ya. Lakacher, Fluctuation in averaging scheme for differential equations with random right hand side, (Russian) Theory of random process, 12 (1984), 8-17.

[3] S.D. Eidel'man, Parabolic system, 1969, North-Holland, Amsterdam.

[4] S.Geman, A method of averaging for random differential equations with applications to stability and stochastic approximations, in Approximate solution of random equations edited by Bharucha-Reid, 1979, 37-85, North-Holland, New York.

[5] R.Z. Khas'minskii, Principle of averaging for parabolic and elliptic differential equations and for Markov processes with small diffusion, Theory of probab. and its appl. 8 (1963), 1-21.

[6] R.Z. Khas'minskii, On the stochastic processes defined by differential equations with a small parameter, Theory of probab. and its appl. 11 (1966), 211-288.

[7] H. Watanabe, Fluctuations in certain dynamical systems with averaging, to appear.

[8] V.V. Zhikov, S,M. Kozlov and O.A. Oleinik, Averaging of parabolic operators, Trans. Moscow Math. Soc. 45 (1982), 189-241.

SEMIMARTINGALE WITH SMOOTH DENSITY

— the problem of "nodes"

ZHENG Weian
Department of Mathematical Statistics
East-China Normal University
Shanghai, China

Given a continuous process $X_t(\omega)$ having smooth density function $p(x,t)$ with respect to the Lebesgue's measure $d\mu$. In which case the following equality holds?

(1) $$P[\inf_{s<t} p(X_s,s)>0]=1 \quad , \quad \forall t\in R_+ \quad .$$

It is a problem raised in the stochastic mechanics. Some authors have already obtained some results in the case where X is a diffusion process (see Nagasawa[1], Nelson[1], Meyer and Zheng[1]).

In this paper, we will prove (1) for a class of semimartingales of type (5), which appeared often in stochastic mechanics. Furthermore, we will prove also the following fact: given $r\in R^d$ $(d\geq 3)$, if $X_t(\omega)$ is a semimartingale of type (5), having a continuous density function, then X never reach r in finite time a.s., which is a well known property for the brownian motion.

We introduce at first a lemma. Suppose X is a continuous process defined on R_+ and taking values in R^d. Suppose X has a continuous density function $p(x,t)$. We call a closed set $A\subset R^d\times R_+$ "negligible for X" if there is a sequence of positive numbers $\varepsilon_m\downarrow 0$ such that

(2) $$\int I_{A(\varepsilon_m,k)}(x,t)p(x,t)d\mu dt=o(\varepsilon_m^{\,2}), \quad \forall \text{fixed } k\in R_+,$$

where $A(\varepsilon_m,k)=\{(x,t);(x,t)\in R^d\times R_+,\ \gamma(0,(x,t))<k, \text{and } \gamma((x,t),A)\leq\varepsilon_m\}$(if A and B are two closed sets in $R^d\times R_+$, we denote by $\gamma(A,B)$ the Euclidean distance between A and B).

LEMMA 1. Given a filtration space $(\Omega,\underline{F},F_t,P)$. Suppose that X is a (F_t)-adapted continuous process such that there are a sequence of stopping times $K_n\uparrow\infty$ and a sequence of constants C_n such that for any pair of stopping times $S\leq T\leq K_n$,

(3) $$E|X_T-X_S|^2\leq C_nE|T-S| \quad .$$

Suppose X has a continuous density function $p(x,y)$.

Then, for every negligible closed set A, we have

(4) $$\sup_{t\in R_+} I_A(X_t,t)=0 \text{ a.s.}$$

PROOF. Without loosing generality, we suppose $K_n \leq n$. We define

$$S^n(\omega)=\inf\left\{t,(X_t(\omega),t)\in A, \gamma(0,(X_t(\omega),t))<n\right\}\wedge K_n.$$

If (4) is false, then we have $P[S^n<K_n]>0$ when n is sufficiently large. we fix such a n and define also

$$T^n_\varepsilon(\omega)=\inf\left\{t\geq S^n(\omega), \gamma((X_t,t),A)>\varepsilon,\ \gamma(0,(X_t(\omega),t))<n\right\}\wedge K_n .$$

There is $\hat{\varepsilon}>0$ such that $P[S^n<T^n_\varepsilon<K_n]>\hat{\varepsilon}>0$ for $\forall\varepsilon<\hat{\varepsilon}$, because the continuity of X_t and the fact $\int I_A(X_t,t)dt=0$ a.s., which is a corollary to (2).

Thus, from the hypotheses of the theorem, we choice a sequence $\varepsilon_m\downarrow 0$ as in (2), when $\varepsilon_m<\hat{\varepsilon}$,

$$\hat{\varepsilon}\,\varepsilon_m^2 = E[(|X_{T^n_{\varepsilon_m}}-X_{S^n}|^2+|T^n_{\varepsilon_m}-S^n|^2)] \leq C_n E[T^n_{\varepsilon_m}-S^n]+nE[T^n_{\varepsilon_m}-S^n]$$

$$=(C_n+n)E[T^n_{\varepsilon_m}-S^n]\leq(C_n+n)E\left[\int I_{A(\varepsilon_m,n)}(X_t,t)dt\right]$$

$$=(C_n+n)\left[\int I_{A(\varepsilon_m,n)}(x,t)p(x,t)d\mu dt\right]=o(\varepsilon_m^2) .$$

It is impossible. Thus, we prove (4).

We introduce an another lemma as the following:

LEMMA 2. Let X_t be a (F_t)-semimartingale of dimension d such that

(5) $$X^i_t=X^i_0+\int_0^t H^i_s ds+M^i_t ,\quad 0\leq i\leq d ,$$

where H^i_s be a (F_t)-adapted process such that $E\int_0^{T_n}|H^i_s|^2ds<\infty$, and M^i_t be a continuous local (F_t)-martingale such that $\langle M^i,M^i\rangle=\int_0^t Q^i_s ds$ and $\sup_{s\leq T_n}\sum_i|Q^i_s|\leq n$, where $T_n\uparrow\infty$ is a sequence of (F_t)-stopping times. Then, there is a sequence of stopping times (K_n) satisfying (3).

PROOF. Define

$$K_n(\omega)=\inf\left\{t;\ \sum_i\int_0^t|H^i_s|^2ds>n\right\}\wedge T_n,$$

then we can easily verify (3). In fact, from Schwarz's inequality, if $S\leq T\leq K_n$,

$$\left|\int_S^T H^i_s ds\right|^2\leq(T-S)\left(\int_S^T|H^i_s|^2ds\right)\leq n(T-S),\quad \int_S^T Q^i_s ds\leq n(T-S) .$$

But, according to Davis' inequality, we have

$$\sum_i E|M^i_T\ M^i_S|^2\leq\sum_i E\int_S^T Q^i_s ds\leq nE[T-S] .$$

Thus, there is a constant C_n such that for $S \leq T \leq K_n$,

$$E|X_T - X_S|^2 = \sum_i E|\int_S^T H_s^i ds|^2 + \sum_i E|M_T - M_S|^2 \leq C_n E|T-S| .$$

In stochastic mechanics, we consider the diffusion processes of the type

(6) $$dX_t = b(X_t, t)dt + a(X_t, t)dW_t ,$$

and assume that X has a density function $p(x,t) = |\psi(x,t)|^2$ where $\psi(.,.)$ is a derivable complex valued function.

LEMMA 3. Suppose $\psi(x,t)$ is a complex valued function such that $\frac{\partial}{\partial x}\psi(x,t)$ and $\frac{\partial}{\partial t}\psi(x,t)$ exist and are continuous. Suppose X is a continuous process having a continuous density function $p(x,t) = |\psi(x,t)|^2$. Then, the set $A = \{(x,t);\ p(x,t) = 0\}$ is negligible for X.

PROOF. We write

(7) $$\int I_{A(\varepsilon,k)}(x,t)p(x,t)d\mu dt = \int I_{[A(\varepsilon,k)\setminus A]}(x,t)p(x,t)d\mu dt .$$

But on the set $A(\varepsilon,k)$, it holds that

(8) $$p(x,t) = |\psi(x,t)|^2 \leq (\varepsilon^2) \sup_{(x,t)\in A(\varepsilon,k)} [|\tfrac{\partial}{\partial x}\psi(x,t)| + |\tfrac{\partial}{\partial t}\psi(x,t)|]^2,$$

and since that $[A(\varepsilon,k)\setminus A]\downarrow\emptyset$, we have

(9) $$\int I_{[A(\varepsilon,k)\setminus A]}(x,t)d\mu dt \xrightarrow{\varepsilon\to 0} 0 .$$

Thus, from (7), (8) and (9), we obtain

$$\int I_{A(\varepsilon,k)}(x,t)p(x,t)d\mu dt = o(\varepsilon^2) .$$

We have then a theorem as follows:

THEOREM 4. Suppose X is a semimartingale satisfying the hypothesis of lemma 2, and suppose X has a density function $p(x,t) = |\psi(x,t)|^2$ where $\psi(x,t)$ satisfies the hypothesis of lemma 3. If we write

$$\tau(\omega) = \inf\{t;\ p(X_t(\omega),t) = 0\},$$

then $P[\tau(\omega)<\infty] = 0$.

PROOF. In fact, it is a corollary to the three preceding lemmas.

COROLLARY. Suppose X is a semimartingale satisfying the hypothesis of lemma 2, and suppose X has a density function $p(x,t)$ which is twice continuously derivable. Then, $P[\tau(\omega)<\infty] = 0$.

PROOF. Because $p(x,t)$ is non-negative, every zero point is at least of degree two. Utilizing the same argument as in the proof of lemma 3, we deduce the closed set $A = \{(x,t);\ p(x,t) = 0\}$ is negligible for X. By lemma 1 and 2, we arrive at the conclusion.

We consider now the second problem of this paper.

LEMMA 5. Suppose X is a continuous process of dimension $d \geq 3$, and suppose X has a continuous density function $p(x,t)$. If x^* is a fixed point in R^d, then the closed $A^* = \{(x^*,t),\ \forall t \in R_+\}$ is negligible for X.

PROOF. Let $B_{\varepsilon,k} = \{(x,t);\ \gamma(x,x^*) \leq \varepsilon,\ 0 \leq t \leq k+\varepsilon\}$, then $A(\varepsilon,k) \subset B_{\varepsilon,k}$. Thus,

$$\int I_{A(\varepsilon,k)}(x,t)p(x,t)d\mu dt \leq \int I_{B_{\varepsilon,k}}(x,t)p(x,t)d\mu dt$$

$$\leq \Big[\sup_{(x,t)\in B_{\varepsilon,k}} p(x,t)\Big] \int I_{B_{\varepsilon,k}}(x,t)d\mu dt$$

$$= k\Big[\sup_{(x,t)\in B_{\varepsilon,k}} p(x,t)\Big] \int I_{\{x;\ \gamma(x,x^*)\leq\varepsilon\}}(x)d\mu .$$

But, since $\int I_{\{x;\ \gamma(x,x^*)\leq\varepsilon\}}(x)d\mu = o(\varepsilon^d)$, we obtain (2) from (10) under the hypothesis $d \geq 3$.

THEOREM 6. Suppose X is a semimartingale satisfying the hypothesis of lemma 2, and suppose X has a continuous density function $p(x,t)$. If x^* is a fixed point in R^d, and if $\zeta(\omega) = \inf\{t;\ X_t = x^*\}$, then $P[\zeta(\omega) < \infty] = 0$.

PROOF. It is a corollary to lemma 2 and lemma 3.

The readers can easily verify that the hypotheses of theorems 4 and 6 are satisfied by a "good" diffusion process with a smooth initial density function $p(x,0)$.

REFERENCE

Nagasawa, M. [1] "Segregation of a population in an environment", Journal of Mathematical Biology, 9, pp. 213-235 (1980);

Nelson, E. [1] Quantum Fluctuations, Princeton, 1984;

Meyer, P.A. and Zheng, W.A. [1] "Construction de processus de Nelson réversibles", Séminaire de Probabilité XVIII, 1984.